Practical Terahertz Electronics: Devices and Applications, Volume 2

Optical devices and applications

Practical Terahertz Electronics: Devices and Applications, Volume 2

Optical devices and applications

Vinod Kumar Khanna
Independent Researcher, Chandigarh, India;
Retired Chief Scientist, CSIR-CEERI, Pilani, India

IOP Publishing, Bristol, UK

ISBN 978-0-7503-4886-7 (ebook)
ISBN 978-0-7503-4884-3 (print)
ISBN 978-0-7503-4887-4 (myPrint)
ISBN 978-0-7503-4885-0 (mobi)

DOI 10.1088/978-0-7503-4886-7

Version: 20211201

IOP ebooks

British Library Cataloguing-in-Publication Data: A catalogue record for this book is available from the British Library.

Published by IOP Publishing, wholly owned by The Institute of Physics, London

IOP Publishing, Temple Circus, Temple Way, Bristol, BS1 6HG, UK

US Office: IOP Publishing, Inc., 190 North Independence Mall West, Suite 601, Philadelphia, PA 19106, USA

To the memory of my respected parents

The late Shri Amarnath Khanna and Shrimati Pushpa Khanna

For their sacrifices and inspiration,

To my beloved grandson Hansh, daughter Aloka and wife Amita

For their affection and whole-hearted support

Contents

Preface

We are very fortunate to be living in an era where we have witnessed several technological breakthroughs and experienced revolutionary changes in lifestyle and comforts. Notwithstanding our achievements, there still remains and very surprisingly too, a chunk of the electromagnetic spectrum which is full of promises but whose potential is still awaiting widespread technological utilization and commercialization for the benefit of civilization.

'Terahertz' is a very big unit of frequency equal to 10^{12} times the base unit called hertz. So when we talk about terahertz electronics we are concerned with very high frequencies in the parlance of electronics. Terahertz electronics is the electronics of extremely high frequencies. In electronics, we deal with two entities, charge carriers and electromagnetic waves. Oscillations in terahertz waves are so fast that electrons cannot follow them and therefore lag behind. So electronic devices become inefficient here.

The frequencies, which appear so high for electronics are too low for photonics. Frequency is linked to an important parameter 'energy'. A quantum of energy is called a photon. The photons have energies in accordance with their frequencies. Naturally so, because the faster a particle moves, the more is its kinetic energy. But here the situation is different compared to electronics. Energies of photons of terahertz frequencies are lower than desired values because photons of these energies produce effects of the same intensity or severity as thermal energy at room temperature. So the effects of terahertz photons become indistinguishable from those of thermal energy, creating chaos and confusion in existing photonic devices, which therefore become erratic and lose their meaningfulness in the terahertz range of frequencies. Hence, this range represents a bothersome zone both for electronics and photonics. Furthermore, these arguments lead us to think about frequency and energy of electrons/photons as the key parameters at the center of all discussions in terahertz electronics and optoelectronics.

The technologically awkward portion of electromagnetic spectrum known as terahertz gap, exhibits amazing properties. The electromagnetic waves of these frequencies are strongly attenuated by absorption by oxygen and water vapor in the atmosphere disallowing their use for medium and long distance communication but at the same time offering the capability to transmit vast amounts of data over short distances of a few meters to share the ever-increasing loads of expanding communication networks. High data rates are achievable with availability of wider bandwidths.

Terahertz radiation offers the opportunity to see through various non-metallic and nonpolar materials in dry state to reveal any prohibited object being transported past entry points at airports with malicious intentions or at crowded public places. Paper, cardboard, plastics, foam and clothing are transparent to terahertz waves. Unlike x-rays, this non-ionizing radiation is harmless to human tissue and DNA, encouraging its use in medical diagnostic imaging such as for epithelial cancer and tooth decay. It opens up immense opportunities in analytical chemistry allowing facile analysis of semiconductors, liquids and biological specimens. Prolific applications of terahertz radiation are expected in three premier technologies viz., communication, spectroscopy and imaging in the forthcoming years.

Realizing the immense potentialities of terahertz electronics, this field has of late captured the attention of scientists worldwide and some strides have been made during the last few decades. At the foundation of availing the fascinating prospects of terahertz waves, lie the various electronic devices that have been developed in the recent past to generate terahertz waves, control their properties and make them work in the designed plan to execute desired tasks. The devices of solid-state category cater mostly to low power applications. The high power zone is captured by vacuum tubes. The vacuum electronic devices benefit from the scattering-free medium of propagation of electrons, 'the vacuum', as opposed to collision-prone medium available to electrons in solid-state devices. This is the primary reason, which is at the root of all differences in properties of vacuum electronic and solid-state devices. The most remarkable characteristic of terahertz electronics is the vital contribution from photothermal and optoelectronic devices. Particularly interesting are the non-linear optical devices. It is this complete team of solid-state and vacuum electronic as well as photothermal and optoelectronic devices whose working principles, fabrication and operation constitute the theme of this book.

The 17 chapters of this book are subdivided into four parts. After introducing the subject in chapter 1, the ensuing six chapters (chapter 2–7) of Part I concentrate on solid-state devices. Then the next three chapters (chapters 8–10) of Part II treat vacuum electronic devices. For long the paucity of suitable optoelectronic devices has hindered progress of terahertz technology partly because of inadequate optical materials. Four chapters (chapters 11–14) of Part III deal at length with photothermal and optoelectronic devices. No field of science is complete without applications because applications determine the utility of the efforts and tell us how closely and deeply the scientific advances have touched our lives and improved our living styles. The closing chapters of the book (chapters 15–17) dwell at length on what terahertz electronics has done and is likely to do in the near future for betterment of human life.

Parts I and II constitute volume 1 of the book while parts III and IV are covered in volume II. Parts I and II of Volume 1 contain chapters 1–10. Parts III and IV of Volume 2 contain chapters 11–17 numbered as chapters 1–7 of this volume.

Two faces of any science are recognized, theoretical, concerning the knowledge gained about the laws of nature by an evidence-based systematic methodology of observation and experimentation, and practical, dealing with application of that knowledge for real-world uses. So our focus in this book is not only to know the truth about the behavior and properties of terahertz waves (theoretical aspects) but about how the knowledge of terahertz waves is beneficially used for their generation, modulation, transmission, detection, or performing any conceivable operation on them so that they can be controlled and manipulated as desired (practical aspects). Greater stress will be laid on the latter part. Numerical worked examples, conceptual and thought-provoking types, and those dealing with practical problems are therefore sprinkled all through the text.

The book is written from a practical engineering perspective. Therefore, dependency on detailed theoretical proofs or higher-level mathematics is avoided to prioritize the practical side of terahertz electronics. All through the book mathematical formulation has been restricted to closed form solutions. Device and process design involve extensive computer simulations involving complicated numerical modeling and optimization techniques. The analysis presented will serve as the first steps towards performing design of devices.

As this book primarily deals with operation of devices at high frequencies, it will be presumed that the reader has prior knowledge of fundamentals of device operation such as for Schottky diode, MOSFET, etc, and they will not be treated at length to make room for inclusion of relevant matter related to high frequencies. Assuming this familiarity with rudimentary knowledge regarding devices, we shall make only short digressions to succinctly touch upon the principles of these devices for sake of revision, and quickly move on to terahertz electronics.

Some flowers are late bloomers. Like these flowers, the terahertz part of the electromagnetic spectrum is a late blooming flower. But now that it has started to bloom, its fragrance will soon pervade the garden of electronics.

I hope the book will fulfill its goal of acquainting the reader with the up-to-date developments in this rapidly evolving, progressive discipline where invigorating innovations are expected every now and then. The author will deem his efforts amply rewarded if readers are able to get abreast with the latest information in the field and are able to tune up with the contemporary scientific advances.

Acknowledgements

I received generous help and support from all quarters during the course of book writing. I feel deeply overwhelmed with humility and gratefulness as I record this support.

Most importantly and above all, I am indebted to God, The Almighty for giving me the strength and wisdom to undertake, pursue and accomplish this task from chapter to chapter. Words become insufficient to convey gratitude for the benevolent blessings showered by God for a smooth journey towards execution of a task.

I acknowledge the valuable contributions of all authors and scientists whose work forms the subject matter of this book. These stalwarts are enlisted in the 'References sections' at the end of each chapter. The references are pointers to the original works and I urge the readers to read them thoroughly to appreciate the full potential of terahertz electronics. The book is intended to make the reading and understanding of original research papers in the field easier and enjoyable. Three cheers for the inventors!

The publishers, acquisitions and production editors and staff at IOP ebooks deserve my thanks for giving me the opportunity to write this book. Their keen enthusiasm and kind co-operation in every stage of this work is worthy of appreciation.

Last but not the least, 'a special thank-you' to my family. They have tolerated with great patience my pre-occupation with this work. Despite my busy schedule encroaching on the time to be spent with them, they have maintained their unfailing support all through the project.

Vinod Kumar Khanna
Chandigarh, India
Retired Chief Scientist, CSIR-Central Electronics Engineering Research Institute, Pilani and Professor, AcSIR, India

Author biography

Vinod Kumar Khanna

Introduction

Vinod Kumar Khanna is an independent researcher at Chandigarh, India. He is a retired Chief Scientist from Council of Scientific & Industrial Research (CSIR)-Central Electronics Engineering Research Institute (CEERI), Pilani-India, and retired Professor from Academy of Scientific & Innovative Research (AcSIR), Ghaziabad, India. He is a former Emeritus Scientist, CSIR and Professor Emeritus, AcSIR, India. His broad areas of research were the design, fabrication and characterization of power semiconductor devices, micro- and nanosensors.

Academic qualifications

He received the MSc Degree in Physics with specialization in Electronics from University of Lucknow in 1975 and PhD degree in Physics from Kurukshetra University in 1988 for the thesis entitled, 'Development, Characterization and Modeling of the Porous Alumina Humidity Sensor'.

Work experience and accomplishments

His research experience spans over a period of 40 years from 1977 to 2017. He started his career as a Research Assistant in the Department of Physics, University of Lucknow from 1977 to 1980. He joined CSIR-Central Electronics Engineering Research Institute, Pilani (Rajasthan) in April 1980. At CSIR-CEERI he worked on several CSIR-funded as well as sponsored research and development projects. His major fields of research included power semiconductor devices and microlectronics/MEMS and nanotechnology-based sensors and dosimeters.

In the power semiconductor devices area, he worked on the high-voltage and high-current rectifier (600 A, 4300 V) for railway traction, high voltage TV deflection transistor (5 A, 1600 V), power Darlington transistor for AC motor drives (100A, 500V), fast-switching thyristor (1300 A, 1700 V), power DMOSFET and IGBT. He contributed towards the development of sealed tube Ga/Al diffusion for deep junctions, surface electric field control techniques using edge beveling and contouring of large-area devices, and floating field limiting ring design, and characterization of minority-carrier lifetime as a function of process steps. He also contributed towards the development of P–I–N diode neutron dosimeter and PMOSFET-based gamma ray dosimeter.

In the area of sensor technology, he worked on the nanoporous aluminum oxide humidity sensor, ion-sensitive field-effect transistor-based microsensors for biomedical, food and environmental applications; microheater embedded gas sensor for automotive electronics, MEMS acoustic sensor for launch vehicles and capacitive MEMS ultrasonic transducer for medical applications.

Semiconductor facility creation and maintenance

He was responsible for setting up and looking after diffusion/oxidation facilities, edge beveling and contouring, reactive sputtering and carrier lifetime measurement facilities. As the Head of MEMS and microsensors group, he looked after the maintenance of six-inch MEMS fabrication facility for R&D projects as well as augmentation of processing equipment under this facility at CSIR-CEERI.

Scientific positions held

During his tenure of service at CSIR-CEERI from April 1980 till superannuation in November 2014, he was promoted to various positions including one merit promotion. He retired as Chief Scientist and Professor (AcSIR: Academy of Scientific & Innovative Research), and Head of MEMS & Microsensors Group. Subsequently, he worked for three years as Emeritus Scientist, CSIR and Emeritus Professor, AcSIR from November 2014 to November 2017. After completion of the emeritus scientist scheme, he now lives at Chandigarh. He is a passionate author, and enjoys reading and writing.

Membership of professional societies

He is a Fellow and life member of the Institution of Electronics & Telecommunication Engineers (IETE), India. He is a life member of Indian Physics Association (IPA), Semiconductor Society, India (SSI) and Indo-French Technical Association (IFTA).

Foreign travel

He is widely travelled. He participated and presented research papers in IEEE Industry Application Society (IEEE-IAS) Annual Meeting at Denver Colorado, USA, in September–October, 1986. His short-term research assignments include deputations to Technische Universität Darmstadt, Germany in 1999, at Kurt-Schwabe-Institut fur Mess-und Sensortechnik e.V., Meinsberg, Germany in 2008 and at Fondazione Bruno Kessler, Trento, Italy in 2011 under collaborative programmes. He was a member of the Indian Delegation to Institute of Chemical Physics, Novosibirsk, Russia in 2009.

Scholarships and awards

Awarded National Scholarship by the Ministry of Education & Social Welfare, Govt of India, on the basis of Higher Secondary result, 1970; CEERI Foundation

Day Merit Team Award for projects on fast-switching thyristor (1986); power Darlington transistor for transportation (1988), P–I–N diode neutron dosimeter (1992); and high-voltage TV deflection transistor (1994); Dr N G Patel Prize for best poster presentation in 12th National Seminar on Physics and Technology of Sensors 2007, BARC, Mumbai; CSIR-DAAD Fellowship in 2008 under Indo-German bilateral exchange programme of Senior scientists, 2008.

Research publications and books

He has published 192 research papers in leading peer-reviewed national/international journals and conference proceedings. Prior to the present book, he has authored 16 books, and has also contributed six chapters in edited books. He has five patents to his credit including two US patents.

About the Book

Practical Terahertz Electronics: Devices and Applications, Volume 2: Optical devices and applications brings forth a refreshing change from the conventional style of presentation of subject matter. Rather than introducing the subject in the conventional descriptive style, it asks questions in the form of numerical problems, and tries to seek the solutions to these problems. These problem-solving examples integrated in the body of the text keep the readers' interest alive all throughout and make the learning process smoother and deeper in knowledge acquisition. The concepts are appreciated and assimilated with little effort and also in a long-lasting manner.

Perched on the crossroads of electronics and photonics, the enormous edifice of terahertz electronics has two distinct appearances depending upon the contributions from electronics and photonics. Further, two sprouts of the electronics component include solid-state and vacuum or gas-filled devices. In addition, the developing application areas of terahertz electronics impart utility to these activities for humanity. Accordingly, the book is subdivided into two parts devoted to electronic devices, a third part on photonic and optothermal devices, and a fourth part dealing with the applications resulting from these endeavors. The four parts of the book containing seventeen chapters present a seamless and cohesive treatment of this expansive technological field. Volume 2 of the book comprising the third and fourth parts is devoted to optical devices and applications of terahertz electronics. The first and second parts of the book were presented in Volume 1 subtitled: Solid-state and vacuum electronic devices.

In line with the adage, 'a picture is worth a thousand words', the book is profusely illustrated with clear line diagrams for easy visualization of device structures and following their operational mechanisms.

The book will be immensely beneficial to both students and professional segments. The book will fulfill the requirements of graduate students in electronics and electrical engineering. Scientists engaged in research in this field as well as practicing engineers will find the book a useful resource providing up-to-date information in a concise format.

Abbreviations and acronyms

A	Ampere
A°	Angstrom unit
AC	Alternating current
ADP	Ammonium dihydrogen phosphate
AE	Antenna emitter
AFB1	Aflatoxin B1
AM	Amplitude modulation
AWGN	Additive white Gaussian noise
BCC	Basal cell carcinoma
BJT	Bipolar junction transistor
BWO	Backward wave oscillator
C	Coulomb
CB	Conduction band
CCD	Charge coupled device
CHIRP	Coherent heterointerfaces for reflection and penetration
cm	Centimeter
CMC	Ceramic matrix composite
CMOS	Complementary metal-oxide-semiconductor (FET)
CNL	Charge neutrality level
COM	Center of mass
C-RAN	Cloud or centralized radio access network
CRM	Cyclotron resonance maser
CT	Computed tomography
CW	Continuous wave (power)
2-D	Two-dimensional
3-D	Three-dimensional
DAST	4-dimethylamino-N-methyl-4-stilbazolium tosylate
dB	Decibel
dBi	Decibel isotropic
dBm	Decibel-milliwatt
DBQW	Double barrier quantum well
DBR	Distributed-Bragg-reflector (laser)
DC	Direct current
2DEG	Two-dimensional electron gas
DFB	Distributed-feedback (laser)
DFG	Difference frequency generation
DFT	Discrete Fourier transform
DHBT	Double heterojunction bipolar transistor
DIBL	Drain-induced barrier lowering
DNA	Deoxyribonucleic acid
DRIE	Deep reactive ion etching
EHF	Extremely high frequency
ELF	Extremely low frequency
eV	Electron volt
F	Farad
FEL	Free electron laser
FET	Field-effect transistor
fF	Femtofarad

FFT	Fast Fourier transform
FIR	Far infrared
FM	Frequency modulation
FMCW	Frequency modulated continuous wave (radar)
FOMs	Figures of merit
FPA	Focal plane array
fs	Femtosecond
FSPL	Free space path loss
FT	Fourier transform
FT-IR	Fourier-transform infrared (spectroscopy)
g	Gram
Gbps	Gigabits per second
GEO	Geosynchronous equatorial orbit
GeV	Giga electron volt
GHz	Gigahertz
GW	Gigawatt
HBT	Heterojunction bipolar transistor
HDPE	High-density polyethylene
HEMT	High-electron-mobility transistor
HF	High frequency
hPa	Hectopascal (I 1hPa = 100Pa)
HRFZ	High resistivity float zone (silicon)
Hz	Hertz
I	Intrinsic (region)
IC	Integrated circuit
IMPATT	IMPact ionization Avalanche Transit-Time (diode)
IR	Infrared
ISAR	Inverse synthetic-aperture radar
ITU	International Telecommunication Union
ITU-R	International Telecommunication Union Radiocommunication (Sector)
J	Joule
K	Kelvin
kA	Kiloampere
kg	Kilogram
kHz	Kilohertz
km	Kilometer
LAE	Large-area emitter
LED	Light-emitting diode
LINAC	Linear accelerator
LF	Low frequency
LO	Longitudinal optical (phonon)
LOS	Line of sight
LT-GaAs	Low temperature gallium arsenide
m	Meter
mA	Milli ampere
Mbps	Megabits per second
MEMS	Microelectromechanical systems
MESFET	Metal–semiconductor-field-effect transistor
MeV	Mega electronvolt
meV	Milli electronvolt

MF	Medium frequency
MHz	Megahertz
MIG	Magnetron injection gun
MIM	Metal–insulator–metal
MIR	Mid infrared
mol	Mole
MOSFET	Metal-oxide-semiconductor field-effect transistor
mS	Milli Siemens
ms	Milli seconds
mW	Milli Watt
mΩ	Milli ohm
N	Newton
NDR	Negative differential resistance
NDT	Non-destructive testing
NEP	Noise equivalent power
NETD	Noise equivalent temperature difference
NIR	Near infrared
nm	Nanometer
NMOS	N-channel metal-oxide–semiconductor (FET)
O/E	Optical-to-electrical (conversion)
OFHC	Oxygen-free high conductivity (copper)
OR	Optical rectification
Pa	Pascal
PCA	Photoconductive antenna
PD	Photodiode
PIN	P-type region/intrinsic region/N-type region (diode)
ps	Picosecond
PVDF	Polyvinylidene fluoride
pW	Picowatt
QCL	Quantum cascade laser
Q.O.	Quasi-optical (mode converter)
QoS	Quality of service
QW	Quantum well
QWITT	Quantum well injection transit time (diode)
RAN	Radio access network
RF	Radio frequency
RMS	Root-mean-square (current/voltage)
RMS SNR	Root-mean-square signal-to-noise ratio
RTD	Resonant tunneling diode
RX	Receiver
S	Siemen
s	Second
SASE	Self-amplified-spontaneous-emission (FEL)
SBD	Schottky barrier diode
SCEs	Short channel effects
SFG	Sum frequency generation
SHBT	Single heterojunction bipolar transistor
SHF	Super high frequency
SHG	Second harmonic generation
SHM	Simple harmonic motion

SI-GaAs	Semi-insulating gallium arsenide
SIN	Superconducting metal film/Insulating layer/ normal metal film
SIS	Superconducting metal film/insulating layer/ superconducting metal film
SLF	Super low frequency
SLM	Spatial light modulator
SNR	Signal-to-noise ratio
SOI	Silicon-on-insulator
sq	Square
SQUID	Superconducting quantum interference device
SS	Subthreshold swing
SSD	Self-switching diode
SWS	Slow wave structure
T	Tesla
Tbps	Terabits per second
TDS	Time-domain spectroscopy
TEA	Transversely excited atmosphere (laser)
TEM_{00} mode	Transverse electromagnetic mode of lowest order (fundamental mode having the same form as a Gaussian beam)
TE_{mn}	Transverse electric mode, m = number of half-wave patterns across the width of the rectangular waveguide and n = number of half-wave patterns across its height
TeraFET	Terahertz field-effect transistor
THz	Terahertz
THz-CW	Terahertz continuous wave (spectroscopy)
THz-TDS	Terahertz-time domain spectroscopy
TWT	Travelling wave tube
TX	Transmitter
UHF	Ultra high frequency
ULF	Ultra low frequency
UTC-PD	Unitravelling-carrier photodiode
UV	Ultraviolet
UV-LIGA	Ultraviolet-lithographie, galvanoformung, abformung (German): ultra-violet-lithography, electroplating, and molding
UV–vis	Ultraviolet–visible (spectroscopy)
V	Volt
VB	Valence band
VBM	Valence band maximum
VF	Voice frequency
VHF	Very high frequency
ViSAR	Video Synthetic aperture radar
VLF	Very low frequency
VUV	Vacuum ultraviolet (wavelength:100–300nm)
W	Watt
WiFi	Not an acronym, but trademarked term for a wireless local area networking (WLAN) technology, which is commonly thought as the shortened form of 'wireless fidelity'
XCOM	X ray communication
μK	Micro Kelvin
μm	Micron, micrometer
μs	Microsecond
μW	Microwatt

Chemical symbols

ADP:$NH_4H_2PO_4$	Ammonium dihydrogen phosphate
Al	Aluminum
AlAs	Aluminum arsenide
AlGaAs	Aluminum gallium arsenide
AlGaN	Aluminum gallium nitride
AlInAs	Aluminum indium arsenide
AlN	Aluminum nitride
Al_2O_3, AlO_x	Aluminum oxide
AO	Auramine O, a diarylmethane dye
AlSb	Aluminum antimonide
Au	Gold (Aurum)
Ba	Barium
$BaTiO_3$	Barium titanate
BCA	Barium calcium aluminate
BeO	Beryllium oxide
CCH	Chlortetracycline hydrochloride
CdTe	Cadmium telluride
CdZnTe	Cadmium zinc telluride
$C_{32}H_{16}CoN_8$	Cobalt phthalocyanine
CO_2	Carbon dioxide
$CsNO_3$	Caesium nitrate
CF_4	Carbon tetrafluoride
CH_3F	Fluoromethane
$^{13}CH_3F$	Fluoromethane-13C
CH_3OD	Methanol-OD or mono-deuteromethanol
CH_3OH	Methanol
Cl	Chlorine
C–O	Carbon–oxygen (bond)
CO_2	Carbon dioxide
Cr_2O_3	Chromium oxide
Cu	Copper
CuO	Copper (II) oxide or cupric oxide
DAST	4-dimethylamino-N-methyl-4-stilbazolium tosylate
GaAs	Gallium arsenide
GaInAs	Gallium indium arsenide
GaN	Gallium nitride
GaP	Gallium phosphide
GaSe	Gallium selenide
Ge	Germanium
H	Hydrogen
HCl	Hydrogen chloride gas, hydrochloric acid
HDPE	High-density polyethylene
HMX	High melting explosive: cyclotetramethylene-tetranitramine
InAlAs	Indium aluminum arsenide
InAs	Indium arsenide
InGaAs	Indium gallium arsenide
InGaAsP	Indium gallium arsenide phosphide

InGaN	Indium gallium nitride
InP	Indium phosphide
Ir	Iridium
KDP: KH_2PO_4	Potassium dihydrogen phosphate
$LiNbO_3$	Lithium niobate
$LiTaO_3$	Lithium tantalate
LT-GaAs	Low temperature gallium arsenide
MDMA	3,4-Methylenedioxymethamphetamine
Mo	Molybdenum
Nb	Niobium
NbN	Niobium nitride
NbTi	Niobium titanium
NbTiN	Niobium titanium nitride
Ni	Nickel
NiCr	Nickel–chromium (nichrome)
NiO	Nickel oxide
OFHC	Oxygen free high conductive (copper)
OH	Hydroxide
Os	Osmium
$PbTiO_3$	Lead titanate
Pd	Palladium
PETN	Pentaerythritol tetranitrate
PolySi	Polysilicon
Pt	Platinum
PVDF	Polyvinylidene fluoride
Re	Rhenium
RDX	Royal demolition explosive: cyclotrimethylenetrinitramine
Ru	Ruthenium
Si	Silicon
SI-GaAs	Semi-insulating GaAs
SiGe	Silicon Germanium
Si_3N_4	Silicon nitride
SiO_2	Silicon dioxide
SiON	Silicon-oxide-nitride, silicon oxynitride
TCsH	Tetracyclines hydrochloride
Ti	Titanium
$TiCl_4$	Titanium tetrachloride
TiO_2	Titanium oxide
TNT	Trinitrotoluene
Vox	Vanadium oxide
W	Tungsten (Wolfram)
ZnO	Zinc oxide
ZnSe	Zinc selenide
ZnTe	Zinc telluride

Mathematical notation

Roman alphabet

A	Junction area, Area, Voltage gain, A constant, Amplitude
a	Radius of the circular contact, Anode radius
A_E	Effective aperture area or aperture
$A_{\text{effective}}$	Effective Richardson constant for thermionic emission
A_i	Short-circuit current gain
A_l	The amplitudes
A_p	Power gain of a helix
$\lvert A_1(\omega)\rvert$, $\lvert A_2(\omega)\rvert$	Amplitudes of sine waves (angular frequency)
B	Magnetic field, Bandwidth, Rotational constant
$\mathbf{B}$	Magnetic field vector, Magnetic induction
b	Cathode radius
B_0	Peak magnetic field, Magnetic field at the cathode, Magnetic pole field
$B_{\text{Effective}}$	Effective bandwidth
B_{Peak}	Peak magnetic field at the resonant cavity
BW	Bandwidth
B_y	Magnetic field component in the Y-direction
C	Pierce gain parameter, Capacitance, Electronic heat capacity, Shannon information capacity
c	Velocity of light in vacuum
C_0	Capacitance in the resonant tunneling diode model, Geometrical depletion capacitance
c_1, c_2, c_3,….	First, second and third derivatives of polarization P with respect to electric field E
C_{BE}	Emitter-base capacitance
C, D	Constants
$C_{\text{Depletion(BC)}}$	Depletion capacitance across the collector-base junction
$C_{\text{Depletion (BE)}}$	Depletion capacitance across the emitter-base junction
$C_{\text{Diff(BE)}}$	Emitter-base diffusion capacitance
$C_{\text{Gate-to-channel}}$	Gate-to-channel capacitance
C_{GD}	Gate-drain capacitance
$C_{\text{GD(Extrinsic)}}$	Extrinsic gate-drain capacitance
$C_{\text{GD(Intrinsic)}}$	Intrinsic gate-drain capacitance
C_{GS}	Gate-source capacitance
C_{gSD}	Capacitance originating from the leakage of charging current for the gate capacitance $C_{GS}+C_{GD}$ through the output source-drain conductance g_{SD}
$C_{\text{GS(Extrinsic)}}$	Extrinsic gate-source capacitance
$C_{\text{GS(Intrinsic)}}$	Intrinsic gate-source capacitance
C_j	Junction capacitance
C_{j0}	Zero-bias junction capacitance
C_{jBE}	Base-emitter junction capacitance
C_{Miller}	Miller capacitance
C_{overlap}	Overlap capacitance
C_{ox}	Oxide capacitance per unit area

C_P	Parallel capacitance
$C_{Parasitic}$	Parasitic capacitance
C_Q	Quantum capacitance
C_S	Displacement capacitance
C_W	Quantum well capacitance
$\mathbf{D}$	Displacement vector
D	Diffusion coefficient, Length of the dipole, Distance between the transmitter and receiver, Diameter, Diameter of the aperture
d	Thickness of film, thickness of the dielectric, Diameter, Period of the planar periodic slow wave structure, Second-order nonlinear coefficient, Separation between the detectors, Distance between telescopes
D_1, D_2	Apertures of telescopes
d_1	Optical path length of the pump arm
d_2	Optical path length of the probe arm
D_e	Diffusion coefficient of electrons
d_{GC}, $d_{Gate\text{-}Channel}$	Separation between the gate and the channel
dI_{Wave}/dt	Wave intensity growth rate
D_n	Diffusion coefficient of electrons
dq	Infinitesimal charge
dW/dt	Rate of change of energy of an electron
E	Electric field, Energy
$\mathbf{E}$	Electric field vector
e	Electronic charge = 1.602×10^{-19}C
E_0	Peak electric field, Resonant energy level, Maximum value or amplitude of the wave, Ground state energy level
E_1, E_2, E_3	Energy levels in the quantum well
E_C	Energy of conduction band edge
E_{EB}	Electron beam energy
E_F	Fermi level, Fermi energy
E_G, E_g	Energy gap
E_j, $E_{J'}$	Rotational energy levels
$E_{Kinetic}$	Kinetic energy
E_{Peak}	Peak electric field
$E(t)$	Near IR pulse
E_{THz0}	Peak value of terahertz field
$E_{THz}(t)$	Electric field of the terahertz radiation pulse, at time t, Continuous-wave terahertz field
$E_{THz}(t')$	Electric field of the terahertz radiation pulse at time t'
$e(t_n)$	A signal; amplitude of the input electric field at an instant of time t_n
E_x	Electric field component in the X-direction
$E(x, t)$	Electric field as a function of distance in X-direction and time t
$E(x, y, z)$	A periodic function of z with period L
$E(\omega)$	Fourier transform of a function of time $E(t)$
$\|E(\omega)\|$	Continuous amplitude spectrum of the wave
$E(\omega_k)$	Discrete Fourier transform of a signal $e(t_n)$
F	Force, RF noise figure
f	Frequency
F_{12}	Force acting on particle of mass m_2 due to particle of mass m_1
F_{21}	Force acting on particle of mass m_1 due to particle of mass m_2

$f_C, f_{Cut\text{-}off}$	Linear cut-off frequency
f_H	Upper cut-off or corner frequency
f_L	Lower cut-off or corner frequency
F_m	Magnetic field compression
f_{Max}, f_{max}	Maximum cut-off frequency, Maximum frequency of oscillation of a transistor
f_{Osc}	Linear oscillation frequency
f_p	Linear plasma frequency
f_r	Resonant frequency
$f_{Roll\text{-}off}$	Roll-off linear frequency
f_T	Tansition frequency or Cut-off frequency of a transistor
$f(x)$	Function of x
G	Electrical conductance, Differential conductance across the active region of RTD, Thermal conductance, Gap length, Gain of the antenna
GB	Gain-bandwidth product
G_D	Differential conductance
g_{DS}	Drain-source conductance
g_m	Transconductance
$g_{mintrinsic}$	Intrinsic transconductance
$g_{oi}, g_{oIntrinsic}$	Intrinsic output conductance
G_{RX}	Gain of receiver antenna
g_{SD}	Output source-drain conductance
g_{ss}	Small-signal gain of the planar undulator
G_{TX}	Gain of transmitter antenna
H	Irradiance, Magnetic field, Hamiltonian
h	Planck's constant = $6.626\ 07 \times 10^{-34}$ J.s = $4.135\ 6677 \times 10^{-15}$ eV.s
$\hbar$	Reduced Planck's constant = $1.054\ 5718 \times 10^{-34}$ J.s = $6.582\ 119\ 57 \times 10^{-16}$ eV. s
h_{FE}	DC common – emitter current gain of a transistor
h_{fe}	Small – signal common – emitter current gain of a transistor
I	Total current flowing through the diode, Moment of inertia, Intensity of terahertz radiation after traversing a distance x through the material
I_0	Reverse saturation current (dark current) of the diode, Beam current, Intensity of the incident terahertz radiation
I_A	Alfvén current
$I_{Average}$	Average intensity of electromagnetic wave
I_B	Base current of a transistor
I_{DS}	Drain–source current
I_E	Emitter current
I_{EB}	Average electron beam current
$I_{EB(Peak)}$	Peak electron beam current
I_C	Collector current of a transistor
i_i, i_{IN}	Input current
i_L	Output current
I_{Max}	Maximum current
$I_{Modulated}$	Intensity of modulated terahertz wave
I_n	Electron current
$I_{Opt}(t)$	Intensity of the optical beam at time t
I^0_{Opt}	Amplitude of the optical beam

i_{OUT}	Output current
$I_{\mathbf{P}}$, I_{Peak}	Peak current
i_p	Pyroelectric current
$I_{Probe}(t)$	Intensity profile of the IR probe beam
I_{RMS}	Root-mean-square current
I_S	Saturation current
I_{THz}	Terahertz component of the photocurrent
$I_{Unmodulated}$	Intensity of unmodulated terahertz wave
I_V	Valley current
I_{Wave}	Intensity of the wave
$I_{Wave(0)}$	Initial intensity of the wave at time t=0
$\mathbf{J}$	Current density vector
J	Rotational energy level
j	An operator which prefixed before a number identifies it as an imaginary number or imaginary part of a complex number
J_0, J_1	Bessel functions of the first kind of zeroth and first order
$JJ(\xi)$	Reduction factor of FEL interaction in the undulator
$J'_m(x)$	Bessel function of the first kind of order m
J_n	Current density of electrons
J_W	Instantaneous current density flowing through the quantum well
K	A dimensionless number called the undulator parameter
k	Momentum, Wavevector, Wavenumber, Force constant or spring constant
k_B	Boltzmann constant=1.380 649×10^{-23}JK^{-1}=8.617 333 26×10^{-5}eVK^{-1}
k_{OH}	Force constant for strong (covalent) oxygen-to-hydrogen bonding
$k_{O.....H}$	Force constant for weak oxygen-to-hydrogen bonding
k_z, k_{zl}	Axial wave number/wave vector
k_{z0}	Wave number of the zeroth order spatial harmonic in the axial direction
$k_{\perp l}$	Transverse wave vector/wave number
$k(\Omega)$	Wave vector of the terahertz signal produced at angular frequency Ω
L	Length, Length of the quantum well, Drift length, Channel length, Gate length, Length of the microbridge, Angular momentum
l	Harmonic number, Length
$\hat{L}^2$	A quantum-mechanical operator (angular momentum operator)
$L_{Atmospheric}$	Atmospheric loss
L_B	Width of the barrier layer
L–C	Inductance–capacitance
$L_{Collector}$	Length of collector region
L_D	Width of the depletion region
L_{Dead}	Dead space
$L_{Diffusion}$	Thermal diffusion length
$L_{Emitter}$	Emitter length
L_G	Gate length, Power gain length
L_Q	Quantum inductance
L_{RX}	Receiver losses
L_S	Inertial inductance
L_{Sat}	Power saturation length along the undulator
$L_{\mathbf{TX}}$	Transmitter losses
L_W	Width of the quantum well region

L_x	Length in the *X*-direction
L_y	Length in the *Y*-direction
L_x, L_y, L_z	Projections of angular momentum on the three axes
$\mathbf{M}$	Magnetization of the medium
M	Link margin
m	Mass of the electron, Modulation depth or modulation index, A quantum number, Number of times that the measurements are done
m_0	Rest mass of electron = 9.109×10^{-31} kg
m_1, m_2	Masses of particles
m^*, m_e^*	Effective mass of electron
M_A	Atomic mass
$m_{\mathrm{Effective}}$	Effective mass of electron
m_H	Mass of hydrogen atom
m_h	Effective mass of holes
m_O	Mass of oxygen atom
$m_{\mathrm{Relativistic}}$	Relativistic mass of electron
$m_{\mathrm{Rest\ mass}}$	Rest mass of electron
N	Circuit length, Total number of free electrons, Number *N* of quantized levels used for data representation, Number of time samples
n	An ideality factor ~1–2 (also called quality factor or emission coefficient), Refractive index of a medium, Carrier concentration, An integer, Harmonic number, Number of detectors
N_A	Acceptor concentration in the semiconductor, Avogadro's number = 6.022×10^{23} mol^{-1}
N_C	Effective density of states in the conduction band
N_D	Donor concentration in the semiconductor
$n_{\mathrm{Extraordinary\ ray}}$	Extraordinary refractive index
n_i	Intrinsic carrier concentration in the semiconductor
$n_{\mathrm{Ordinary\ ray}}$	Ordinary refractive index
$\mathrm{Noise}_{\mathrm{RX}}$	Noise at receiver
$n_p(0)$	Electron concentration at the emitter-base boundary (x=0)
n_{p0}	Electron concentration in thermal equilibrium in the P-base
N_{Pump}	Carrier density at the semiconductor surface generated by optical pumping with laser
$n_p(x)$	Electron concentration in P-region as a function of distance x
$n_{\mathrm{Slow\ wave}}$	Refractive index of slow wave structure
$n(t)$	Noise signal
N_u	Number of periods in the undulator
N_V	Effective density of states in the valence band, Number of electrons per atom
$N(\delta)$	Delta doping density
$n(\Omega)$	Refractive index of terahertz pulse at frequency Ω
$n_g(\omega)$	Group velocity refractive index of near IR pulse of frequency ω
$\mathbf{P}$	Polarization density of the medium
P	Power, Electromagnetic power, Total path length for the electromagnetic wave travelling along the helical path, Vapor pressure
p	Distance between the two adjoining turns of a helix (pitch of the helix), Momentum
$\mathbf{p}$	Momentum of the electron

P_0	Amplitude of the modulated terahertz radiation
Pa	Pascal
P_{AC}	AC power
P_{THz}^{AE}	Terahertz power of antenna emitter
P_{CW}	Continuous-wave output power
P_{dBm}	Power in decibel-milliwatt
P_{DC}	DC operating power, DC polarization
P_{Del}	Power delivered
P_{EB}	Electron beam power
$P_{Feedpoint}$	Input or feedpoint power
P_{FEL}	Power of free electron laser
P_L	Power delivered to the load
P_{Laser}	Power of the laser
$\mathbf{P}_{Linear}$	Linear polarization
P_{Max}, $P_{Maximum}$	Maximum power
P_{mW}	Power in milliwatt
P_n	Shot noise power of spontaneous radiation
$P_{Non\text{-}linear}$	Non-linear polarization
P_R	Received power
P_S	Power received from signal source
P_s	Spontaneous polarization
P_{Sat}	Power saturation level
$P_{Straight}$	Path length for the electromagnetic wave propagating along the straight line path
P_T	Transmitted power
P_{TX}	Power transmitted
P_{Watts}	Power in Watts
Q	Charge, Quality factor
q	Electronic charge = 1.602×10^{-19}C
Q_B	Base charge
R	Resistance, Radius, Bond length
r, r_0	Radius of a circle, Larmor radius
$\mathbf{r}_1$, $\mathbf{r}_2$	Position vectors of particles
$\mathbf{r}_{12}$	Vector sum of $\mathbf{r}_1$ and $\mathbf{r}_2$
r_1, r_2	Radii of circles
R_{bb}	Base resistance of a transistor
$R_{b,\ Contact}$	Contact resistance of base
$R_{b,\ gap}$	Emitter-base gap resistance
$R_{b,\ Sheet}$	Base sheet resistance
$R_{b,Spreading}$	Spreading resistance under the emitter
R–C	Resistance–capacitance
R_C, $R_{Collector}$	Collector resistance of a transistor
R_D	Series resistance of drain terminal
r_d	Dynamic resistance
R_E, $R_{Emitter}$	Emitter resistance of a transistor
$R_{Feedpoint}$	Feedpoint resistance
R_L	Load resistance
R_{Losses}	Resistance arising from losses
r_{max}	Channel capacity (data rate)
R_R	Radiation resistance of the antenna

$R_{\mathrm{Radiation\ (dipole)}}$	Dipole radiation resistance
R_{S}	Series resistance, DC spreading resistance, series resistance of source terminal
R_{s}	Sheet resistance of the superconducting film
R_{Spr}	Spreading resistance
$r(t)$	Sum of transmitted signal and noise signal
R_{tt}	Transit-time negative resistance
R_{W}	Specific resistance of quantum well region
r, θ, ϕ	Radial distance, polar angle, and azimuthal angle
S	Instantaneous intensity of electromagnetic waves
s	Velocity of plasma waves, Velocity of electron, Harmonic number, Distance travelled by the wave front, Latency, length of the path
S_{Average}	Average intensity of electromagnetic waves
$\mathrm{Signal}_{\mathrm{RX}}$	Signal at receiver
S_{T}	Transmitted power density
$S(t)$	An instantaneous signal
$s(t)$	Transmitted signal
$S(t')$	Signal at a single time point t'
$\mathrm{Sensitivity}_{\mathrm{RX}}$	Receive sensitivity
T	Absolute temperature, Time period of the wave, Kinetic energy of a particle, Sampling interval or sampling period
$1/T$	Sampling rate
t	Time
t_1	Time during which the crest of a wave shifts from one point to another
t_2	Time during which the wave moves from one point to another
T_{C}	Transition (critical) temperature of the superconductor
t_{C}	Channel thickness
$T_{\mathrm{Collector}}$	Thickness of collector region
T_{Contact}	Thickness of contact layer
$t_{\mathrm{Depletion\ (BC)tt}}$	Time taken by electronic current to cross the collector-base depletion region
T_{F}	Fermi temperature
t_{n}	An instant of time
U	Unilateral power gain of a transistor
U_0	Positive DC gate-source voltage which can induce strong inversion
U_1	Amplitude or peak value of gate-source voltage
U_{GS}	Gate-source voltage
v, υ	Velocity of electrons
$\mathbf{v}$	Velocity vector
V_0	DC bias, Peak voltage, Beam voltage
V, V_{a}	Applied bias, Energy, Volume, Potential energy
v_1, v_2	Linear velocities of particles
V_{BE}	Base-emitter voltage
V_{bi}	Built-in potential of a diode/metal–semiconductor contact
$V_{bi}\|_{BC}$	Built-in potential of collector-base junction
$V_{bi}\|_{BE}$	Built-in potential of base-emitter junction
v_{C}	Rate of escape of electrons from the quantum well to the collector
V_{CE}	Collector–emitter voltage of a transistor
V_{D}	Voltage drop across the diode, Drain voltage

v_d	Electron drift velocity
V_{DC}	DC bias
V_{DS}	Drain-source voltage
$V_{D(Sat)}$	Saturation drain voltage
v_E	Rate of escape of electrons from the quantum well into the emitter
v_e	Velocity of electron beam
$v_e\|_{Approximate}$	Approximate velocity of electron
$v_e\|_{Relativistic}$	Relativistically obtained velocity of electron
V_G	Gate voltage
v_g	Group velocity
V_{GD}	Gate-drain voltage
V_{GS}	Gate-source voltage
$v_{High\ field\ domain}$	High field domain velocity
$v_{Incoming}$	Velocity of incoming waves
V_{Max}	Maximum voltage
V_{Noise}	Root mean square noise voltage
$v_{Outgoing}$	Velocity of outgoing waves
V_P, V_{Peak}	Peak voltage
v_p	Phase velocity
V_R	Voltage drop across the resistor R
V_{RMS}	Root-mean-square voltage
V_S	Source voltage
v_S	Saturated electron velocity in the drift region
v_{Sat}	Electron saturation velocity
V_{Signal}	Root-mean-square signal voltage
V_{Th}	Threshold voltage
V_V	Valley voltage
V_W	Instantaneous voltage across the quantum well
v_x	Component of electron velocity in the X-direction
v_z	Axial translational velocity of electron beam, component of electron velocity in the Z-direction
$v_\perp$	Perpendicular component of electron velocity **v** directed perpendicular to the magnetic field
$v_{\|\|}$	Component of electron velocity **v** directed parallel to the magnetic field
$v(\Omega)$	Velocity of terahertz waves of frequency Ω
W	Width of the depletion region, Thickness, Channel width, Work done
$W_{b,\ Contact}$	Base contact width
$W_{Collector}$	Width of collector region
$W_{Depletion\ (BC)}$, $W_{Depletion}\|_{BC}$	Width of the depletion region across the collector-base junction
$W_{Electrically\ active\ emitter}$	Width of electrically active emitter
$W_{Emitter} = W_{Emitter\ contact}$	Width of the emitter
$W_{Emitter\text{-}Base\ spacing}$	Emitter-base spacing
$W_{Emitter\ junction}$	Width of emitter junction
$W_{Optimum}$	Optimum thickness of depletion region
x	Aluminum fraction in $Al_xGa_{1-x}As$, extension/shortening of the spring

x_0	A particular x position
X_C	Capacitive reactance
x_d	Zero-bias depletion region width
$x_{Depletion}\|_{BE}$	Width of depletion region across the base-emitter junction
$x(\omega)$	Real part of the Fourier transform $E(\omega)$
$Y(\theta, \phi)$	Spherical harmonics
$y(\omega)$	Imaginary part of the Fourier transform $E(\omega)$
Z	Impedance
z	Distance in the Z-direction
Z_0	Impedance of the slow wave structure, Impedance or characteristic impedance of free space (wave impedance of a plane wave in free space)
$Z_{Bulk\text{-}Spreading}$	Bulk spreading impedance
Z_{in}	Input impedance
Z_S	Bulk spreading impedance
$Z_{Skin\text{-}effect}$	Skin-effect impedance
Z_{tt}	Transit-time specific impedance
Z_W	Specific impedance of the quantum well

Greek alphabet

α	Real part of the complex number $\alpha + j\beta$, Ratio of radial-to-axial velocity (pitch factor), Coupling coefficient
α_T	Base transport factor
β	Imaginary part of the complex number $\alpha + j\beta$
β_{AC}	Small signal common-emitter current gain of a transistor
β, β_{DC}	Large signal DC common-emitter current gain of a transistor
β_x	Twiss factor
β_z	Relative average velocity in the forward Z-direction
Γ	Pyroelectric coefficient
γ	A coefficient quantifying the influence of surface states, Lorentz factor
δ, δ_S	Skin depth
ΔE_C	Conduction band offset
ΔE_G	Bandgap difference
ΔE_V	Valence band offset
ΔI	(Peak current–valley current) of resonant tunneling diode
ΔI_B	Change in base current
ΔI_C	Change in collector current
$\Delta I_{EO}(t)$	Effective time-varying electro-optic signal
$\langle \Delta I_{EO}(t) \rangle$	Time-averaged signal
ΔJ	Difference between rotational energy levels J and J'
Δk	Change in wave number
Δn	Refractive index change
$\Delta n\|_{Linear}$	Refractive index change in linear electro-optic effect
$\Delta n\|_{Quadratic}$	Refractive index change in quadratic electro-optic effect
Δt	Interval of time, Time taken by the electron to traverse a distance from a point A to an equivalent point B
$\delta\ (t)$	Delta function
$\Delta t'$	Time taken by the wave to traverse a distance from a point A to an equivalent point B (one undulator wavelength)
ΔV	(Peak voltage–valley voltage) of resonant tunneling diode
$\Delta \nu$	Measurement bandwidth, Frequency separation between any two consecutive absorption lines
$\Delta \nu_{THz}$	Frequency difference in terahertz range
$\Delta \omega$	Change in angular frequency
ε_0	Permittivity of free space = 8.854×10^{-12} Fm^{-1}
ε, ε', ε_r	Relative permittivity of a material
ε_B, $\varepsilon_{Barrier\ layer}$	Relative permittivity of barrier layer
ε_D	Relative permittivity of depletion region
$\varepsilon_{Effective}$	Effective permittivity
$\varepsilon_{n,x}$	Emittance
ε_{ox}	Dielectric constant of oxide film
ε_S	Dielectric constant of the semiconductor
$\varepsilon_{Semiconductor}$	Permittivity of the semiconductor
ε_W	Relative permittivity of quantum well region
η	Efficiency, Absorbance of the radiation-sensitive film, $\sqrt{\frac{\mu_0}{\varepsilon_0}} / n$, Mass attenuation or absorption coefficient

η_{FEL}	Optical conversion efficiency of a free electron laser
η_{OPT}	Output coupling efficiency of the optical cavity
θ	An angle, Angle of emission of the wavefront with respect to the axis of the electron beam, Polar angle, Angular resolution
$\theta_{\mathrm{Telescope}}$	Angular resolution of the telescope
λ	Wavelength
λ_{R}	Material-specific correction factor
λ_{u}	Undulator wavelength in a laboratory frame of reference
λ_u^*	Undulator wavelength after length contraction, Wavelength in a moving frame of reference
μ	Mobility of charge carriers, Refractive index, Effective mass of carriers in the semiconductor, Reduced mass of particles of masses m_1 and m_2, Linear attenuation constant
μ_0	Permeability of free space or permeability constant = $4\pi \times 10^{-7}$ H/m = $1.256\ 637 \times 10^{-6}$ N A^{-2}
$\mu_{\mathrm{Effective}}$	Effective permeability
μ_{n}	Electron mobility
$\mu_{\mathrm{O.....H}}$	Reduced mass for hydrogen bonded oxygen and hydrogen atoms (O…H)
μ_{r}	Relative permeability of the material
ν	Frequency
$\overline{\nu}$	Wavenumber
ν_+	Sum frequency
ν_-	Difference frequency
ν_0	Central frequency
$\nu_1, \nu_2, \nu_3, \ldots$	Frequencies of absorption lines
ν_{Beats}	Beat frequency
$\nu_{\mathrm{Cut\text{-}off}}$	Cut–off frequency
ν_{RC}	Cut–off frequency determined by resistance–capacitance time constant
ξ	A variable = $2\pi z/\lambda_{\mathrm{u}}$
$\xi_{\mathrm{Roll\text{-}off}}$	Roll–off factor
ρ	Resistivity, Pierce or FEL parameter, Density of a material, Free charge density, Mass density
$\rho_{\mathrm{b,Contact}}$	Base contact resistivity
$\rho_{\mathrm{Collector}}$	Resistivity of collector region
$\rho_{\mathrm{Contact(Specific)}}$	Specific contact resistivity
σ	Conductivity, Conductance
σ_{DC}	DC electrical conductivity
$\sigma_{\mathrm{Interband}}(\omega)$	Optical conductivity at frequency ω for transitions from one energy band to another band
$\sigma_{\mathrm{Intraband}}(\omega)$	Optical conductivity at frequency ω for transitions in the same band
$\sigma_{\mathrm{Optimum}}$	Optimum injection conductance
σ_{x}	Electron beam RMS transverse dimension
$\sigma(\omega)$	Optical conductivity due to absorption of waves of frequency ω
τ	Time, Time constant, Electron lifetime in the quantum well, Total time spent by plasma waves in the round trip of the channel, Momentum relaxation time for scattering of carriers
τ_{Backward}	Time taken by plasma wave to travel from drain to source

τ_{Btt}	Base transit time
τ_{C}	Electron lifetime in the collector, Time spent in charging the depletion region of the collector
τ_{E}	Emitter charging time
τ_{ee}	Electron-electron interaction time
τ_{escape}	Escape time of the phonon
$\tau_{\mathrm{Extrinsic}}$	Extrinsic delay
τ_{Forward}	Time taken by plasma wave to travel from source to drain
$\tau_{\mathrm{Intrinsic}}$	Intrinsic delay
$\tau_{\mathrm{Parasitic}}$	Parasitic delay
τ_{pe}	Phonon-electron interaction time
τ_{RC}	Resistance-capacitance time constant
$\tau_{\mathrm{Relaxation}}$	Momentum relaxation time
τ_{t}	Transit time of carriers
τ_{Thermal}	Thermal response time
ϑ	Velocity of electromagnetic waves
ϕ	Phase angle, Azimuthal angle
ϕ_0	Energy of charge neutrality level reckoned from valence band maximum
ϕ_1	A phase term linked to the optical path length of the pump arm
ϕ_2	A phase term linked to the optical path length of the probe arm
ϕ_{B}	Zero-bias Schottky barrier height
ϕ_{bi}	Built-in potential at the metal-semiconductor interface
ϕ_{M}	Work function of the metal
ϕ_{S}	Work function of the semiconductor
$\phi\ (x,\ y,\ z)$	A periodic function of z with period L
$\phi(\omega)$	Continuous phase spectrum of the wave
$\phi_1(\omega),\ \phi_2(\omega)$	Phases of the waves
$\chi,\ \chi_{\mathrm{S}}$	Electron affinity of the semiconductor
$\chi^{(1)}$	Linear susceptibility
$\chi^{(1)}, \chi^{(2)}, \chi^{(3)}, \ldots..$	First-order, second-order and third-order susceptibilities
χ_{mn}	nth zero of the derivative of the mth order Bessel function of the first kind
ψ	Opening angle for the radiation emitted by electron beam, Wave function
Ω	Ohm, Frequency of terahertz radiation pulse, Radian-frequency sampling interval, Beat frequency
ω	Angular frequency, Angular velocity
ω_0	Cut-off frequency of a waveguide
ω^*	Angular frequency after length contraction
Ω_{C}	Cyclotron frequency
$\omega_{\mathrm{Cut\text{-}off}}$	Cut-off angular frequency
ω_{D}	Dielectric relaxation frequency
ω_{osc}	Angular oscillation frequency
ω_{P}	Plasma frequency
$\omega_{\mathrm{Roll\text{-}off}}$	Roll-off angular frequency
ω_{S}	Carrier scattering frequency
ω_{T}	Angular transition or cut-off frequency

Part I

Photothermal and optoelectronic devices

IOP Publishing

Practical Terahertz Electronics: Devices and Applications, Volume 2

Optical devices and applications

Vinod Kumar Khanna

Chapter 1

Bolometers, Golay cells and pyroelectric detectors

In the room-temperature bolometer category, antenna coupled-, nanostructured thin film integrated- and microelectromechanical systems (MEMS)-based devices are described. Limitations of superconductor-insulator-superconductor bolometers are discussed. Amongst the available choices of bolometers for operation at terahertz frequencies, hot electron bolometers are the favourite devices. The functioning of these bolometers does not depend on the energy gap of the superconducting film. Two types of hot-electron bolometers are delved into, namely the diffusion-cooled and electron–phonon cooled versions. The operation of Golay cell working on the photoacoustic effect is explained. Its advantages and shortcomings are indicated. Pyroelectricity and related terms are defined, and the pyroelectric detector is introduced. The optical and electrical responsivities of detectors are differentiated, as also are their optical and electrical noise equivalent powers. The response time of the detector is defined. Typical values of these figures of merit of the micro-bolometers, Golay cells and pyroelectric detectors are compared to assess their relative performance capabilities.

After learning about solid-state electronic devices in part I of the book and vacuum electronic devices in part II, we begin exploration of thermal and optical aspects of the subject in part III, and examine how the effects of heat and light on materials are utilized in terahertz electronics for generation and detection of terahertz waves and performing other operations on them.

In this chapter, we take up the devices based on heating effects of electromagnetic waves, which are widely used as radiation detectors. The heating effects cause resistance changes in microbolometers and pyroelectric detectors. They are also applied to make a device working on optical microphone principle, viz., the Golay cell.

1.1 Bolometers

A bolometer is a thermal detector comprising an array of minuscule sensing elements which get heated on illumination with electromagnetic radiation, and the resultant change in their electrical resistances is processed into temperature values to generate a thermal image of the radiation source.

Bolometers are categorized into two main types, viz., uncooled and cooled systems. The uncooled bolometers operate at room temperature. Non-requirement of cryocooling accessories provides long-lasting maintenance-free operation with low power consumption. Additionally, they are cheaper to manufacture and assemble. Cooled bolometers are more expensive, consume more power and suffer from more downtime due to frequent wear and tear than the uncooled ones. But they are very sensitive and are capable of detecting exceedingly low temperature excursions.

1.2 Uncooled bolometers

1.2.1 Antenna-coupled bolometer

Mature infrared bolometer technology comes to the rescue in developing suitable bolometers for terahertz radiation. The traditional IR bolometers using the microbridge structure are less sensitive to terahertz waves so that a noise equivalent power (section 1.6(ii)) of 14 $pW/Hz^{½}$ for IR absorption becomes 300 pW $Hz^{-½}$ for terahertz absorption, rendering reliable measurements difficult (Gou *et al* 2015). Therefore, an efficient approach to enhance absorption of these waves uses the antenna coupled microbridge structure where the antenna provides high-sensitivity terahertz absorption and the microbridge gives high-performance thermal detection. Several spiral-type antenna coupled microbridge bolometer designs are described for room-temperature terahertz sensing (Gou *et al* 2018).

The bolometer is fabricated on a 2 μm thick silicon-oxide-nitride (SiON membrane), figure 1.1. The titanium thermistor film is formed by electro-deposition. The titanium film is suspended in air for minimizing thermal losses by etching the underlying SiON film using micromachining technology. Thus, a free-standing titanium bolometer is obtained. Metal contacts are formed on the titanium film for connection. The double dipole or log-periodic type terahertz antenna is made of aluminum. A metallized bottom plate under the SiON membrane reflects any radiation reaching it. The geometry of the dipole antenna and the resonant cavity are designed to achieve selectivity to radiation. The reflector is positioned at a distance of a quarter wavelength of the radiation.

The bolometers are characterized by a quasi-optical setup in the 0.1–1 THz range (Kašalynas *et al* 2013, 2016). The responsivity of titanium bolometers is 1000 V W^{-1}, and the noise equivalent power (NEP) is 5 pW $Hz^{-1/2}$ (Trontelj *et al* 2014).

1.2.2 Nanostructured thin film absorber integrated bolometer

In this scheme, terahertz absorption is considerably improved by integrating a nanostructured absorber layer with the bolometer. Gou *et al* (2015) reported a focal plane array comprising 320×240 vanadium oxide pixels (pitch = 35 μm) integrated

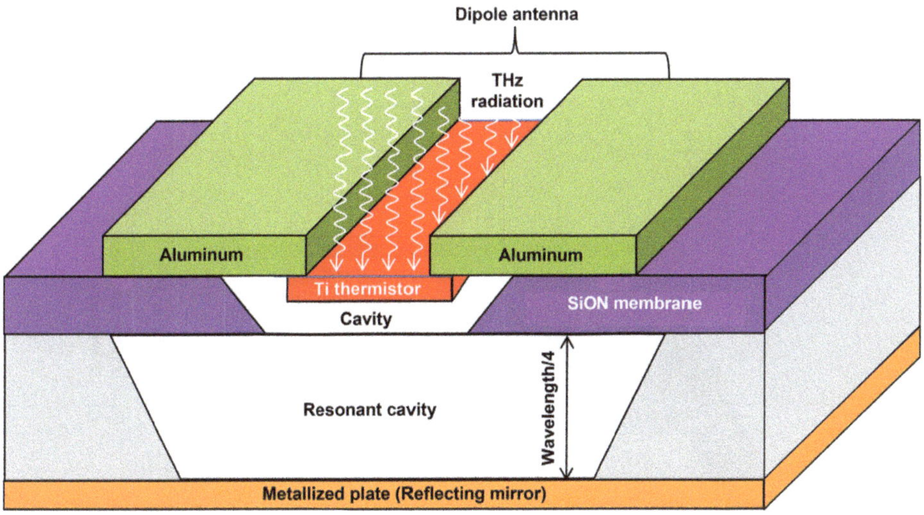

Figure 1.1. Bolometer with titanium thermistor suspended in air in a cavity overlying a resonant cavity with reflector plate. The aluminum dipole antenna is also shown.

with a nanostructured titanium thin film for absorption of terahertz radiation (figure 1.2). The titanium film is deposited by magnetron sputtering and thinned down by reactive ion etching to obtain optimized terahertz reflection and transmission curves. Since

$$\text{Transmission} + \text{Reflection} + \text{Absorption} = 1 \tag{1.1}$$

and transmission + reflection for 9 nm Ti film is smaller than the same for 5 nm thick Ti film, it is evident that the absorption of terahertz radiation is greater for 9 nm Ti film than 5 nm film. The diaphragm of each pixel contains four layers from the bottom upwards: silicon nitride support layer (250 nm), vanadium oxide thermal sensitive layer (50 nm), silicon nitride passivation layer (100 nm) and Ti thin film terahertz absorption layer (thinned down to 9 nm from 20 nm deposited thickness). The reflector is made of NiCr film (250 nm). The focal plane array is packaged in vacuum $>10^{-5}$ Pa.

The readout integrated circuit is fabricated under the focal plane array on the substrate. The change in current of each pixel is amplified and the resistance change is obtained from the current change. The resistance change in accordance with the terahertz radiation absorbed by the pixel provides the terahertz imaging.

A 2.52 THz CO_2 laser is used in the experimental setup for terahertz imaging. The measured responsivity of the bolometer is 2186 V W^{-1}, and the noise equivalent power is 45.7 pW $Hz^{-1/2}$ (Gou *et al* 2015).

1.2.3 MEMS bolometer

As the name of this bolometer indicates, it is based on a thermomechanical transduction principle (Zhang *et al* 2019). The thermistor used is a doubly clamped

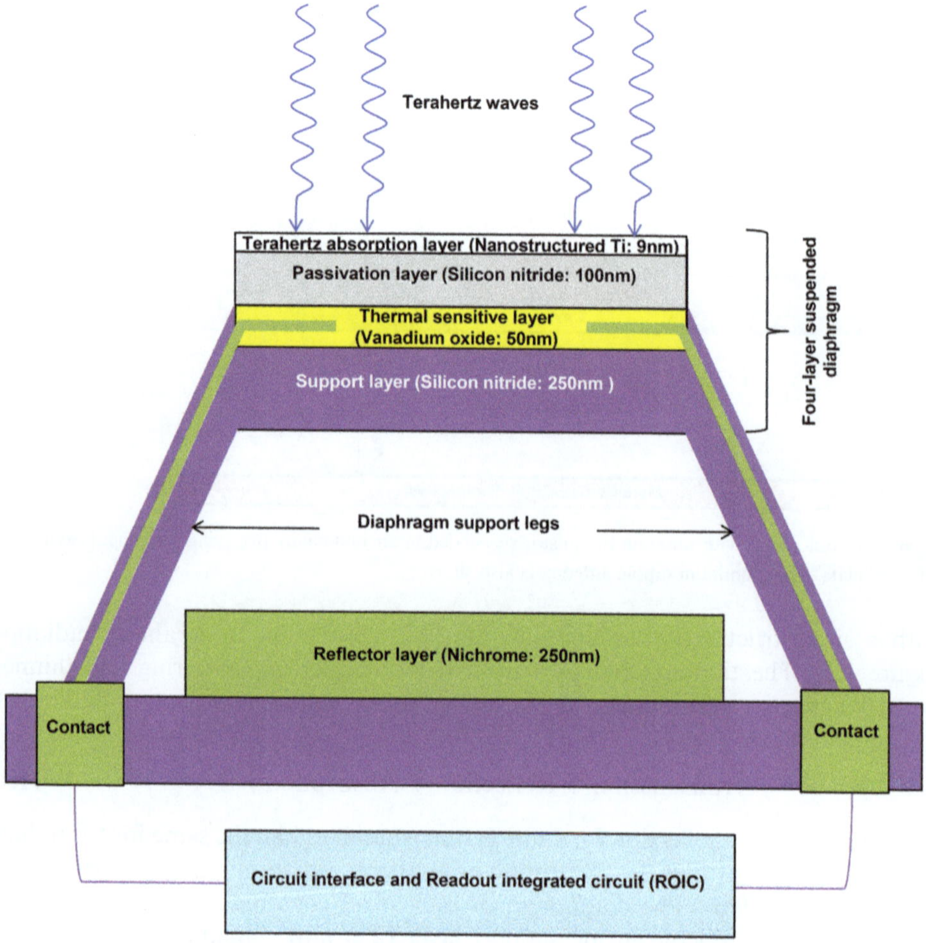

Figure 1.2. Pixel of the vanadium oxide microbolometer focal plane array.

gallium arsenide beam resonator (figure 1.3). The temperature changes are indicated by the variation in resonance frequency of the device.

For measurements, the terahertz radiation is generated by difference frequency mixing of photons. The arrangement consists of two tunable lasers feeding a unitravelling-carrier diode (chapter 3, section 3.6). The terahertz beam produced is focused on the resonator by mounting it on a hemispherical lens. The frequency is scanned in the range from 0.25 to 3.0 THz. Temperature changes as small as 1 μK are detectable. The NETD (noise equivalent temperature difference) value at room temperature is 1 μK $Hz^{-1/2}$. The speed of detection is of the order of several kHz. The electrical noise equivalent power is 90 pW $Hz^{-1/2}$ and the dynamic range is 3×10^7.

The MEMS bolometer is fabricated by standard semiconductor technology using selective etching which is well suited to realization of large area multielement arrays. The constituent layers of the structure are shown in the diagram. An NiCr thin film

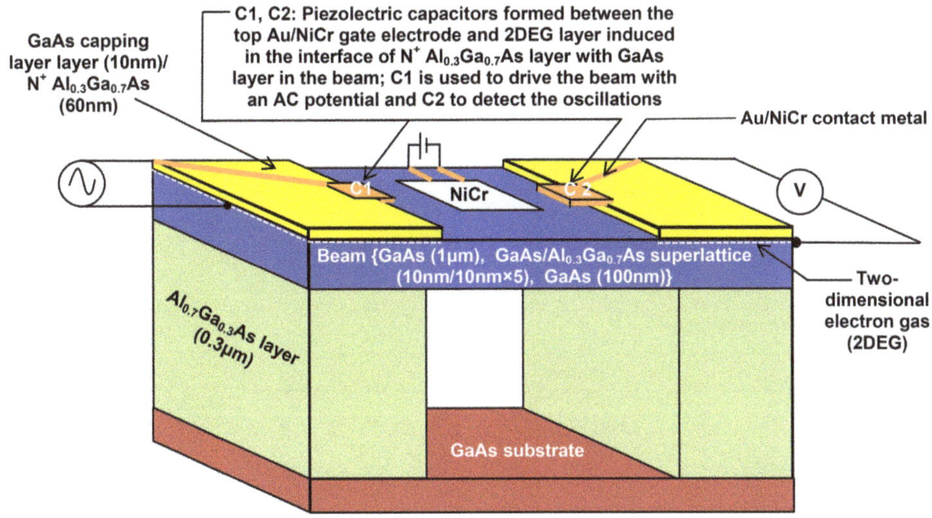

Figure 1.3. Doubly damped GaAs beam resonator working on thermomechanical transduction scheme.

deposited on the top surface of the beam is used as a terahertz absorber and heater for calibration of responsivity of the bolometer (Zhang *et al* 2019).

1.3 Cooled superconducting bolometers

Superconducting bolometers derive advantage of high sensitivity from the fact that the resistance of superconductor is zero below a critical or transition temperature and jumps up to a high value immediately when the temperature exceeds this transition temperature. In this condition, the superconductor no longer acts as a superconductor but becomes a normal metallic conductor. Keeping this in view, the superconductor is kept at a temperature in the center of the transition region width. Then with a slight increase of its temperature, the resistance of the superconductor drastically increases, enabling the small temperature rise to be measured exactly.

1.3.1 Limitations of SIS tunnel junction devices

Superconducting tunnel junction bolometers work on the variation of tunneling current of a tunnel junction with temperature rise produced by incident terahertz radiation (figure 1.4). A tunnel junction consists of a thin insulating film sandwiched between two metallic electrodes (Clarke *et al* 1974). The thickness of the insulating film should be sufficiently small to ensure that quantum mechanical tunneling can take place through it. One configuration has the structural arrangement: superconducting metal film/insulating layer/normal metal film (SIN), figure 1.4(a). The successive layers in the other configuration are: superconducting metal film/insulating layer/superconducting metal film (SIS), figure 1.4(b). The SIS junction has extremely low impedance. Consequently, the superconducting quantum interference device, famous by the name SQUID, can act as a low noise amplifier for the SIS bolometer. So the SQUID is utilized for accurate

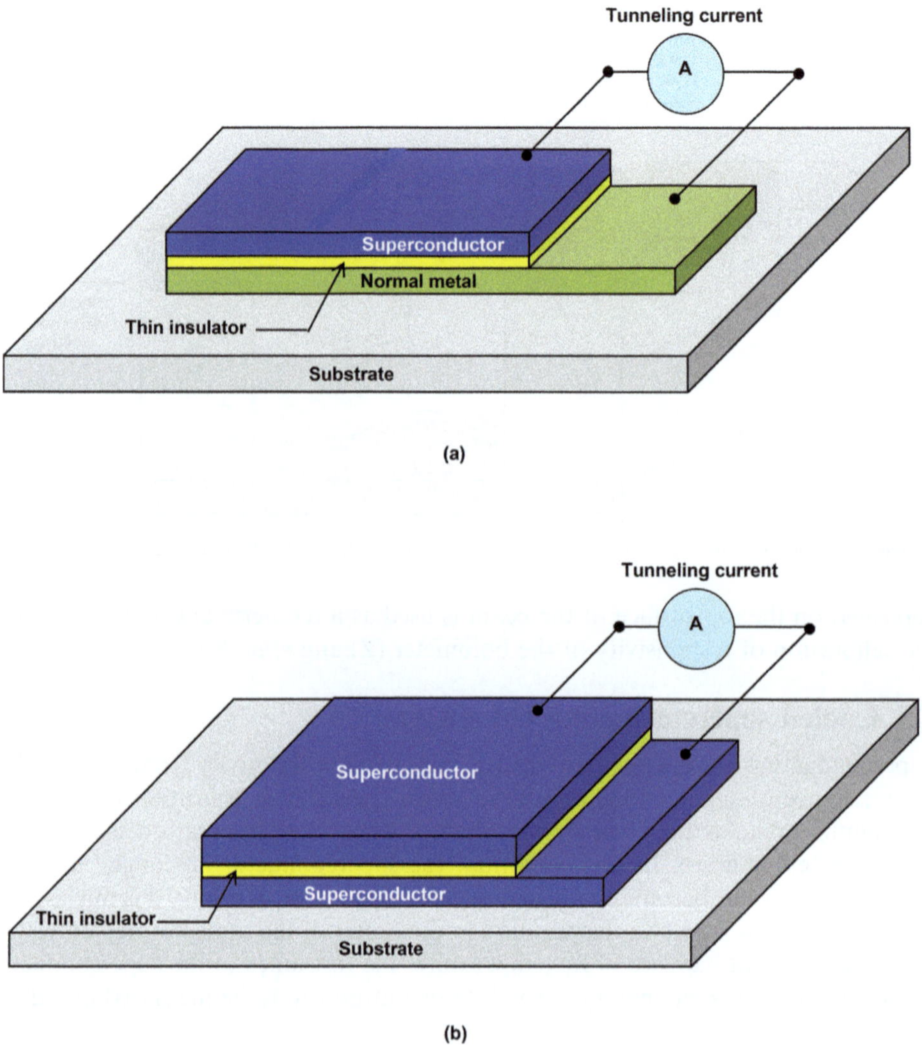

Figure 1.4. Tunnel junctions: (a) superconductor/insulator/normal metal (SIN) and (b) superconductor/insulator/superconductor (SIS).

measurements, and the temperature dependent Josephson current flowing through the bolometer is recorded with a SQUID galvanometer. During measurements with the SIN bolometer, it is biased at a potential difference less than the superconducting energy gap, and the temperature dependent tunneling current is measured with an FET amplifier.

Bolometers employing the superconductor–insulator–superconductor (SIS) tunnel junctions, e.g., niobium–AlO_x–niobium, nibium–AlN–niobium, niobium nitride–AlN–niobium nitride, can be operated up to a certain frequency determined by the superconducting energy gap. In a superconductor, the electrons near the

Fermi level pair into Cooper pairs. The Cooper pairs have a slightly lower energy leaving an energy gap above them which prevents backscattering. The energy gap is the energy required to break up a Cooper pair. Therefore, above the gap frequency, the noise temperature of the bolometer increases profoundly. These frequencies are ~ 0.68 THz for Nb, 1.16 THz for NbTiN and 1.35 THz for NbN films. Although SIS junction bolometers offer low noise and high sensitivity values at frequencies ~1 THz, new research avenues must be explored for higher frequency operation. At frequencies >1 THz, the available options include superconducting bolometers which operate independently of the energy gap. One such bolometer is the superconducting hot electron bolometer.

Example 1.1 The superconducting energy gap of niobium is 2.8 meV (Lodewijk *et al* 2009) and that of niobium nitride is 5.6 meV (Uzawa *et al* 2020). What are their gap frequencies?

Solution: From the formula

$$E = h\nu \tag{1.2}$$

we have

$$\nu = \frac{E}{h}. \tag{1.3}$$

Hence,

$$\begin{aligned}\nu_{\mathrm{Nb}} &= \frac{E_{\mathrm{Nb}}}{h} = \frac{2.8 \times 10^{-3}\mathrm{eV}}{4.135\,668 \times 10^{-15}\,\mathrm{eVs}} = \frac{2.8}{4.135\,668} \times 10^{-3+15} \\ &= 0.677 \times 10^{12}\mathrm{s}^{-1} \approx 0.68\ \mathrm{THz}\end{aligned} \tag{1.4}$$

and

$$\begin{aligned}\nu_{\mathrm{NbN}} &= \frac{E_{\mathrm{NbN}}}{h} = \frac{5.6 \times 10^{-3}\,\mathrm{eV}}{4.135\,668 \times 10^{-15}\,\mathrm{eVs}} = \frac{5.6}{4.135\,668} \times 10^{-3+15} \\ &= 1.3541 \times 10^{12}\mathrm{s}^{-1} \approx 1.35\ \mathrm{THz}\end{aligned} \tag{1.5}$$

Hence the gap frequencies are in the ratio of the superconducting energy gaps, as expected.

Example 1.2 The theoretical energy gap of niobium–titanium–nitride (NbTiN) superconducting film is 1.7 times that of Nb which is used up to 0.68 THz (Cecil *et al* 2007). What is the expected frequency of low loss circuits made with this film?

Solution:

$$\nu_{\mathrm{NbTiN}} = 1.7 \times \nu_{\mathrm{Nb}} = 1.7 \times 0.68 = 1.156\ \mathrm{TH\,z} \approx 1.16\ \mathrm{THz} \tag{1.6}$$

1.3.2 Superconducting hot electron bolometer

This bolometer applies two basic principles:

(i) Utilization of temperature dependence of a superconducting thin film near its critical temperature as a thermometer: The transition edge sensor is a superconducting thin film which is maintained at a temperature close to its critical temperature. A small change in temperature T produced by an incoming photon causes the superconductor to change into a normal metal and consequently its resistance R jumps from a small value to a large value resulting in a high dR/dT value. As the transition width, i.e., the temperature range over which this drastic resistance change occurs is only a small fraction of a Kelvin, small temperature changes can be detected and the function of an accurate thermometer is realized. However, the sensitivity of transition edge sensor is restricted only to transition region because the moment the superconductor changes to the normal metal state, the dR/dT value decreases and hence responsivity is lost.

(ii) Dependence of resistance of a very thin film on electron temperature: The thickness of Nb or NbN films is taken as ~10 nm. In these thin films, the mean free path of electrons is extremely small ~1–10 nm owing to intensified scattering at the surface. The electron–electron interaction increases but the electron–phonon interaction decreases. When the incoming photons are absorbed by electrons, the electron temperature at thermal equilibrium can differ from lattice temperature. This means that the electrons warm up to a temperature relative to the lattice temperature. These are hot electrons, and the bolometer in which the electrical resistance of the absorbing film depends on the electron temperature is called a hot-electron bolometer. Since only the electrons are heated, the thermal capacity of the absorbing film is very small leading to lowering of the thermal response time, i.e., the time in which the bolometer responds to changes in the power of incident terahertz waves. This is particularly true when the size of the device is of submicron dimensions. Thus the frequency restriction imposed by the energy gap frequency on an SIS superconducting bolometer no longer holds. From the viewpoint of cooling mechanism, there are two geometries of the hot-electron bolometers using electron diffusion-cooled and electron–phonon (substrate) cooling mechanisms (McGrath 1995).

Example 1.3 The electron–electron scattering time is 10^{-15} s. What is the frequency limit of the hot-electron bolometer?

Solution: The real high-frequency impedance of the hot-electron bolometer does not depend on the frequency from approximately the energy gap frequency up to a higher frequency governed by inverse electron–electron scattering time.

$$\begin{aligned}\text{Limiting frequency} &= \frac{1}{2\,\pi \times \text{Electron} - \text{electron scattering time}} \\ &= \frac{1}{2 \times 3.14159 \times 10^{-15}} = 159.155 \times 10^{12}\ \text{Hz} \approx 159\ \text{THz}\end{aligned} \tag{1.7}$$

1.3.3 Diffusion-cooled hot electron bolometer

The bolometer consists of a strip of superconductor of submicron length called the microbridge terminated on its two sides with contact regions made of non-superconducting or normal metal (figure 1.5). The hot electrons in this bolometer are cooled by diffusion from the microbridge to the normal metal contacts. Diffusion-aided electron–electron interaction is the preponderant mechanism responsible for cooling over electron–phonon interactions. This can happen only when the length L of the microbridge is kept below a certain defined value. Let us find this value. Consider the absorption of terahertz wave energy by an electron in the center of the microbridge. This electron can diffuse a distance $L/2$ to reach the left contact or a distance $L/2$ to reach the right contact. Diffusion length is the average distance traversed by the electron before it interacts with another electron and shares its

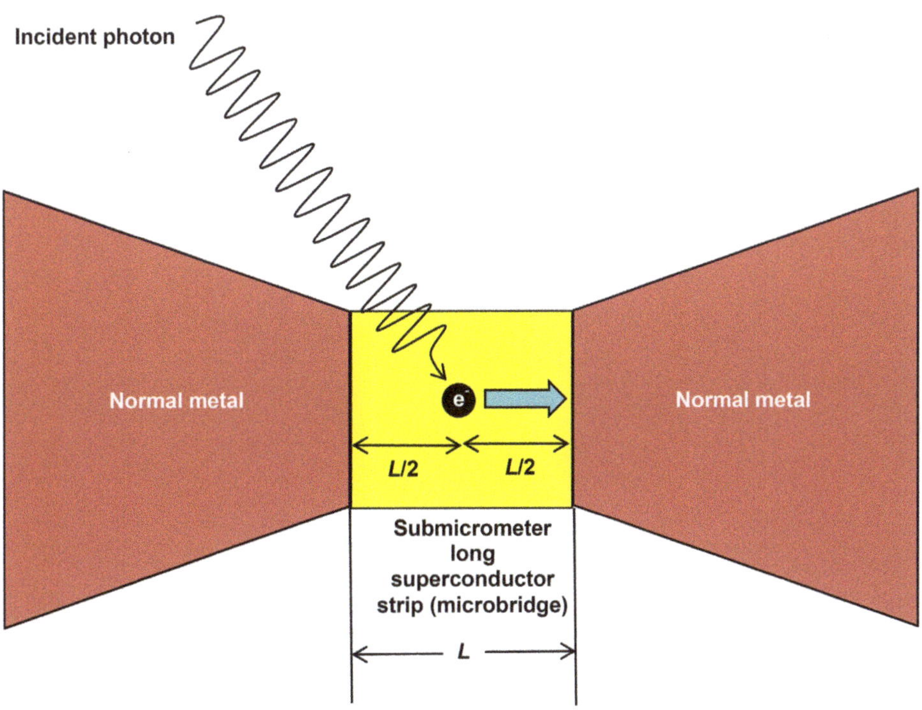

Figure 1.5. Diffusion-cooled hot electron bolometer.

energy with it. Let us denote the electron–electron interaction time by τ_{ee}. This is an inelastic interaction in which the energy of the incident electron is not conserved.

Diffusion is a random phenomenon by which the particles spread out from regions of high concentration to low concentration. The diffusion coefficient D is a constant in the Fick's first law equation

$$\text{Diffusion flux (number of particles m}^{-2}\,\text{s}^{-1}) = -\text{Diffusion coefficient} \times \text{Concentration gradient}\left(\frac{\text{number of particles m}^{-3}}{\text{m}}\right) \tag{1.8}$$

from which we get

$$\text{Unit of diffusion coefficient} = \frac{\text{number of particles m}^{-2}\,\text{s}^{-1}}{\text{number of particles m}^{-3}} \times \text{m} = \text{m}^2\,\text{s}^{-1} \tag{1.9}$$

From the underlying mathematical theory, we note that the diffusion coefficient for the diffusion of particles across a defined boundary is given by

$$\begin{aligned} D = {} & \text{Probability or transfer rate per unit time} \\ & \text{of a particle to move across the boundary with unit } [T]^{-1} \times \\ & \quad (\text{Average distance travelled by the particle in the time taken} \\ & \quad \text{to reach the boundary})^2 \end{aligned} \tag{1.10}$$

Here the time taken by the electron must be greater than τ_{ee} in order that the electron diffuses from superconducting absorbing microbridge region into the normal metal contact pad without interacting with another electron. So the minimum value of this time is τ_{ee} and the distance covered is $L/2$. Taking these time and distance values, we write

$$D = \frac{1}{\tau_{ee}} \times \left(\frac{L}{2}\right)^2 \tag{1.11}$$

or,

$$\left(\frac{L}{2}\right)^2 = D\tau_{ee} \tag{1.12}$$

or,

$$\frac{L}{2} = \sqrt{D\tau_{ee}} \tag{1.13}$$

or,

$$L = 2\sqrt{D\tau_{ee}} \tag{1.14}$$

The hot electron from the microbridge diffusing to normal metal contact is cooled in this region which serves as a heat sink. The electron–electron inelastic interaction time τ_{ee} is given by (Santhanam and Prober 1984)

$$\tau_{ee} = \frac{1}{13.4 \times 10^7 \times R_s T_C} \tag{1.15}$$

where R_s is the sheet resistance of the superconducting film and T_C is the transition temperature of the superconductor.

Equations (1.14) and (1.15) help us in the geometrical design of the bolometer because the appropriate length of the microbridge can be chosen to confine the bolometer in the hot electron regime.

Example 1.4 A diffusion-cooled hot electron bolometer is designed with niobium film of thickness 9.5 nm. The sheet resistance of the niobium film is 19 Ω sq^{-1}. The transition temperature of niobium thin film is found to be lowered with thickness (Kodama *et al* 1983). The transition temperature of Nb thin film is 5 K and transition width is 0.5 K (McGrath 1995). The diffusion coefficient of electrons in niobium is 1 $cm^2 s^{-1}$ (Dalrymple *et al* 1986). Calculate the maximum length of the superconductor microbridge.

Solution: The length of the superconductor microbridge separating the two normal metal contacts must be shorter than the thermal diffusion length given in equation (1.14)

$$L_{\text{Diffusion}} = 2\sqrt{D_e \tau_{ee}} \tag{1.16}$$

where D_e denotes the diffusion coefficient of electrons in the superconductor and τ_{ee} stands for the electron–electron inelastic interaction time.

From equation (1.15),

$$\begin{aligned} \tau_{ee} &= \frac{1}{13.4 \times 10^7 \times R_s T_C} = \frac{1}{13.4 \times 10^7 \times R_s T_C} = \frac{1}{13.4 \times 10^7 \times 19 \times 5} \\ &= \frac{10^{-7}}{13.4 \times 19 \times 5} = 7.855 \times 10^{-4} \times 10^{-7} = 7.855 \times 10^{-11}\ \text{s} \end{aligned} \tag{1.17}$$

Substituting the value of τ_{ee} obtained from equation (1.17) in equation (1.16)

$$\begin{aligned} L_{\text{Diffusion}} = 2\sqrt{D_e \tau_{ee}} &= 2\sqrt{1 \times 10^{-4} \times 7.855 \times 10^{-11}} \\ &= 2\sqrt{7.855 \times 10^{-15}} \\ &= 2 \times 8.8628 \times 10^{-8} = 1.7726 \times 10^{-7}\ \text{m} \\ &= \frac{1.773 \times 10^{-7}}{10^{-9}}\text{nm} = 177.3\ \text{nm} \approx 177\ \text{nm} \end{aligned} \tag{1.18}$$

Therefore, the length of superconductor microbridge must be smaller than 177 nm.

Example 1.5 In a Nb superconductor diffusion-cooled hot electron bolometer, the dimensions of the superconductor bridge are: length = 130 nm, width = 1 μm and thickness = 9.5 nm. Calculate the (a) electronic heat capacity, (b) thermal conductance of the superconductor and (c) the thermal response time. (d) Find the 3 dB roll-off frequency and comment on its application to terahertz detection.

The critical temperature of the superconductor is 5 K and the sheet resistance of the film is 19 Ω sq^{-1}.

Solution: (a) We shall apply the following equations of the free electron Fermi gas model of metals (Kittel 2005):

Electronic heat capacity is

$$C = \frac{\pi^2}{2} N k_B \frac{T}{T_F} \tag{1.19}$$

where N is the total number of free electrons, k_B is Boltzmann constant = 1.380 65 × 10^{-23} J K^{-1}, T is the temperature in Kelvin scale, and T_F is the Fermi temperature

$$T_F = \frac{E_F}{k_B} \tag{1.20}$$

The Fermi energy E_F is given by

$$E_F = \frac{\hbar^2}{2m}\left(\frac{3\pi^2 N}{V}\right)^{2/3} \tag{1.21}$$

where

$$\hbar = \frac{\text{Planck's constant}}{2\pi} = \frac{h}{2\pi} = 1.054\ 57 \times 10^{-34}\ \text{J. s} \tag{1.22}$$

$$m = \text{the mass of electron} = 9.109 \times 10^{-31}\ \text{kg} \tag{1.23}$$

$$\begin{aligned}\frac{N}{V} &= \text{Number density of electrons}\\ &= \frac{\text{Density }(\rho) \times \text{Avogadro's number }(N_A) \times \text{Number of electrons per atom }(N_V)}{\text{Atomic mass }(M_A)}\end{aligned} \tag{1.24}$$

For niobium,

$$\begin{aligned}\rho &= 8.57\ \text{g cm}^{-3} = \frac{8.57 \times 10^{-3}}{(10^{-2})^3} = 8.57 \times 10^{-3+6} \text{kg m}^{-3}\\ &= 8.57 \times 10^3\ \text{kg m}^{-3},\ N_V = 1,\ M_A\\ &= 92.906\ 38\ \text{g mol}^{-1} = 92.906\ 38 \times 10^{-3}\ \text{kg mol}^{-1}\end{aligned} \tag{1.25}$$

Niobium has atomic number 41. The electronic configuration of Nb is $1s^2\ 2s^2\ p^6\ 3s^2\ p^6\ d^{10}\ 4s^2\ p^6\ d^4\ 5s^1$, and valence electrons are $4d^4\ 5s^1$. We have taken the number of electrons in outermost shell =1 as the free electron number in this calculation.

Also,

$$N_A = 6.022 \times 10^{23}\ \text{mol}^{-1} \tag{1.26}$$

From equations (1.24)–(1.26)

$$\begin{aligned}\left|\frac{N}{V}\right|_{Nb} &= \frac{8.57 \times 10^3\ \text{kgm}^{-3} \times 6.022 \times 10^{23}\ \text{mol}^{-1} \times 1}{92.906\ 38 \times 10^{-3}\ \text{kg mol}^{-1}} = 0.555\ 49 \times 10^{3+23+3}\\ &= 5.56 \times 10^{28}\ \text{m}^{-3}\end{aligned} \tag{1.27}$$

Applying equations (1.21) and (1.27), we get

$$\begin{aligned} E_F\,|_{Nb} &= \frac{\hbar^2}{2m}\left(3\pi^2 \left|\frac{N}{V}\right|_{Nb}\right)^{2/3} \\ &= \frac{(1.054\,57 \times 10^{-34})^2}{2 \times 9.109 \times 10^{-31}}\{3(3.141\,59)^2 \times 5.56 \times 10^{28}\}^{2/3} \\ &= \frac{111.212 \times 10^{-70}}{18.218 \times 10^{-31}}\{164.6247 \times 10^{28}\}^{2/3} \\ &= 6.1045 \times 10^{-70+31} \times 1.394 \times 10^{20} = 8.5097 \times 10^{-70+31+20} \\ &= 8.5097 \times 10^{-19}\ \mathrm{J} = \frac{8.51 \times 10^{-19}}{1.602 \times 10^{-19}}\mathrm{eV} = 5.312\ \mathrm{eV} \end{aligned} \tag{1.28}$$

From equation (1.20),

$$\begin{aligned} T_F\,|_{Nb} &= \frac{E_F\,|_{Nb}}{k_B} = \frac{5.312\ \mathrm{eV}}{8.617\,33 \times 10^{-5}\ \mathrm{eV\ K}} = 0.616\,432\,2 \times 10^5\ \mathrm{K} \\ &= 61\,643\ \mathrm{K} \end{aligned} \tag{1.29}$$

Using equation (1.27) and putting the given dimensions of the superconductor microbridge we get

$$\begin{aligned} N\,|_{Nb} &= \left|\frac{N}{V}\right|_{Nb} \times V\,|_{Nb} \\ &= 5.56 \times 10^{28}\mathrm{m}^{-3} \times 9.5 \times 10^{-9}\ \mathrm{m} \times 130 \times 10^{-9}\ \mathrm{m} \times 1 \times 10^{-6}\ \mathrm{m} \\ &= 5.56 \times 9.5 \times 130 \times 10^{28-9-9-6} = 6866.6 \times 10^4 \\ &= 6.8666 \times 10^7 \approx 6.87 \times 10^7 \end{aligned} \tag{1.30}$$

By equations (1.19), (1.29), and (1.30)

$$\begin{aligned} C &= \frac{\pi^2}{2} N|_{Nb} k_B \frac{T_C|_{Nb}}{T_F|_{Nb}} \\ &= \frac{(3.14159)^2}{2} \times 6.87 \times 10^7 \times 1.380\,65 \times 10^{-23}\ \mathrm{J\ K^{-1}} \times \frac{5\ \mathrm{K}}{61\,643\ \mathrm{K}} \\ &= \frac{(3.14159)^2 \times 6.87 \times 1.380\,65 \times 10^{7-23}}{2 \times 61\,643}\mathrm{J\ K^{-1}} \\ &= 0.000\,759\,3 \times 10^{-16}\ \mathrm{J\ K^{-1}} = 7.593 \times 10^{-20}\ \mathrm{J\ K^{-1}} \end{aligned} \tag{1.31}$$

(b) Here the microbridge itself serves as the thermal link to the constant-temperature reservoir.

By Wiedmann–Franz law

$$\frac{\text{Thermal conductivity}}{\text{Electrical conductivity}} = \frac{\pi^2}{3}\left(\frac{k_B}{q}\right)^2 T \tag{1.32}$$

where k_B is Boltzmann constant, q is electronic charge and $T = T_C$, the critical temperature of superconductor.

Hence,

$$\text{Thermal conductivity of the microbridge} = \frac{\pi^2}{3}\left(\frac{k_B}{q}\right)^2 T_C \times \text{Electrical conductivity of the microbridge} \tag{1.33}$$

or,

$$\begin{aligned}&\text{Thermal conductance of the microbridge}\frac{\text{Length of the microbridge}}{\text{Its cross}-\text{sectional area}}\\ &= \frac{\pi^2}{3}\left(\frac{k_B}{q}\right)^2 T_C \times \frac{1}{\text{Electrical resistivity of the microbridge}}\\ &= \frac{\pi^2}{3}\left(\frac{k_B}{q}\right)^2 T_C\\ &\times \frac{1}{\text{Electrical resistance of the microbridge}\frac{\text{Cross}-\text{sectional area of the microbridge}}{\text{Length of the microbridge}}}\end{aligned} \tag{1.34}$$

or,

$$\text{Thermal conductance of the microbridge} = \frac{\pi^2}{3}\left(\frac{k_B}{q}\right)^2 T_C \times \frac{1}{\text{Electrical resistance of the microbridge}} \tag{1.35}$$

But

$$\begin{aligned}&\text{Electrical resistance of the microbridge} = \left(\frac{1}{12}\right)\text{Sheet resistance of superconducting film}\\ &\times \frac{\text{Length of the microbridge}}{\text{Width of the microbridge}} = \left(\frac{1}{12}\right) R_S \frac{L}{W}\end{aligned} \tag{1.36}$$

where the factor of 12 takes account of the spreading of the current symmetrically from both the ends of the microbridge. Combining equations (1.35) and (1.36) and letting G denote the thermal conductance, we get

$$G = \frac{\pi^2}{3}\left(\frac{k_B}{q}\right)^2 T_C \times \frac{12W}{R_S L} = 12\frac{\pi^2}{3}\left(\frac{k_B}{q}\right)^2 T_C \frac{W}{R_S L} \tag{1.37}$$

Putting the values of the variables in equation (1.37),

$$
\begin{aligned}
G &= 12\frac{\pi^2}{3}\left(\frac{k_B}{q}\right)^2 T_C \frac{W}{R_S L} \\
&= 12\frac{(3.14159)^2}{3}\left(\frac{1.38065\times10^{-23}\,\mathrm{JK^{-1}}}{1.602\times10^{-19}\,\mathrm{C}}\right)^2 5\mathrm{K}\frac{1\times10^{-6}}{19\times130\times10^{-9}} \\
&= \frac{12\times(3.14159)^2\times(1.38065\times10^{-23})^2\times5\times1\times10^{-6}}{3\times(1.602\times10^{-19})^2\times19\times130\times10^{-9}} \\
&\quad \frac{\mathrm{J^2\,K^{-2}\times K\times m}}{\mathrm{C^2\times\Omega\times m}} \\
&= \frac{12\times9.8696\times1.9062\times10^{-46}\times5\times1\times10^{-6}}{3\times2.5664\times10^{-38}\times19\times130\times10^{-9}} \\
&\quad \frac{(\mathrm{m^2\times kg\times s^{-2}})^2\mathrm{K^{-2}\times K\times m}}{\mathrm{C^2\times(kg\times m^2\times s^{-3}\times A^{-2})\times m}} \\
&= \frac{12\times9.8696\times1.9062\times10^{-46-6}\times5\times1}{3\times2.5664\times10^{-38-9}\times19\times130} \\
&\quad \frac{(\mathrm{m^2\times kg\times s^{-2}})^2\mathrm{K^{-2}\times K\times m}}{\mathrm{C^2}\times\left(\mathrm{kg\times m^2\times s^{-3}}\times\frac{\mathrm{C^{-2}}}{\mathrm{s^{-2}}}\right)\times\mathrm{m}} \\
&= 0.05936\times10^{-52+47}\frac{\mathrm{m^4\times kg^2\times s^{-4}\times K^{-2}\times K\times m\times s^{-2}}}{\mathrm{C^2\times kg\times m^2\times s^{-3}\times C^{-2}\times m}} \\
&= 0.05936\times10^{-5}\mathrm{m^2\times kg\times s^{-3}\times K^{-1}} \\
&= 5.936\times10^{-7}(\mathrm{m^2\times kg\times s^{-3}})\times\mathrm{K^{-1}} \\
&= 5.94\times10^{-7}\,\mathrm{WK^{-1}}
\end{aligned}
\tag{1.38}
$$

(c) The thermal response time is

$$
\begin{aligned}
\tau_{\text{Thermal}} &= \frac{C}{G} = \frac{7.593\times10^{-20}\,\mathrm{JK^{-1}}}{5.94\times10^{-7}\,\mathrm{WK^{-1}}} = \frac{7.593}{5.94}\times10^{-20+7}\frac{\mathrm{JK^{-1}}}{\mathrm{Js^{-1}K^{-1}}} \\
&= 1.2783\times10^{-13}\,\mathrm{s}
\end{aligned}
\tag{1.39}
$$

where equations (1.31) and (1.38) are applied.

(d) The increase in temperature ΔT of the bolometer is

$$
\Delta T = \frac{\eta P_0}{G\sqrt{1+\omega^2\tau^2_{\text{Thermal}}}}
\tag{1.40}
$$

where η is the absorbance of the radiation-sensitive film, P_0 is the amplitude of the modulated terahertz radiation, ω is the angular frequency of modulation, and G is the thermal conductance to the reservoir. At a frequency where $\omega\tau_{\text{Thermal}} = 1$, the thermal response ΔT becomes

$$\Delta T = \frac{\eta P_0}{G\sqrt{1+1}} = \frac{1}{\sqrt{2}}\frac{\eta P_0}{G} = 0.7071\frac{\eta P_0}{G} \quad (1.41)$$

which represents the 3 dB or roll-off frequency at which the output temperature increase has dropped to 70.71% of its maximum value. Hence,

$$\omega_{\text{Roll-off}}\tau_{\text{Thermal}} = 1 \quad (1.42)$$

or,

$$2\pi f_{\text{Roll-off}}\tau_{\text{Thermal}} = 1 \quad (1.43)$$

$$\begin{aligned} \therefore f_{\text{Roll-off}} &= \frac{1}{2\pi\tau_{\text{Thermal}}} = \frac{1}{2 \times 3.14159 \times 1.2783 \times 10^{-13}} \\ &= \frac{10^{13}}{2 \times 3.14159 \times 1.2783} = 0.1245 \times 10^{13}\ \text{Hz} \approx 1.2\ \text{THz} \end{aligned} \quad (1.44)$$

using equation (1.39)

The bolometer is capable of detection of terahertz waves by mixing the incoming high-frequency signal with a signal from a local oscillator (heterodyne detection).

1.3.4 Electron–phonon cooled hot-electron bolometer

A strip of superconductor, the microbridge, terminates at its two ends in superconducting contacts (figure 1.6). The film lies above a substrate, and the hot electrons in the film are cooled by electron–phonon interactions to the substrate instead of electron diffusion from the superconductor microbridge to normal metal contacts at the two ends.

The bolometer works as follows: Any photon falling on the superconducting film delivers its energy to an electron. In an interval of time τ_{ee}, the electron–electron interaction time, the energy received by the electron is shared with other electrons. The combined effect of the terahertz energy absorption and energy sharing amongst electrons produces a distribution of hot electrons in the superconducting thin film. The hot electron creates a phonon in time τ_{ep}. As the superconducting film is extremely thin, this phonon escapes without any collision, i.e., ballistically, to the substrate in a time τ_{escape}. In order for the device to operate as a hot-electron bolometer, it is necessary that:

(i) The electron–electron interaction time τ_{ee} must be much smaller than the time τ_{ep} required for production of a phonon. Then only the electrons can warm up from the received energy. This condition is fulfilled when a thin film is used at temperatures below 10 K.

(ii) The escape time τ_{escape} of the phonon must be much smaller than the phonon–electron interaction time τ_{pe} in order that the phonon escapes to the substrate without undergoing any interaction with other electrons. This happens when the film is very thin. Further, the substrate material should have a high thermal conductivity so that it works as constant-temperature bath.

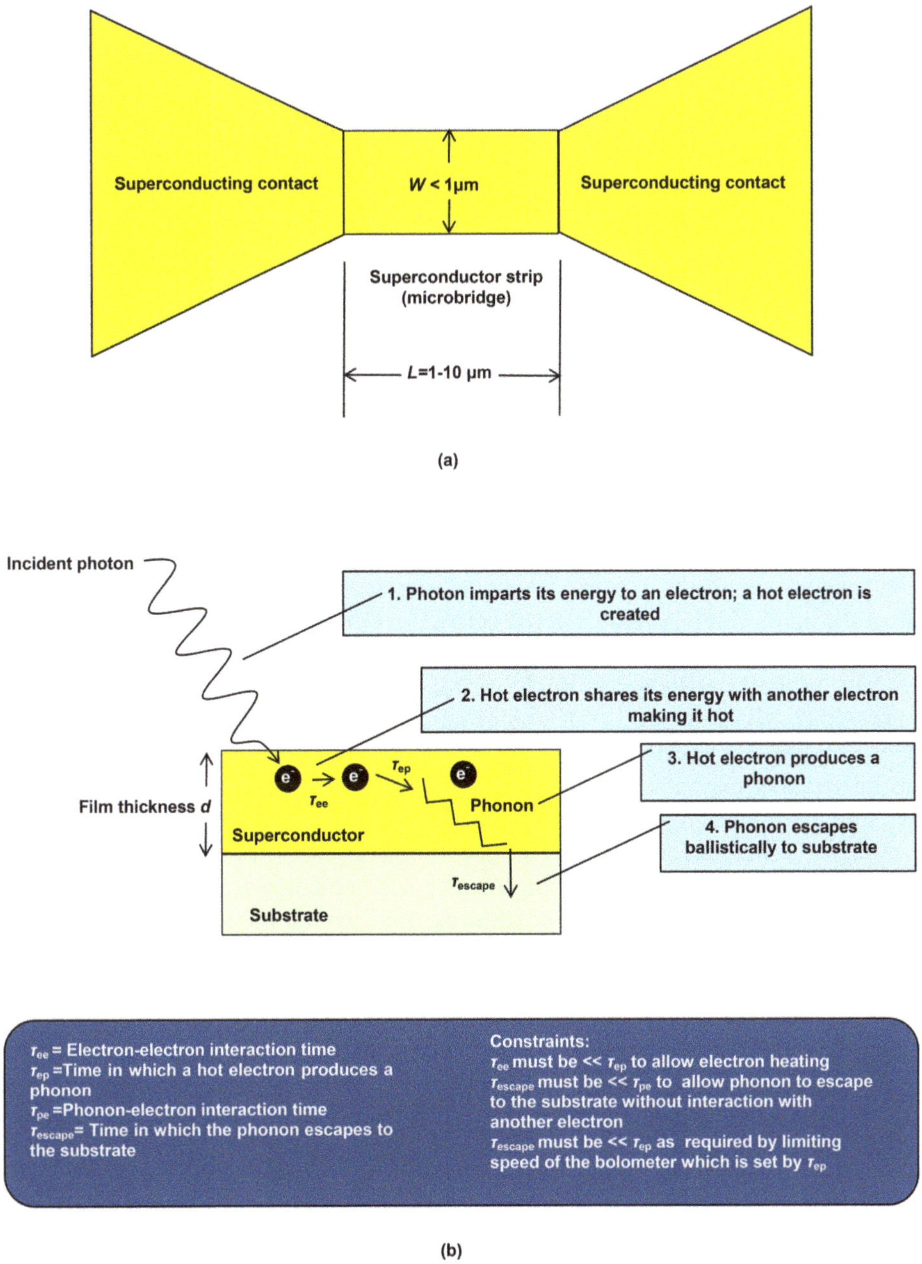

Figure 1.6. Electron–phonon-cooled hot electron bolometer.

1.4 Golay cell

Invented by M J E Golay in 1947 (Golay 1947a, 1947b), this cell can detect electromagnetic radiation in a broad range of wavelengths, typically 5–8000 μm (frequency ~ 6×10^{13}–3.75×10^{10} Hz).

It is a photo-acoustic detector working on pneumatic control. The photo-acoustic or opto-acoustic effect is the phenomenon involving the production of sound waves by a material on absorption of light. The sound is a mechanical disturbance in the form of vibrations. It propagates through a material medium, and can be converted by a transducer such as microphone into an electrical signal. So, the photo-acoustic effect produces mechanical vibrations of a medium. Pneumatics is concerned with the use of pressurized air or gas, i.e., air or gas under pressure. So let us see the constructional details of the Golay cell and examine how it works.

The Golay cell consists of an enclosure filled with an inert gas such as xenon (figure 1.7). On one side of this sealed enclosure is a window made of a material through which the terahertz waves can enter the enclosure. This material could be diamond. High-density polyethylene (HDPE) is sometimes used. This material is a thermoplastic polymer. A flexible diaphragm or membrane is fixed across the opposite side of the enclosure. The outer surface of this membrane is carefully mirrored. The mirroring ensures that any light falling on it is reflected back. Inside the enclosure and at its center is a semitransparent film. It is this film which absorbs the energy in the incident radiation.

When performing the experiment, light from an LED is focused on the mirrored side of the diaphragm. A photoelectric cell detects the light reflected from the surface of the mirror.

When terahertz radiation is made to fall on the window, it passes through the window and strikes the semi-transparent film. The film absorbs the energy from the waves. Consequent upon the absorption of energy, the gas in the enclosure is heated. The heated gas expands and so has a greater volume. The pressure inside the enclosure rises and the increased gas pressure distorts the diaphragm. As the diaphragm undergoes distortion in accordance with variations in strength of terahertz waves incident on the window, the light reflected from the mirror on the diaphragm varies in the same proportion. So the varying intensities of illumination of the photoelectric cell replicate the instantaneous shapes of the mirror and

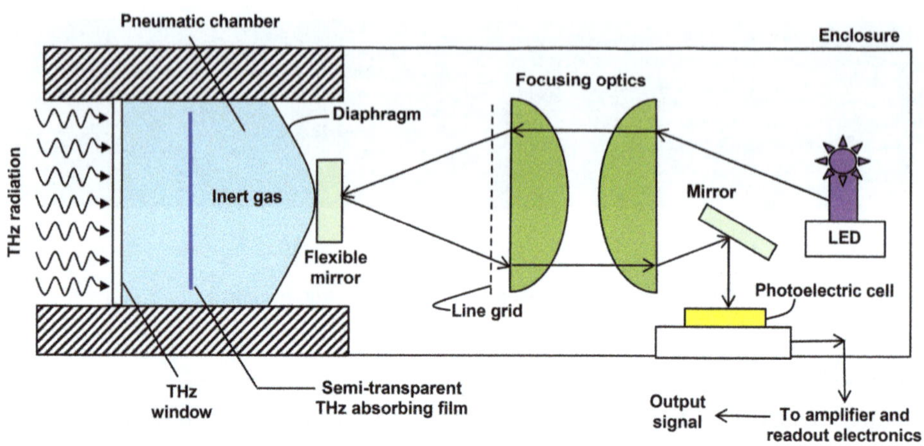

Figure 1.7. Sketch of the construction and optical arrangement of a Golay cell.

therefore provide a display of the changes in terahertz radiation. A preamplifier circuit consisting of an operational amplifier and field-effect transistors converts the intensity variations into a voltage waveform. The instantaneous output voltage of the photodiode is proportional to the displacement of the mirrored outer surface of the diaphragm. A reference source of known power is used to calibrate the instrument in Volts/Watt of incident power. Then the incident terahertz power can be determined from the graph generated from terahertz radiation.

It is noted that in the above experiment, we could record the vibrations of the diaphragm produced by the expansion of gas caused by the absorption of electromagnetic waves and resultant heat generation. So we can say that an optical microphone has been implemented.

It is easy to see that any type of vibration of the platform on which the Golay cell is mounted will interfere with the vibrations produced in the diaphragm by terahertz radiation and thereby vitiate the measurements. Looking at the detrimental influence of platform vibrations on Golay cell performance, it is mandatory that the Golay cell be stationed on a vibration-isolated base. The base provides a facility for adjusting the position of the Golay cell to receive the incoming radiation. The essential precondition of a vibration-free platform along with bulky structure of the Golay cell and its exorbitant cost sometimes make it inconvenient for field measurements. Other disadvantages are its fragility and slow response with typical rise time of 25 ms. Nonetheless, it is a reliable detector used in astronomical observations. It requires no cooling and offers the advantage of reliable operation at room temperature.

1.5 Pyroelectric detector

The pyroelectric detector is a thermal sensor. It caters to an extensive spectral range covering a major chunk of the electromagnetic spectrum inclusive of the terahertz region. Like the Golay cell, it works at room temperature. Let us first understand pyroelectricity and related terms.

1.5.1 Pyroelectricity

Pyroelectricity (Pyr = fire) is the electricity produced in certain non-centrosymmetric crystals, i.e., crystals that do not have a center of symmetry. Examples of crystals showing pyroelectricity are tourmaline, gallium nitride, lithium sulfate monohydrate, polyvinylidene fluoride (PVDF), lithium tantalate crystal ($LiTaO_3$), barium titanate ($BaTiO_3$), caesium nitrate ($CsNO_3$) and cobalt phthalocyanine ($C_{32}H_{16}CoN_8$). Pyroelectricity is observed when the crystals are either heated or cooled.

These non-centrosymmetric crystals are naturally electrically polarized because they contain a dipole in each unit cell and the molecules with fixed dipole moment are aligned to an appreciable degree. Hence they are said to exhibit spontaneous polarization. The crystals themselves are referred to as polar crystals. Any dielectric crystal will show polarization under the action of an applied electric field but that does not necessarily imply that it is polar. A polar crystal will show polarization even when there is no applied field.

A polar crystal has a polar axis, which is the direction that is not transformed in the opposite direction when any symmetry operation is performed on the crystal.

1.5.2 Pyroelectric current

If metal contacts are made on two opposite faces of a pyroelectric crystal along planes orthogonal to its polar axis, and the temperature of the crystal is continuously varied, a potential difference is recorded between the metal contacts. This potential difference causes the flow of a current across the crystal faces, which is the pyroelectric current. The events leading to this current flow are: firstly, change in temperature; secondly, change in polarization and hence surface charge density; thirdly, voltage creation and fourthly, current flow. The pyroelectric current i_p is given by

$$i_{\mathrm{P}} = \Gamma A \frac{d\Delta T}{dt} \tag{1.45}$$

where A is the area of the crystal and Γ is the pyroelectric coefficient:

$$\Gamma = \frac{\text{Change in spontaneous polarization}}{\text{Change in temperature}} = \frac{\partial P_{\mathrm{s}}}{\partial T} \tag{1.46}$$

The behavior of these crystals under different conditions is examined in figure 1.8.

1.5.3 Impermanency of pyroelectric voltage

The voltage produced across the crystal faces is purely temporary. If the change in temperature of the crystal is stopped, the charge leaks away either due to the electrons flowing in the crystal or through the ions in the air or through the contacts. Therefore, for continuous current flow, the variation $\Delta T/\Delta t$ of temperature T with time t called temporal temperature gradient must be maintained.

Note that under normal unheated/uncooled conditions, the built-in or permanent polarization of a pyroelectric crystal does not produce any potential difference across the faces of the crystal because any free charge carriers migrating to the surfaces of the crystal as well as those in the surroundings neutralize this potential difference. However, on altering the temperature of the crystal, the built-in polarization changes. Consequently pyroelectricity is manifested.

1.5.4 Difference between pyroelectricity and thermoelectricity

Pyroelectricity must not be confused with thermoelectricity where one end of the device such as a thermocouple is heated (hot junction) and the other end is cooled (cold junction), i.e., thermoelectricity is a consequence of a spatial temperature gradient $\Delta T/\Delta l$ from the hot to cold junction along the length l of the device. Pyroelectricity does not require any spatial thermal gradient because it is concerned with heating or cooling of the whole crystal, meaning that the temperature is the same everywhere in the crystal. Pyroelectricity disappears when the temperature variation of the crystal with time ceases ($\Delta T/\Delta t = 0$) while thermoelectricity vanishes when the temperature difference with respect to space = temperatures of the hot and cold junctions are equalized ($\Delta T/\Delta l = 0$).

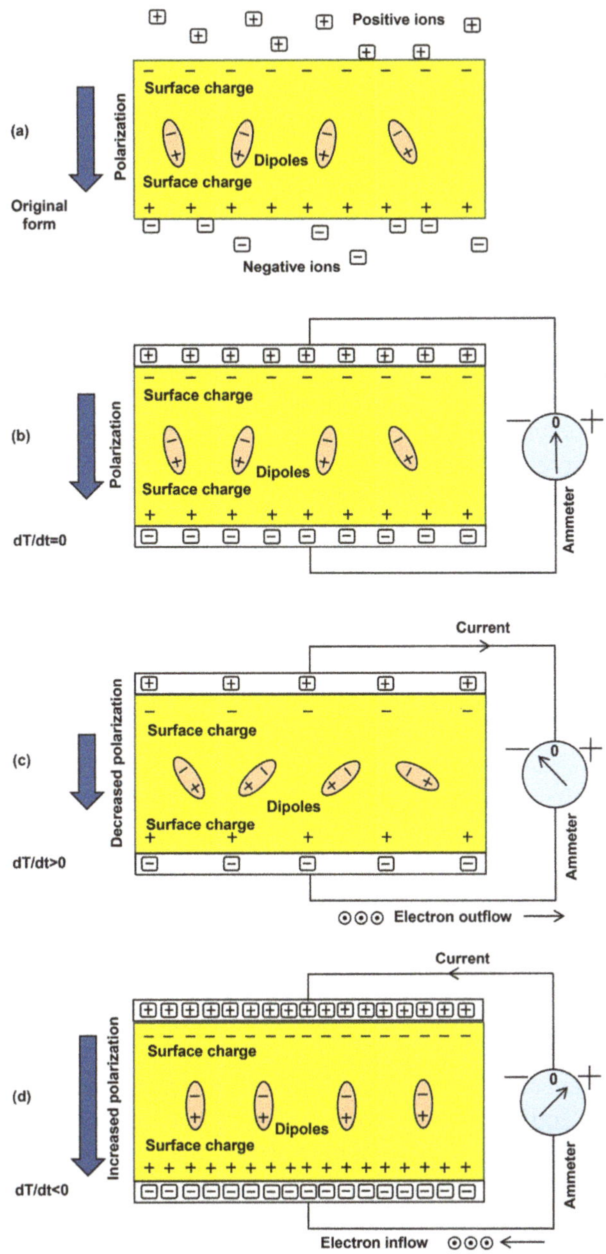

Figure 1.8. Behavior of pyroelectric material: (a) in original form without electrodes when the dipole moments sum up to show spontaneous polarization with ionic charges from surroundings attracted towards surfaces, (b) in original form with electrodes where the temperature is maintained constant, so there is no current flow, (c) subjected to heating gradient when the spontaneous polarization decreases due to disturbance of dipole alignment resulting in current flow in negative direction, and (d) exposed to cooling gradient when the spontaneous polarization increases due to proper alignment of dipoles resulting in current flow in a positive direction.

1.5.5 Pyroelectric, piezoelectric and piezocaloric effects

Pyroelectric effect must also be distinguished from piezoelectric and piezocaloric effects. While pyroelectric effect deals with the generation of electric current by uniform heating or cooling, piezoelectric effect is related to production of electric current by an applied stress and piezocaloric effect is production of heat from pressure or stress.

1.5.6 Ferroelectric materials

There is a specific class of materials known as ferroelectrics which exhibit spontaneous polarization and in which the polarization can be reversed by changing the direction of the applied electric field. This implies that the direction of polar axis can be changed by applying an external field. Not all but some of the pyroelectric materials show ferroelectric nature, e.g., tourmaline, a pyroelectric material is not ferroelectric. $LiTaO_3$ is both pyroelectric and ferroelectric. So are $BaTiO_3$ and $PbTiO_3$. All ferroelectric materials are also pyroelectric. So ferroelectrics constitute a subcategory of pyroelectrics. All pyroelectric materials are piezoelectric but the converse is untrue, e.g., quartz, a piezoelectric, is not pyroelectric. Clearly, ferroelectrics are pyroelectric as well as piezoelectric.

1.5.7 Crystal classes and crystal behavior

As we know, external symmetry of crystals is defined by 32 possible combinations of symmetry operations resulting in a total of 32 classes of crystals. Twenty one of these 32 crystal classes are non-centrosymmetric. Direct piezoelectricity (electric charge generation by mechanical stress) is found in 20 of these 21 classes. Ten classes of the remaining 20 classes are polar in nature. These are the classes, which display pyroelectricity. All the polar classes provide pyroelectricity. As already stated, not all pyroelectrics are ferroelectrics.

1.5.8 Function of the pyroelectric detector

The pyroelectric detector is fabricated in a capacitor-like structure (figure 1.9). It consists of a pyroelectric crystal sandwiched between two metal contacts. One of the metal contacts is coated with a black film to absorb the incoming terahertz radiation. A commonly used crystal is $LiTaO_3$, which is ferroelectric also. The crystal is poled by applying a high intensity electric field whereby all the dipoles are properly aligned in one direction providing a strong polarization.

When the terahertz radiation falls on the blackend face of the detector, it is absorbed by the black film. The consequent heating of the film raises the temperature of the $LiTaO_3$ pyroelectric layer so that its polarization changes. The charge developed on the electrodes is coupled out to the external circuit. As the current produced is very small, it needs to be amplified. Therefore, the detected current signal is fed to a current or voltage amplifier and readout circuit. The detector is calibrated in μA/Watt or Volts/Watt or Volts/Joule.

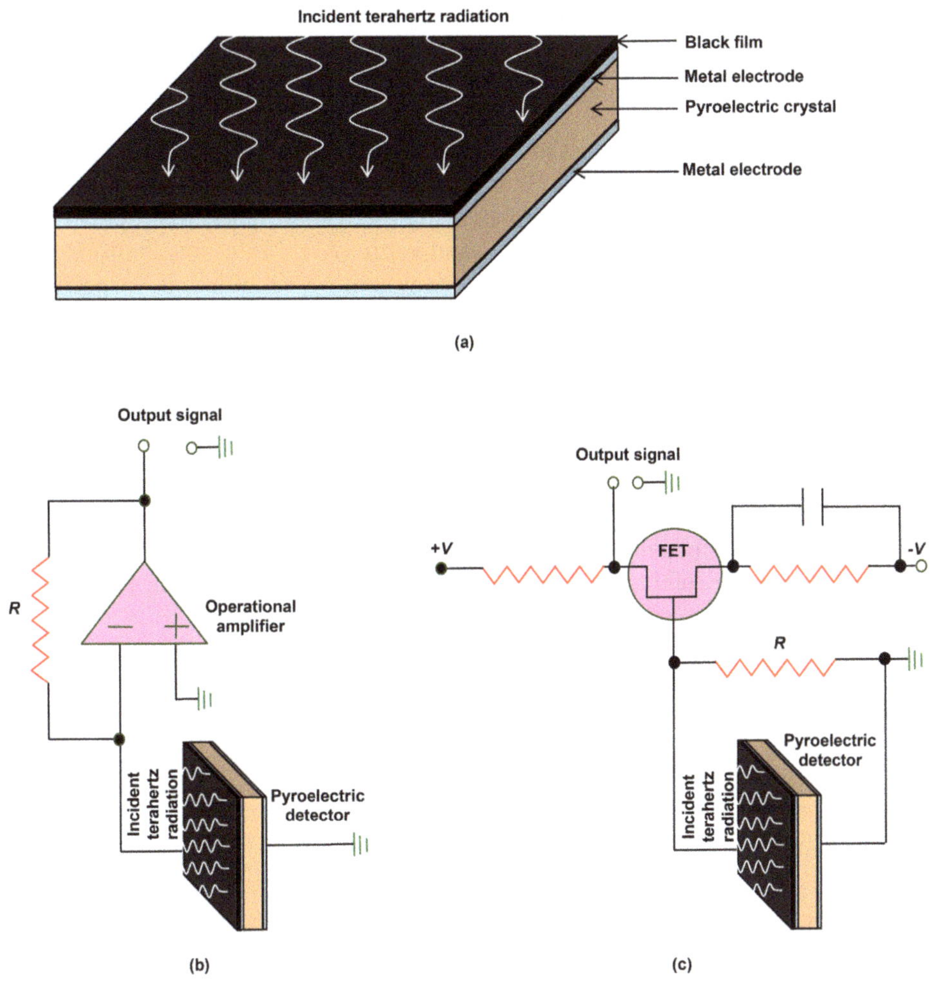

Figure 1.9. Pyroelectric detector: (a) schematic diagram, (b) coupled to a current amplifier and (c) coupled to a voltage amplifier.

1.6 Discussion and conclusions

To compare the performance characteristics of thermal detectors, let us define the chief metrics serving as a yardstick for ranking:

(i) Responsivity of a detector is the ratio of electrical output produced in volts or amperes to the input power in watts. It measures the input–output gain of a detector system:

(ia) Optical responsivity refers to the responsivity calculated using the power incident on the detector.

(ib) Electrical responsivity is determined by using the power absorbed by the detector.

The power absorbed by the detector is less than the power incident upon it due to losses incurred in optical coupling of the radiation source to the detector. So, when considering electrical responsitivity, the input power is smaller by an amount equal to the losses. Hence electrical responsivity is higher than optical responsivity depending on the optical coupling efficiency.

(ii) Noise equivalent power of a detector, acronym NEP, is defined as the minimum optical input power that produces a signal-to-noise ratio = 1 for a bandwidth = 1 Hz at the output. It is the optical power under the condition of equality of signal and noise levels representing the frailest detectable optical signal power. Further, note that an output bandwidth of 1 Hz corresponds to 0.5 s integration time. When the measurement bandwidth is included, it is specified in watts per square root of Hertz ($W/Hz^{1/2}$). When it is not included, then the specification is done in watts only.

The NEP gives an appraisal of the sensitivity of the detector. A smaller NEP value indicates a more sensitive detector, e.g., a detector of NEP = 1×10^{-14} W $Hz^{-1/2}$ is more sensitive than a detector of NEP = 1×10^{-12} W $Hz^{-1/2}$ because the former can detect a signal power of 0.01 pW with SNR = 1 after 0.5 s averaging while the latter detects only 1 pW at the same SNR and averaging time values. The inverse of NEP is detectivity.

(iia) Optical NEP refers to the power incident on the detector, i.e., power at the detector input.

(iib) Electrical NEP pertains to the power absorbed by the detector, i.e., the power at detector output.

Electrical NEP equals the optical NEP multiplied by the optical coupling efficiency of the detector.

Mathematically, NEP can be defined by the equation

$$\mathrm{NEP} = \frac{HA}{\frac{V_{\mathrm{Signal}}}{V_{\mathrm{Noise}}}(\Delta\nu)^{1/2}} = \frac{HAV_{\mathrm{Noise}}}{V_{\mathrm{Signal}}(\Delta\nu)^{1/2}} \tag{1.47}$$

where H is the irradiance (incident power per unit area) in Wm^{-2}, A is the area of the detector, $\Delta\nu$ is the measurement bandwidth, V_{Signal} is root mean square signal voltage, and V_{Noise} is root mean square noise voltage.

(iii) Response time of a detector is the time taken by the detector output signal (current or voltage) to reach 63.2% of its final equilibrium value. It equals the time required to reach the value {1– (1/e)} final equilibrium value where e is the base of the natural logarithm.

Comparison charts showing the performance metrics, and relative merits and demerits of the different thermal detectors are compiled in tables 1.1–1.4.

Example 1.6 A Golay cell window has a window of diameter 6 mm. The NEP is 3×10^{-10} W $Hz^{-1/2}$. The bandwidth is 10 Hz. The ratio of noise voltage to signal voltage is 10^{-5}. Find the power of radiation incident per unit area on the cell?

Solution: Here, area of detector is

$$A = \pi\frac{d^2}{4} = 3.141\,59\frac{(6 \times 10^{-3})^2}{4} = 2.8274 \times 10^{-5}\,\mathrm{m}^2 \tag{1.48}$$

Table 1.1. Bolometers, Golay cells and pyroelectric detectors.

Name of detector	Physical property	Voltage responsivity (V/W)	NEP	Reference
Uncooled VO_x Bolometer	Electrical resistance	4.97×10^3	$1–3 \times 10^{-10}$ W $Hz^{-1/2}$	Rogalski and Sizov (2011)
		9.32×10^5	2.44×10^{-10} W	Yeh *et al* (2020)
Hot Electron Bolometer at 1.5 K		1.6×10^5 (Optical) 6×10^6 (Electrical)	4.4×10^{-14} W $Hz^{-1/2}$ (Optical) 1.2×10^{-14} W $Hz^{-1/2}$ (Electrical)	Han *et al* (2013)
			3×10^{-19} W $Hz^{-1/2}$	Karasik and Cantor (2011)
Golay cell (Room temperature)	Gas volume	$0.1–45 \times 10^3$	$2–4 \times 10^{-10}$ W $Hz^{-1/2}$	Rogalski and Sizov (2011)
Pyroelectric (Room temperature) detector	Electrical polarization	150×10^3	4×10^{-10} W $Hz^{-1/2}$	Rogalski and Sizov (2011)

Table 1.2. Microbolometer.

Advantages	Gives flat terahertz response over a wide frequency range; uncooled versions are small, light in weight, consume less power and are less expensive than cryogenically cooled versions; long life; fast response time ~10 ms; response times of hot electron bolometers are in the range of picoseconds to nanoseconds
Disadvantages	Cooled versions are costly and consume more power

Table 1.3. Golay cell.

Advantages	Broad spectral response, operates at room temperature
Disadvantages	Less sensitive than microbolometer, modest response time ~25 ms, fragile, vulnerable to shock and vibration, large cumbersome housing, expensive

Table 1.4. Pyroelectric detector.

Advantages	Wide spectral response, compact, rugged, inexpensive, works at room temperature, does not require any window like Golay cell
Disadvantages	Less sensitive than Golay cell, long response time ~150 ms

Also, NEP = 3×10^{-10} W Hz$^{-1/2}$, $\Delta\nu = 10$ Hz, $V_{\text{Noise}}/V_{\text{Signal}} = 10^{-5}$. From equation (1.47)

$$\text{NEP} = \frac{HAV_{\text{Noise}}}{V_{\text{Signal}}(\Delta\nu)^{1/2}} \tag{1.49}$$

Rearranging and putting the values

$$H = \frac{\text{NEP} \times V_{\text{Signal}}(\Delta\nu)^{1/2}}{AV_{\text{Noise}}} = \frac{3 \times 10^{-10}(10)^{1/2}}{2.8274 \times 10^{-5} \times 10^{-5}}$$
$$= \frac{3}{2.8274} \times 10^{-10+5+5+1/2} = 3.355 \text{ W m}^{-2} \tag{1.50}$$

Example 1.7 What is the power of terahertz radiation per unit area for an SNR = 1 if the radiation has a bandwidth of 20 kHz in a pyroelectric detector of NEP 4×10^{-10} W Hz$^{-1/2}$?

Solution: The power per unit area (H) is given by equation (1.50)

$$H = \frac{\text{NEP} \times V_{\text{Signal}}(\Delta\nu)^{1/2}}{AV_{\text{Noise}}} = \frac{4 \times 10^{-10} \times (20 \times 10^3)^{1/2}}{1}$$
$$= 4 \times 10^{-10} \times (20 \times 10^3)^{1/2} = 565.685 \times 10^{-10} \text{ W m}^{-2}$$
$$= 5.66 \times 10^{-8} \text{ W m}^{-2} \tag{1.51}$$

References

Cecil T W, Weikle R M, Kerr A R and Lichtenberger A W 2007 Investigation of NbTiN thin films and AlN tunnel barriers with ellipsometry for superconducting device applications *IEEE Trans. Appl. Supercond.* **17** 3525–8

Clarke J, Hoffer G I and Richards P L 1974 Superconducting tunnel junction bolometers *Rev. Phys. Appl.* **9** 69–71

Dalrymple B J, Wolf S A, Ehrlich A C and Gillespie D J 1986 Inelastic electron lifetime in niobium films *Phys. Rev.* **B 33** 7514–9

Golay M J E 1947a Radiation detecting device *US Patent* 2,557,096, patented June 19, 1951, pp 1–16

Golay M J 1947b Theoretical consideration in heat and infrared detection, with particular reference to the pneumatic detector *Rev. Sci. Instrum.* **18** 347–56

Gou J, Wang J, Zheng X, Gu D, Yu H and Jiang Y 2015 Detection of terahertz radiation from 2.52 THz CO_2 laser using a 320 × 240 vanadium oxide microbolometer focal plane array *RSC Adv.* **5** 84252–6

Gou J, Niu Q, Liang K, Wang J and Jiang Y 2018 Frequency modulation and absorption improvement of THz micro-bolometer with micro-bridge structure by spiral-type antennas *Nanoscale Res. Lett.* **13** 1–9

Han Q, Gao T, Zhang R, Chen Y, Chen J, Liu G, Zhang Y, Liu Z, Wu X and Yu D 2013 Highly sensitive hot electron bolometer based on disordered graphene *Sci. Rep.* **3** 1–6

Kašalynas Š A, Adamonis J, Kornijcuk V, Adomavičius R, Krotkus A, Valušis G, Sešek A, Pavlovic L and Trontelj J 2013 Performance of the antenna coupled microbolometers characterized by the quasi-optical measurements at frequencies 0.1–1.0 THz *2013 38th International Conference on Infrared, Millimeter, and Terahertz Waves (IRMMW-THz), Mainz, Germany, 1–6 September, 2013* (New York: IEEE) pp 1–2

Kašalynas I *et al* 2016 Spectroscopic terahertz imaging at room temperature employing microbolometer terahertz sensors and its application to the study of carcinoma tissues *Sensors* **16** 1–15

Karasik B S and Cantor R 2011 Demonstration of high optical sensitivity in far-infrared hot-electron bolometer *Appl. Phys. Lett.* **98** 193503-1–3

Kittel C 2005 *Introduction to Solid-State Physics* 8th edn (Hoboken, NJ: Wiley) ch 6 131–59

Kodama J, Itoh M and Hirai H 1983 Superconducting transition temperature versus thickness of Nb film on various substrates *J. Appl. Phys.* **54** 4050

Lodewijk C F J, Zijlstra T, Zhu S, Mena F P, Baryshev A M and Klapwijk T M 2009 Bandwidth limitations of Nb/AlN/Nb SIS mixers around 700 GHz *IEEE Trans. Appl. Supercond.* **19** 395–9

McGrath W R 1995 Hot-electron bolometer mixers for submillimeter wavelengths: An overview of recent developments *Symp. Proc. Sixth Int. Symp. on Space Terahertz Technology (Pasadena, CA, March 21–23)* (Piscataway, NJ: IEEE) 216–28 pp

Rogalski A and Sizov F 2011 Terahertz detectors and focal plane arrays *Opto-Electron. Rev.* **19** 346–404

Santhanam P and Prober D E 1984 Ineleastic electron scatterng mechanisms in aluminum films *Phys. Rev.* B **29** 3733–66

Trontelj J, Valušis G, Venckevičius R, Kašalynas I, Sešek A and Švigelj A 2014 A high performance room temperature THz sensor *Proc. SPIE* **9199** 91990K

Uzawa Y, Saito S, Qiu W, Makise K, Kojima T and Wang Z 2020 Optical and tunneling studies of energy gap in superconducting niobium nitride films *J. Low Temp. Phys.* **199** 143–8

Yeh T-H, Tsai C-K, Chu S-Y, Lee H-Y and Lee C-T 2020 Performance improvement of Y-doped Vox microbolometers with nanomesh antireflection layer *Opt. Express* **28** 6433–42

Zhang Y, Hosono S, Nagai N, Song S-H and Hirakaw K 2019 Fast and sensitive bolometric terahertz detection at room temperature through thermomechanical transduction *J. Appl. Phys.* **125** 151602-1–6

Chapter 2

Optical rectification and electrooptic effect-based components

Differences between linear and non-linear optics are discussed. Non-linear optics deals with the behavior of light in media in which the polarization is not proportional to the applied electric field, a phenomenon commonly observed with high intensity beams such as lasers. The equation for propagation of electromagnetic waves in linear medium is modified for a nonlinear medium by expanding polarization to include second-order and third-order susceptibilities. By examining the second-order responses of a non-linear medium to two-electromagnetic wave input, single electromagnetic wave input and ultrashort single electromagnetic wave input, the phenomena of optical rectification, second harmonic generation, sum- and difference-frequency mixing, and terahertz pulse emission are explained. Two types of electrooptic effects, linear and quadratic, are defined and application of the linear effect for terahertz detection is described.

Non-linearity is a situation in which the change in effect (dependent variable or output) is not proportional to the change in cause (independent variable or input). When it happens with light, the consequences are very unusual and unexpected, as we shall see in this chapter.

2.1 Non-linear optics

Subpicosecond duration terahertz frequency pulses are widely used in time-domain THz spectroscopy and imaging (Wilke and Sengupta 2008). Generation of these terahertz pulses employs optical rectification (OR) technique. Detection of pulses utilizes the electrooptic effect. Both OR and electrooptic effect are phenomena falling under non-linear optics (Laud 1992, Boyd 2020). Most readers are aware of linear optics. For those who are new to non-linear optics, we begin with a brief review of non-linear optics.

doi:10.1088/978-0-7503-4886-7ch2

2.2 Linear versus non-linear optics

Informally, linear optics is the sub-branch of optics dealing with weak light sources, whereas non-linear optics is concerned with intense light sources. Formally, by intense light sources we mean lasers.

Common applications of lenses, mirrors and diffraction gratings come under linear optics. Properties of linear optical materials are:

(i) Basic optical properties of materials such as refractive index and hence velocity of light do not depend on the intensity of light.
(ii) The principle of superimposition is valid when several light beams are propagating through the material.
(iii) Frequency of light does not change after passing through the medium. If a blue beam of light strikes a material, the beam emerging out from the material is also blue, i.e., the color of the beam does not change.
(iv) Any two beams of light travelling through a given region of material do not exert any influence on each other. Light cannot be used to control light.

For a non-linear optical material:

(i) The refractive index of the material and therefore velocity of light in it can vary with intensity of the light beam.
(ii) The principle of superimposition does not apply if several light beams are travelling through the material.
(iii) The frequency of light can change after passing through the material so that the light beam coming out from the material may be of a different color from the one that entered the material, e.g., infrared (IR) light can become green when passed through a non-linear optical material.
(iv) Light can be used to control light by virtue of interaction between photons of two different beams.

Many exotic and immensely useful phenomena are witnessed in non-linear optics.

2.3 Electromagnetic wave propagation in linear optical media

2.3.1 Maxwell's equations

Four coupled partial differential equations called Maxwell's equations govern the propagation of electromagnetic waves, and lay down the foundation of classical electrodynamics

$$\text{Faraday's law of induction: } \nabla \times \mathbf{E} + \frac{\partial \mathbf{B}}{\partial t} = 0 \tag{2.1}$$

$$\text{Ampere's circuital law: } \nabla \times \mathbf{H} - \frac{\partial \mathbf{D}}{\partial t} = \mathbf{J} \tag{2.2}$$

$$\text{Gauss law of electricity: } \nabla . \mathbf{D} = \rho \tag{2.3}$$

and

$$\text{Gauss law of magnetism: } \nabla . \mathbf{B} = 0 \tag{2.4}$$

where $\mathbf{E}$ = electric field, $\mathbf{H}$ = magnetic field, $\mathbf{D}$ = electric displacement, $\mathbf{B}$ = magnetic induction, $\mathbf{J}$ = current density and ρ = free charge density.

The constitutive relations are:

$$\mathbf{D} = \varepsilon_0 \mathbf{E} + \mathbf{P} \tag{2.5}$$

and

$$\mathbf{B} = \mu_0 \mathbf{H} + \mathbf{M} \tag{2.6}$$

where ε_0 stands for the permittivity of free space and μ_0 is the free-space permeability. $\mathbf{P}$ is the polarization density of the medium. By polarization density we mean the electric dipole moment per unit volume. Polarization is defined as the displacement between positive and negative charges of a dielectric material in an applied electric field. $\mathbf{M}$ is the magnetization of the medium. It represents the magnetic dipole moment per unit volume.

2.3.2 Derivation of the wave equation for linear media

Two assumptions are made:

(i) free charges are absent:

$$\rho = 0, \therefore \mathbf{J} = 0 \tag{2.7}$$

(ii) the material is unmagnetized; hence

$$\mathbf{M} = 0 \tag{2.8}$$

Taking the curl of equation (2.1)

$$\nabla \times \nabla \times \mathbf{E} + \frac{\partial}{\partial t}(\nabla \times \mathbf{B}) = 0 \tag{2.9}$$

From equations (2.6) and (2.8),

$$\mathbf{B} = \mu_0 \mathbf{H} + \mathbf{M} = \mu_0 \mathbf{H} + 0 = \mu_0 \mathbf{H} \tag{2.10}$$

Substituting for $\mathbf{B}$ from equation (2.10) in equation (2.9)

$$\nabla \times \nabla \times \mathbf{E} + \frac{\partial}{\partial t}\left(\nabla \times \mu_0 \mathbf{H}\right) = 0 \tag{2.11}$$

or,

$$\nabla \times \nabla \times \mathbf{E} + \mu_0 \frac{\partial}{\partial t}(\nabla \times \mathbf{H}) = 0 \tag{2.12}$$

From equations (2.2) and (2.7),

$$\nabla \times \mathbf{H} = \frac{\partial \mathbf{D}}{\partial t} + \mathbf{J} = \frac{\partial \mathbf{D}}{\partial t} + 0 = \frac{\partial \mathbf{D}}{\partial t} \tag{2.13}$$

Combining equations (2.12) and (2.13), we get

$$\nabla \times \nabla \times \mathbf{E} + \mu_0 \frac{\partial}{\partial t}\left(\frac{\partial \mathbf{D}}{\partial t}\right) = 0 \tag{2.14}$$

or

$$\nabla \times \nabla \times \mathbf{E} + \mu_0 \frac{\partial^2 \mathbf{D}}{\partial t^2} = 0 \tag{2.15}$$

From equations (2.5) and (2.15), we have

$$\nabla \times \nabla \times \mathbf{E} + \mu_0 \frac{\partial^2}{\partial t^2}(\varepsilon_0 \mathbf{E} + \mathbf{P}) = 0 \tag{2.16}$$

or

$$\nabla \times \nabla \times \mathbf{E} + \mu_0 \frac{\partial^2}{\partial t^2}(\varepsilon_0 \mathbf{E}) + \mu_0 \frac{\partial^2 \mathbf{P}}{\partial t^2} = 0 \tag{2.17}$$

or

$$\nabla \times \nabla \times \mathbf{E} + \mu_0 \varepsilon_0 \frac{\partial^2 \mathbf{E}}{\partial t^2} + \mu_0 \frac{\partial^2 \mathbf{P}}{\partial t^2} = 0 \tag{2.18}$$

or

$$\nabla \times \nabla \times \mathbf{E} = -\left(\mu_0 \varepsilon_0 \frac{\partial^2 \mathbf{E}}{\partial t^2} + \mu_0 \frac{\partial^2 \mathbf{P}}{\partial t^2}\right) = -\left(\frac{1}{c^2}\frac{\partial^2 \mathbf{E}}{\partial t^2} + \mu_0 \frac{\partial^2 \mathbf{P}}{\partial t^2}\right) \tag{2.19}$$

since

$$c = \frac{1}{\sqrt{\mu_0 \varepsilon_0}} \tag{2.20}$$

or

$$\sqrt{\mu_0 \varepsilon_0} = \frac{1}{c} \tag{2.21}$$

or

$$\mu_0 \varepsilon_0 = \frac{1}{c^2}. \tag{2.22}$$

But we know that

$$\nabla^2 \mathbf{E} = \nabla(\nabla.\, \mathbf{E}) - \nabla \times \nabla \times \mathbf{E} = 0 - \nabla \times \nabla \times \mathbf{E} = -\nabla \times \nabla \times \mathbf{E} \tag{2.23}$$

since divergence of electric field at a point in space is the charge density/permittivity of space; and in a charge-free region of space

$$\nabla . \mathbf{E} = 0 \tag{2.24}$$

Equations (2.19) and (2.23) enable us to write

$$\nabla^2\mathbf{E} = \frac{1}{c^2}\frac{\partial^2\mathbf{E}}{\partial t^2} + \mu_0\frac{\partial^2\mathbf{P}}{\partial t^2} \tag{2.25}$$

But

$$\nabla^2\mathbf{E} = \frac{\partial^2\mathbf{E}}{\partial x^2} + \frac{\partial^2\mathbf{E}}{\partial y^2} + \frac{\partial^2\mathbf{E}}{\partial y^2} \tag{2.26}$$

For x-direction, we can write

$$\frac{\partial^2\mathbf{E}}{\partial x^2} = \frac{1}{c^2}\frac{\partial^2\mathbf{E}}{\partial t^2} + \mu_0\frac{\partial^2\mathbf{P}}{\partial t^2} \tag{2.27}$$

or

$$\frac{\partial^2\mathbf{E}}{\partial x^2} - \frac{1}{c^2}\frac{\partial^2\mathbf{E}}{\partial t^2} = \mu_0\frac{\partial^2\mathbf{P}}{\partial t^2} \tag{2.28}$$

We attach the subscript ‘Linear’ with polarization for emphasizing linearity, and write

$$\frac{\partial^2\mathbf{E}}{\partial x^2} - \frac{1}{c^2}\frac{\partial^2\mathbf{E}}{\partial t^2} = \mu_0\frac{\partial^2\mathbf{P}_{\text{Linear}}}{\partial t^2} \tag{2.29}$$

A parameter called electric susceptibility and denoted by the symbol $\chi^{(1)}$ is used for quantification of the extent of polarization induced in a dielectric by an applied electric field. It is introduced through the equation:

$$\chi^{(1)} = \frac{\mathbf{P}_{\text{Linear}}}{\varepsilon_0\mathbf{E}} \tag{2.30}$$

or

$$\mathbf{P}_{\text{Linear}} = \varepsilon_0\chi^{(1)}\mathbf{E} \tag{2.31}$$

The susceptibility parameter measures of the ability of a material to become polarized when subjected to an electric field. Susceptibility $\chi^{(1)}$ is unitless. So, the unit of polarization is Cm^{-2}, as shown below:

$$\begin{aligned} \text{P}_{\text{Linear}} &= \text{Fm}^{-1}\text{Vm}^{-1} = \text{s}^4\text{A}^2\text{m}^{-2}\text{kg}^{-1}\text{m}^{-1}\text{kgm}^2\text{s}^{-3}\text{A}^{-1}\text{m}^{-1} \\ &= \text{sAm}^{-2} = \text{sCs}^{-1}\text{m}^{-2} = \text{Cm}^{-2} \end{aligned} \tag{2.32}$$

Let us replace $\mathbf{P}_{\text{Linear}}$ in equation (2.29) using equation (2.30). Then

$$\frac{\partial^2 \mathbf{E}}{\partial x^2} - \frac{1}{c^2}\frac{\partial^2 \mathbf{E}}{\partial t^2} = \mu_0 \frac{\partial^2(\varepsilon_0 \chi^{(1)} \mathbf{E})}{\partial t^2} = \mu_0 \varepsilon_0 \chi^{(1)} \frac{\partial^2 \mathbf{E}}{\partial t^2} = \frac{1}{c^2}\chi^{(1)}\frac{\partial^2 \mathbf{E}}{\partial t^2} \tag{2.33}$$

where equation (2.22) has been applied. Equation (2.33) can be rewritten as

$$\frac{\partial^2 \mathbf{E}}{\partial x^2} - \frac{1}{c^2}\frac{\partial^2 \mathbf{E}}{\partial t^2} - \frac{1}{c^2}\chi^{(1)}\frac{\partial^2 \mathbf{E}}{\partial t^2} = 0 \tag{2.34}$$

or,

$$\frac{\partial^2 \mathbf{E}}{\partial x^2} - \frac{1}{c^2}\left(\frac{\partial^2 \mathbf{E}}{\partial t^2} + \chi^{(1)}\frac{\partial^2 \mathbf{E}}{\partial t^2}\right) = 0 \tag{2.35}$$

or,

$$\frac{\partial^2 \mathbf{E}}{\partial x^2} - \frac{1 + \chi^{(1)}}{c^2}\frac{\partial^2 \mathbf{E}}{\partial t^2} = 0 \tag{2.36}$$

or,

$$\frac{\partial^2 \mathbf{E}}{\partial x^2} - \frac{\left\{\sqrt{1 + \chi^{(1)}}\right\}^2}{c^2}\frac{\partial^2 \mathbf{E}}{\partial t^2} = 0 \tag{2.37}$$

or,

$$\frac{\partial^2 \mathbf{E}}{\partial x^2} - \frac{n^2}{c^2}\frac{\partial^2 \mathbf{E}}{\partial t^2} = 0 \tag{2.38}$$

where

$$n = \sqrt{1 + \chi^{(1)}} = \text{Refractive index}. \tag{2.39}$$

2.4 Electromagnetic wave propagation in non-linear optical media

2.4.1 Modification of the equation for induced polarization

At low electric fields, the polarization **P** directly proportional to the field **E**, is said to be linear, and is described by the susceptibility $\chi^{(1)}$. At intense electric fields, the induced polarization becomes a non-linear function of the field, and is expanded in ascending powers of electric field as

$$P = \varepsilon_0\{\chi^{(1)}E + \chi^{(2)}E^2 + \chi^{(3)}E^3 + \ldots\} \tag{2.40}$$

where $\chi^{(1)}, \chi^{(2)}, \chi^{(3)}$,..... are first-order, second-order and third-order susceptibilities. $\chi^{(1)}$, associated with the first power of electric field is linear susceptibility, whereas $\chi^{(2)}, \chi^{(3)}$, accompanying the second and third powers of electric field are non-linear susceptibilities. The magnitudes of $\chi^{(1)}, \chi^{(2)}, \chi^{(3)}$ are in descending order as of $\chi^{(1)}:\chi^{(2)}:\chi^{(3)} = 1:10^{-8}:10^{-16}$. In place of one proportionality constant $\chi^{(1)}$, we now have several constants $\chi^{(2)}$, $\chi^{(3)}$,...... These susceptibilities give rise to different interesting phenomena. A few examples are: refractive index and $\chi^{(1)}$, as already stated in equation (2.39); $\chi^{(2)}$ and second harmonic generation (SHG), $\chi^{(3)}$ and stimulated Raman scattering. Thus the polarization P can be decomposed into linear and non-linear components as:

$$P_{\text{Linear}} = \varepsilon_0 \chi^{(1)} \mathbf{E} \tag{2.41}$$

$$P_{\text{Non-linear}} = \varepsilon_0 \{\chi^{(2)} \mathbf{E}^2 + \chi^{(3)} \mathbf{E}^3 + \ldots\} \tag{2.42}$$

and

$$P = P_{\text{Linear}} + P_{\text{Non-linear}}. \tag{2.43}$$

2.4.2 Modification of the wave equation for non-linear media

Equation (2.38) of electromagnetic wave propagation in a linear medium is modified to write the equation describing wave movement in a non-linear medium. We restart from equation (2.29) and put the expanded version of equation for polarization in equation (2.42) to get

$$\begin{aligned}\frac{\partial^2 \mathbf{E}}{\partial x^2} - \frac{1}{c^2}\frac{\partial^2 \mathbf{E}}{\partial t^2} &= \mu_0 \frac{\partial^2 \left[\varepsilon_0 \{\chi^{(1)} \mathbf{E} + \chi^{(2)} \mathbf{E}^2 + \chi^{(3)} \mathbf{E}^3 + \ldots\}\right]}{\partial t^2} \\ &= \mu_0 \varepsilon_0 \left\{\chi^{(1)} \frac{\partial^2 \mathbf{E}}{\partial t^2}\right\} + \mu_0 \varepsilon_0 \frac{\partial^2}{\partial t^2} \{\chi^{(2)} \mathbf{E}^2 + \chi^{(3)} \mathbf{E}^3 + \ldots\} \\ &= \frac{1}{c^2} \chi^{(1)} \frac{\partial^2 \mathbf{E}}{\partial t^2} + \mu_0 \varepsilon_0 \frac{\partial^2}{\partial t^2} \{\chi^{(2)} \mathbf{E}^2 + \chi^{(3)} \mathbf{E}^3 + \ldots\}\end{aligned} \tag{2.44}$$

or

$$\frac{\partial^2 \mathbf{E}}{\partial x^2} - \frac{1}{c^2}\frac{\partial^2 \mathbf{E}}{\partial t^2} - \frac{1}{c^2} \chi^{(1)} \frac{\partial^2 \mathbf{E}}{\partial t^2} = \mu_0 \varepsilon_0 \frac{\partial^2}{\partial t^2} \{\chi^{(2)} \mathbf{E}^2 + \chi^{(3)} \mathbf{E}^3 + \ldots\} \tag{2.45}$$

or

$$\frac{\partial^2 \mathbf{E}}{\partial x^2} - \frac{1}{c^2}\left(\frac{\partial^2 \mathbf{E}}{\partial t^2} + \chi^{(1)} \frac{\partial^2 \mathbf{E}}{\partial t^2}\right) = \mu_0 \frac{\partial^2}{\partial t^2} \{\varepsilon_0 \chi^{(2)} \mathbf{E}^2 + \varepsilon_0 \chi^{(3)} \mathbf{E}^3 + \ldots\} \tag{2.46}$$

or

$$\frac{\partial^2 \mathbf{E}}{\partial x^2} - \frac{1 + \chi^{(1)}}{c^2}\frac{\partial^2 \mathbf{E}}{\partial t^2} = \mu_0 \frac{\partial^2}{\partial t^2} \{\varepsilon_0 \chi^{(2)} \mathbf{E}^2 + \varepsilon_0 \chi^{(3)} \mathbf{E}^3 + \ldots\} \tag{2.47}$$

or

$$\frac{\partial^2 \mathbf{E}}{\partial x^2} - \frac{n^2}{c^2}\frac{\partial^2 \mathbf{E}}{\partial t^2} = \mu_0 \frac{\partial^2}{\partial t^2} \{\varepsilon_0 \chi^{(2)} \mathbf{E}^2 + \varepsilon_0 \chi^{(3)} \mathbf{E}^3 + \ldots\} = \mu_0 \frac{\partial^2 \mathbf{P}_{\text{Non-linear}}}{\partial t^2} \tag{2.48}$$

again applying equation (2.42). $\mathbf{P}_{\text{Non-linear}}$ is given by equation (2.42). Equations (2.22) and (2.39) are also applied. The susceptibility $\chi^{(2)}$ allows us to study a second-order response and similarly the susceptibility $\chi^{(3)}$ enables studying the third-order response of a non-linear medium to an electromagnetic wave. The unit of susceptibility $\chi^{(2)}$ is $\mathrm{mV^{-1}}$, which gives the unit of polarization due to $\chi^{(2)}$ term as $\mathrm{Cm^{-2}}$:

$$\begin{aligned}\varepsilon_0\chi^{(2)}E^2 &= \mathrm{Fm^{-1}mV^{-1}(Vm^{-1})^2 = FV^{-1}V^2m^{-2} = s^4A^2m^{-2}kg^{-1}Vm^{-2}} \\ &= \mathrm{s^4A^2m^{-2}kg^{-1}kgm^2s^{-3}A^{-1}m^{-2} = sAm^{-2} = sCs^{-1}m^{-2} = Cm^{-2}}\end{aligned} \tag{2.49}$$

As an alternative route, polarization P is expanded in a Taylor series about $E = 0$

$$\begin{aligned}P_{\text{Non-linear}} &= c_1E + \frac{1}{2!}c_2E^2 + \frac{1}{3!}c_3E^3 + \ldots = c_1E + \frac{1}{2}c_2E^2 + \frac{1}{3\times 2}c_3E^3 + \ldots \\ &= c_1E + \frac{1}{2}c_2E^2 + \frac{1}{6}c_3E^3 + \ldots\end{aligned} \tag{2.50}$$

where the coefficients c_1, c_2, c_3,.... are first, second and third derivatives of P with respect to E, noting that these derivatives are evaluated at $E = 0$. The coefficients c_1, c_2, c_3,.... represent constants characteristic of the medium. At small values of E, the first term is dominant. The coefficient c_1 is written as

$$c_1 = \varepsilon_0\chi^{(1)} \tag{2.51}$$

where $\chi^{(1)}$ is linear susceptibility related to the relative permittivity ε and refractive index n of the medium as

$$\chi^{(1)} = \frac{\varepsilon}{\varepsilon_0} - 1 = n^2 - 1 \tag{2.52}$$

The coefficient c_2 signifying the second-order nonlinearity is written as

$$c_2 = 4\varepsilon_0 d \tag{2.53}$$

where d is the second-order nonlinear coefficient. For the coefficient c_3, we write

$$c_3 = 24\varepsilon_0\chi^{(3)} \tag{2.54}$$

Putting the values of c_1, c_2, c_3 from equations (2.51), (2.53) and (2.54) in equation (2.50) we get

$$\begin{aligned}P_{\text{Non-linear}} &= c_1E + \frac{1}{2}c_2E^2 + \frac{1}{6}c_3E^3 + \ldots = \varepsilon_0\chi^{(1)}E + \frac{1}{2}4\varepsilon_0 dE^2 + \frac{1}{6}24\varepsilon_0\chi^{(3)}E^3 + \ldots \\ &= \varepsilon_0\chi^{(1)}E + 2\varepsilon_0 dE^2 + 4\varepsilon_0\chi^{(3)}E^3 + \ldots\end{aligned} \tag{2.55}$$

The nonlinear optical medium can be mathematically characterized using equations (2.40)–(2.42) and (2.55). Note that $\chi^{(2)} = 2d$.

Example 2.1 The electrooptic coefficient of ZnTe is $d = 4.0 \times 10^{-12}$ m $\mathrm{V^{-1}}$ and that of GaP is 1.0×10^{-12} m $\mathrm{V^{-1}}$ (Casalbuoni *et al* 2008). Find the polarization arising from second-order susceptibility $\chi^{(2)}$ for the two materials at an electric field of 5×10^8 V m^{-1}.

Solution: As mentioned in the remark to equation (2.55), the susceptibility

$$\chi^{(2)} = 2d \tag{2.56}$$

$$\therefore |\chi^{(2)}|_{\text{ZnTe}} = |2d|_{\text{ZnTe}} = 2 \times 4.0 \times 10^{-12} = 8 \times 10^{-12}\ \text{mV}^{-1} \tag{2.57}$$

and

$$\therefore |\chi^{(2)}|_{\text{GaP}} = |2d|_{\text{GaP}} = 2 \times 1.0 \times 10^{-12} = 2 \times 10^{-12}\ \text{mV}^{-1} \tag{2.58}$$

Hence polarization produced in ZnTe from the $\chi^{(2)}$ term in equation (2.42) is

$$\begin{aligned}
P_{\text{ZnTe}} &= \varepsilon_0\chi^{(2)}E^2 = 8.854 \times 10^{-12}\text{Fm}^{-1} \times 8 \times 10^{-12}\ \text{mV}^{-1} \times (5 \times 10^8)^2\ \text{V}^2\ \text{m}^{-2} \\
&= 8.854 \times 10^{-12} \times 8 \times 10^{-12} \times (5)^2 \times 10^{16} \times \text{F m}^{-1} \times \text{mV}^{-1} \times \text{V}^2\ \text{m}^{-2} \\
&= 8.854 \times 8 \times (5)^2 \times 10^{-12-12+16} \times \text{F} \times \text{V} \times \text{m}^{-2} \\
&= 1.7708 \times 10^3 \times 10^{-8}\text{s}^4\text{A}^2\ \text{m}^{-2}\ \text{kg}^{-1} \times \text{kg m}^2\ \text{s}^{-3}\ \text{A}^{-1} \times \text{m}^{-2} \\
&= 1.7708 \times 10^{-5}\text{s A m}^{-2} = 1.7708 \times 10^{-5}\text{s C s}^{-1}\ \text{m}^{-2} = 1.7708 \times 10^{-5}\text{C m}^{-2}
\end{aligned} \tag{2.59}$$

and

$$\begin{aligned}
P_{\text{GaP}} &= \varepsilon_0\chi^{(2)}E^2 = 8.854 \times 10^{-12} \times 2 \times 10^{-12} \times (5 \times 10^8)^2\ \text{C m}^{-2} \\
&= 8.854 \times 2 \times (5)^2 \times 10^{16} \times 10^{-12-12}\ \text{C m}^{-2} \\
&= 4.427 \times 10^2 \times 10^{-24+16}\ \text{C m}^{-2} = 4.427 \times 10^{-6}\ \text{C m}^{-2}
\end{aligned} \tag{2.60}$$

Example 2.2 The peak electric field of an electromagnetic wave propagating in a medium of refractive index 1.5 is 125 V m^{-1}. What is the intensity of the wave?

Solution: The average intensity of a sinusoidal electromagnetic wave of electric field strength E is given by

$$I_{\text{Average}} = \frac{E^2}{\eta} \tag{2.61}$$

where

$$\begin{aligned}
\eta &= \sqrt{\frac{\mu_0}{\varepsilon_0}}/n = \frac{1}{n}\sqrt{\frac{4\pi \times 10^{-7}\ \text{N A}^{-2}}{8.854 \times 10^{-12}\ \text{F m}^{-1}}} = \frac{1}{n}\sqrt{\frac{4 \times 3.141\,59 \times 10^{-7}\ \text{N A}^{-2}}{8.854 \times 10^{-12}\ \text{F m}^{-1}}} \\
&= \frac{1}{n}\sqrt{\frac{4 \times 3.141\,59 \times 10^{-7+12}}{8.854}} = \frac{1}{n}\sqrt{1.419\,286 \times 10^5} = 376.73\ \Omega/n
\end{aligned} \tag{2.62}$$

μ_0, ε_0 are permeability and permittivity of vacuum respectively, and n is refractive index of the medium.

If E_0 is the peak electric field strength

$$E = \frac{|E_0|}{\sqrt{2}} \tag{2.63}$$

and from equations (2.61)–(2.63),

$$\begin{aligned} I_{\text{Average}} &= \frac{\left(\frac{|E_0|}{\sqrt{2}}\right)^2}{\eta} = \frac{E_0^2}{2\eta} = \frac{E_0^2}{2\frac{376.73}{n}} = \frac{nE_0^2(\text{V m}^{-1})^2}{753.46\ \Omega} \\ &= \frac{nE_0^2(\text{V}^2\text{m}^{-2})}{753.46\frac{\text{V}}{\text{A}}} = \frac{nE_0^2\ (\text{V A m}^{-2})}{753.46} = \frac{nE_0^2}{753.46}\text{W m}^{-2} \end{aligned} \tag{2.64}$$

Putting the given values ($n = 1.5$, $E_0 = 125$ V m^{-1}) in equation (2.64), we get

$$I_{\text{Average}} = \frac{nE_0^2}{753.46}\text{W m}^{-2} = \frac{1.5 \times (125)^2}{753.46}\text{W m}^{-2} = 31.1065\ \text{W m}^{-2} \tag{2.65}$$

Example 2.3 The second-order nonlinear optical susceptibility of GaAs is 5×10^{-10} m V^{-1} (Park *et al* 1999). Refractive index of GaAs is 3.5. Calculate the intensity of light at which the polarization due to second-order susceptibility equals that due to first-order susceptibility.

Solution: Given

$$n = \sqrt{1 + \chi^{(1)}} = 3.5 \tag{2.66}$$

or,

$$1 + \chi^{(1)} = 3.5^2 = 12.25 \tag{2.67}$$

or,

$$\chi^{(1)} = 12.25 - 1 = 11.25 \tag{2.68}$$

In equation (2.40), the first-order susceptibility term is

$$\varepsilon_0\chi^{(1)}E = 8.854 \times 10^{-12} \times 11.25E = 99.6075 \times 10^{-12}E = 9.96 \times 10^{-11}E \tag{2.69}$$

In equation (2.40), the second-order susceptibility term is

$$\begin{aligned} \varepsilon_0\chi^{(2)}E^2 &= 8.854 \times 10^{-12} \times 5 \times 10^{-10}E^2 = 8.854 \times 5 \times 10^{-10-12}E^2 \\ &= 44.27 \times 10^{-22}E^2 = 4.427 \times 10^{-21}E^2 \end{aligned} \tag{2.70}$$

For equality of first and second terms,

$$9.96 \times 10^{-11}E = 4.427 \times 10^{-21}E^2 \tag{2.71}$$

from equations (2.69) and (2.70).

Since $E \neq 0$, dividing both sides by E we get

$$E = \frac{9.96 \times 10^{-11}}{4.427 \times 10^{-21}} = \frac{9.96 \times 10^{-11+21}}{4.427} = 2.2498 \times 10^{10}\ \mathrm{V\ m^{-1}} \tag{2.72}$$

Applying equations (2.61) and (2.62),

$$I_{\text{Average}} = \frac{E^2}{\eta} = \frac{(2.2498 \times 10^{10})^2}{376.73} = \frac{5.0616 \times 10^{20}}{376.73} = 0.0134 \times 10^{20} \tag{2.73}$$
$$= 1.34 \times 10^{18}\ \mathrm{W\ m^{-2}}$$

Note: This exercise explains why a meaningful role of non-linear optical effects is expected only at very high electric fields and hence high intensities of light, and why these effects are perceptible in the realm of laser beams focused on small areas.

Example 2.4 Find the light intensity in W cm^{-2} at which the ratio of second-order nonlinear component $2dE^2$ to the linear component $\varepsilon_0\chi^{(1)}E$ of polarization P is 1% in an ADP crystal having refractive index $n = 1.5$ and the second-order non-linear coefficient $d = 6.8 \times 10^{-24}$ C V^{-2}.

Solution: From equation (2.55) we can see that the second-order nonlinear component of polarization P is $2\varepsilon_0 dE^2$ while the linear component is $\varepsilon_0\chi^{(1)}E$. Here d is specified in C V^{-2} which is the product of F/m and m/V = (F/m) × (m/V) = F/V = (C/V)/V = C V^{-2}. This means that product of d with vacuum permittivity (having unit F/m) is given. Hence, we can write

$$2dE^2 = 1\% \times \varepsilon_0\chi^{(1)}E = 0.01\varepsilon_0\chi^{(1)}E \tag{2.74}$$

noting that d already contains free-space permittivity as a factor.

Dividing both sides by $E \neq 0$, we have

$$2dE = 0.01\varepsilon_0\chi^{(1)} \tag{2.75}$$

or,

$$E = \frac{0.01\varepsilon_0(n^2 - 1)}{2d} \tag{2.76}$$

using equation (2.52).

Hence applying equations (2.61) and (2.76), we have

$$
\begin{aligned}
I_{\text{Average}} &= \frac{E^2}{\eta} = \frac{\left\{\dfrac{0.01\varepsilon_0(n^2-1)}{2d}\right\}^2}{376.73}\frac{\left(\dfrac{\text{F m}^{-1}}{\text{C/V}^2}\right)^2}{\Omega} \\
&= \frac{10^{-4}\varepsilon_0^2(n^2-1)^2}{4d^2\times 376.73}\frac{\text{F}^2\,\text{m}^{-2}}{\text{C}^2/\text{V}^4}\times\frac{1}{\text{V/A}} \\
&= \frac{10^{-4}\times(8.854\times10^{-12})^2(1.5^2-1)^2}{4(6.8\times10^{-24})^2\times 376.73}\frac{\text{F}^2\,\text{m}^{-2}\,\text{V}^3\text{A}}{\text{C}^2} \\
&= \frac{10^{-4}\times10^{-24}\times 78.393\times(2.25-1)^2}{4\times6.8^2\times10^{-48}\times376.73}\frac{\text{F}^2\,\text{m}^{-2}\,\text{V}^2\,\text{V A}}{(\text{As})^2} \\
&= \frac{10^{-4-24}\times78.393\times1.5625}{4\times46.24\times376.73\times10^{-48}}\frac{(\text{s}^4\text{A}^2\text{m}^{-2}\text{kg}^{-1})^2\text{m}^{-2}(\text{kgm}^2\text{s}^{-3}\text{A}^{-1})^2\text{VA}}{\text{A}^2\text{s}^2} \\
&= 0.001\,758\times10^{-28+48}\frac{\text{s}^8\text{A}^4\,\text{m}^{-4}\,\text{kg}^{-2}\,\text{kg}^2\,\text{m}^4\text{s}^{-6}\text{A}^{-2}\text{V A m}^{-2}}{\text{A}^2\,\text{s}^2} \\
&= 0.001\,758\times10^{20}\frac{\text{s}^2\,\text{A}^2\text{V A m}^{-2}}{\text{A}^2\text{s}^2} = 1.758\times10^{17}\,\text{W m}^{-2} \\
&= \frac{1.758\times10^{17}}{10^4}\text{Wcm}^{-2} = 1.758\times10^{13}\,\text{W cm}^{-2}
\end{aligned}
\tag{2.77}
$$

because $[\text{F}] = [\,\text{s}^4\cdot\text{A}^2\cdot\text{m}^{-2}\cdot\text{kg}^{-1}]$ and $[\text{V}] = [\,\text{kg}\cdot\text{m}^2\cdot\text{s}^{-3}\cdot\text{A}^{-1}]$.

2.5 Second-order response of a non-linear medium to single-electromagnetic wave input

Consider a non-linear medium with second-order susceptibility $\chi^{(2)}$, and suppose that a monochromatic electromagnetic wave of angular frequency ω is incident on it (figure 2.1).

The electric field E of the incident electromagnetic wave is expressed as

$$E = E_0 \cos \omega t \tag{2.78}$$

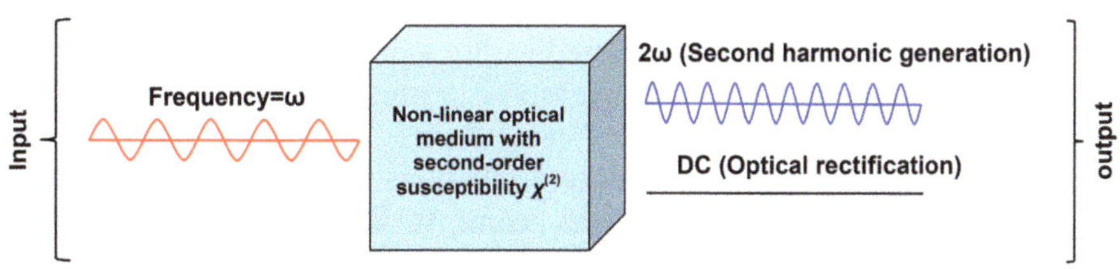

Figure 2.1. Passage of an electromagnetic wave of frequency ω through a non-linear medium and production of an electromagnetic wave at twice the frequency of input electromagnetic wave (2ω) along with a zero-frequency or DC signal.

where E_0 is the amplitude of the electric field. From equation (2.42), and applying equation (2.78), the second-order non-linear polarization is

$$\begin{aligned}P^{(2)}_{\text{Non-linear}} &= \chi^{(2)}E^2 = \chi^{(2)}(E_0\cos\omega t)^2\\ &= \chi^{(2)}E_0^2\cos^2\omega t = \chi^{(2)}E_0^2\frac{1}{2}\times 2\cos^2\omega t\\ &= \frac{1}{2}\chi^{(2)}E_0^2(1+\cos 2\omega t)\\ &= \frac{1}{2}\chi^{(2)}E_0^2 + \frac{1}{2}\chi^{(2)}E_0^2\cos 2\omega t\\ &= \text{Frequency} - \text{independent or DC term}\\ &\quad + \text{Term containing double the frequency of}\\ &\quad\quad \text{electromagnetic wave input}\end{aligned} \tag{2.79}$$

This equation means that the during the propagation of an electromagnetic wave through a non-linear medium, new frequency components are produced which were not supplied as input; one component is a DC polarization and the other component is a polarization at twice the input wave frequency. It may be noted that the input wave was purely a single-frequency wave of frequency ω, and contained neither zero-frequency nor double-frequency polarization components, which are therefore interpreted to be the outcomes of second-order susceptibility only. So a non-linear optical medium can lead to exciting results, not possible in linear optics.

In analogy to electrical rectification by a semiconductor junction diode involving the transformation of an alternating to direct current, the conversion of electromagnetic wave with alternating current or AC polarization to one with direct current or DC polarization is referred to as OR (optical rectification). Further, the production of a component with polarization having double the frequency of input wave is called SHG (second harmonic generation).

The OR is relevant to terahertz production. The SHG is not of interest in the present context. So it will not be discussed further.

2.6 Visualization and observation of optical rectification

Suppose an intense laser beam impinges upon and passes through a non-linear crystal (New 2011). This crystal is enclosed within a pair of parallel plates to form a capacitor. An example of a non-linear crystal is a non-centrosymmetric crystal.

Consequent to the passage of laser beam, electric dipoles are produced in the medium (figure 2.2). The dipoles undergo oscillations, and frequency of oscillation of dipoles equals the frequency of the laser beam. AC linear polarization P_{Linear} and first-order susceptibility $\chi^{(1)}$ describe the creation and behavior of these dipoles. In addition to this AC linear polarization, the crystal suffers from another polarization which is a DC polarization. This polarization takes place due to the relative displacements induced between the positive and negative charges of the crystal by the field of the laser beam. The effect of this DC polarization is observed in the form

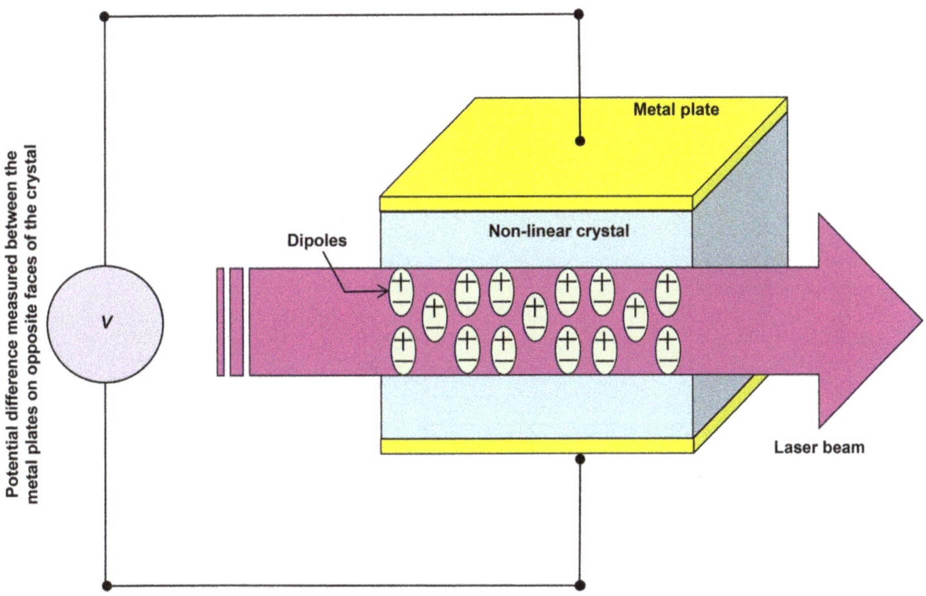

Figure 2.2. Observation of optical rectification by creation of a potential difference between two metal plates separated by a non-linear medium when a laser beam is passed through the medium.

of a potential difference between the two plates of the capacitor enabling the detection of a rectified component of the polarization of the laser beam.

The DC polarization P_{DC} arising from the frequency-independent term in equation (2.79) is written as

$$P_{DC} = \frac{1}{2}\chi^{(2)}E_0^2 \quad (2.80)$$

OR is the generation of a DC or low-frequency polarization produced during the propagation of an intense laser beam through a non-linear medium, e.g., lithium niobate ($LiNbO_3$), lithium tantalate ($LiTaO_3$), cadmium telluride (CdTe), zinc telluride (ZnTe), cadmium zinc telluride (CdZnTe), gallium arsenide (GaAs), gallium phosphide (GaP), gallium selenide (GaSe), and 4-dimethylamino-N-methyl-4-stilbazolium tosylate (DAST). It occurs when the optical field of a laser beam is rectified by the second-order susceptibility $\chi^{(2)}$ of a material.

2.7 Second-order response of a non-linear medium to two-electromagnetic wave input

As in the case of single-electromagnetic wave input, let us now consider the situation in which we have a non-linear medium with second-order susceptibility $\chi^{(2)}$, and two monochromatic electromagnetic waves of angular frequencies ω_1, ω_2 are incident on it (figures 2.3 and 2.4).

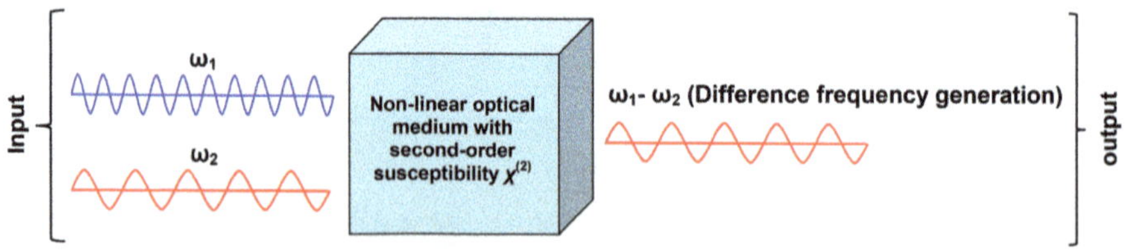

Figure 2.3. Passage of two electromagnetic waves of frequencies ω_1, ω_2 through a non-linear medium and production of an electromagnetic wave having frequency equal to difference of frequencies $\omega_1 - \omega_2$ of the two input electromagnetic waves.

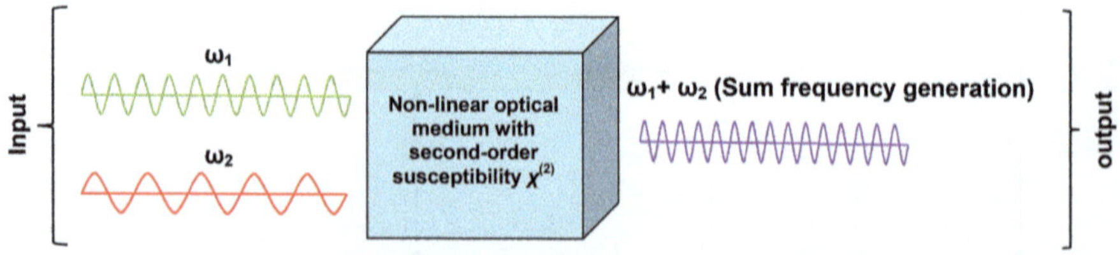

Figure 2.4. Passage of two electromagnetic waves of frequencies ω_1, ω_2 through a non-linear medium and production of an electromagnetic wave having frequency equal to sum of frequencies $\omega_1 + \omega_2$ of the two input electromagnetic waves.

The electric fields of the incident electromagnetic waves are written as

$$E_1 = E_0 \cos \omega_1 t \tag{2.81}$$

for electromagnetic wave 1, and

$$E_2 = E_0 \cos \omega_2 t \tag{2.82}$$

for electromagnetic wave 2. For this case, the second-order nonlinear response of the polarization is obtained by applying equation (2.42), putting $E^2 = E_1 E_2$ and then inputting the expressions for E_1 and E_2 from equations (2.81) and (2.82) as

$$\begin{aligned} P^{(2)}_{\text{Non-linear}} &= \chi^{(2)} E_1 E_2 = \chi^{(2)} (E_0 \cos \omega_1 t)(E_0 \cos \omega_2 t) \\ &= \chi^{(2)} E_0^2 \cos \omega_1 t \cos \omega_2 t = \frac{1}{2} \chi^{(2)} E_0^2 \times 2 \cos \omega_1 t \cos \omega_2 t \\ &= \frac{1}{2} \chi^{(2)} E_0^2 \times \{\cos(\omega_1 - \omega_2)t + \cos(\omega_1 + \omega_2)t\} \end{aligned} \tag{2.83}$$

because

$$\begin{aligned} &\cos(\omega_1 - \omega_2)t + \cos(\omega_1 + \omega_2)t \\ &\quad = \cos\omega_1 t \cos\omega_2 t + \sin\omega_1 t \sin\omega_2 t + \cos\omega_1 t \cos\omega_2 t - \sin\omega_1 t \sin\omega_2 t \\ &\quad = 2\cos\omega_1 t \cos\omega_2 t \end{aligned} \tag{2.84}$$

Since

$$\frac{1}{2}\chi^{(2)}E_0^2 = \text{Constant} \tag{2.85}$$

we can say from equation (2.83) that the second-order response consists of two terms, one of which is proportional to the difference of frequencies of the two input waves and the other is proportional to sum of frequencies of input waves. The former describes difference-frequency generation (DFG), figure 2.3, while the latter represents sum-frequency generation (SFG), figure 2.4. Of these, DFG only is relevant to terahertz wave generation using non-linear optics.

2.8 Second-order response of a non-linear medium to an ultrashort pulse of single-electromagnetic wave input

The DC polarization (sections 2.5–2.6) produced by a laser beam of constant or slowly varying optical power does not cause emission of any radiation. This means that the OR does not lead to any terahertz emission when the laser beam has a constant intensity. But the same does not happen with pulses of laser beams. Consider pulses of IR radiation of femtosecond duration (figure 2.5). When these pulses fall on a non-linear crystalline material, the DC polarization rises and falls in quick succession. As a consequence, a single-cycle pulse of electromagnetic radiation is emitted.

The bandwidth of a near-IR pulse is several terahertz, e.g., a 100 femtosecond near-IR pulse has a bandwidth $\Delta\omega$ = inverse of the pulse duration = $1/(100 \times 10^{-15})$ s = 10 THz. The Fourier transform spectrum $E(\omega)$ of the near-IR pulse $E(t)$ contains a wide range of frequencies from 0.1 to 10 THz. How is a pulse of terahertz radiation produced from femtosecond near-IR pulse? There are a large number of frequencies within the bandwidth $\Delta\omega$ of the femtosecond near-IR pulse. Difference-frequency

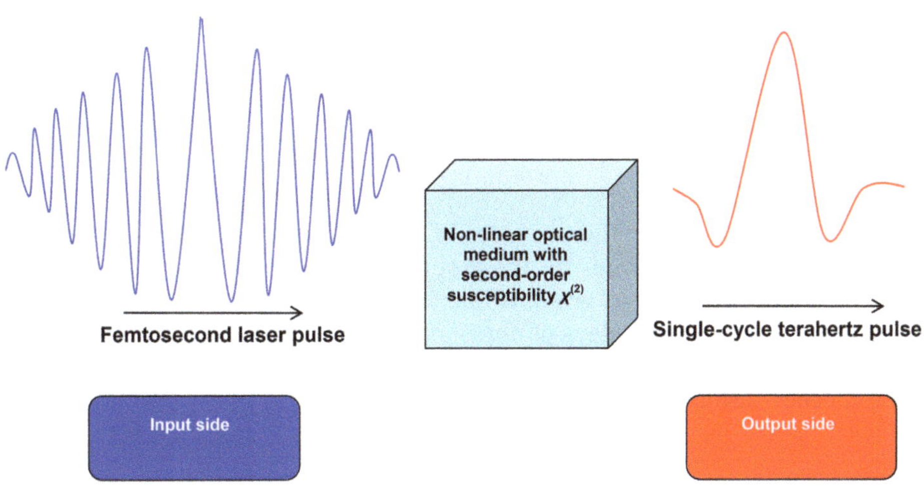

Figure 2.5. Generation of a single cycle of terahertz radiation by shining an ultrashort laser pulse of femtosecond duration on a non-linear medium.

mixing of all these frequencies leads to terahertz pulse emission. Therefore, the bandwidth of the emitted terahertz pulse corresponds to difference-frequency mixing of the frequencies in the near-IR pulse. The second derivative of the non-linear polarization due to $\chi^{(2)}$ term in equation (2.48) is responsible for this emission.

It is obvious that the terahertz emission is obtained at a large number of frequencies. But practically a measurable output is recorded at only one of the many possible frequencies. What is the reason? Understandably, a fixed phase relationship must be maintained during the wave propagation. All the combinations of frequencies cannot fulfill the condition of phase matching at the same time. Only a single frequency shows this compliance.

The phase matching condition is an essential pre-requirement for terahertz pulse generation by OR. What are the two signals between which phase matching is to be done? The two signals are optical pulse (near-IR pulse) and the terahertz radiation pulse. Let the two near-IR waves of angular frequencies $\omega_1 = \omega$ and $\omega_2 = \omega+\Omega$ interact with a non-linear medium. Here Ω is the frequency of terahertz radiation pulse equalling the frequency difference between near-IR pulses $= \omega_2\text{-}\omega_1 = \omega+\Omega\text{-}\omega = \Omega$. For sake of illustration, only two frequencies $\omega_1 = \omega$ and $\omega_2 = \omega + \Omega$ are considered although signals of many terahertz frequencies are produced.

The efficiency of frequency mixing is high when the phase matching is achieved. For this matching, the wave vector change

$$\Delta k = k(\omega + \Omega) - k(\omega) - k(\Omega) = 0 \tag{2.86}$$

In equation (2.86), $k(\Omega)$ is the wave vector of the terahertz signal produced at angular frequency Ω, and

$$k(\omega_2) - k(\omega_1) = k(\omega + \Omega) - k(\omega) \tag{2.87}$$

is the difference between wave vectors of near-IR pulses of frequencies $\omega_1 = \omega$ and $\omega_2 = \omega + \Omega$ used for frequency mixing. Further, k is the wave vector of light at frequency ω given by

$$k(\omega) = \frac{\omega}{c} \tag{2.88}$$

The velocity c of light in vacuum is given by

$$c = n_g(\omega)v_g(\omega) \tag{2.89}$$

Here $n_g(\omega)$ is the group refractive index of the medium. It is also called the group index. The symbol $v_g(\omega)$ stands for the group velocity of light in the medium. The concept of group index is used when we are dealing with a group of frequencies, such as for calculations of time delays of ultrashort pulses of finite bandwidth propagating in a medium. Group index takes into account the change in refractive index n with frequency through the equation

$$n_g(\omega) = n(\omega) + \omega\frac{\partial n}{\partial \omega} \tag{2.90}$$

Equation (2.90) shows that in a pulse of light, the group index $n_g >$ the phase index n.

A pulse of electromagnetic waves of a given bandwidth travels with the group velocity

$$v_g(\omega) = \frac{c}{n_g(\omega)} = \frac{c}{n(\omega) + \omega\dfrac{\partial n}{\partial \omega}} \tag{2.91}$$

where equation (2.90) is applied.

So the velocity of the complete pulse is less than that of light of a given frequency.

Similar to equation (2.88), $k(\Omega)$, the wave vector of terahertz radiation at frequency Ω can be expressed in terms of frequency Ω and velocity $v(\Omega)$ as

$$k(\Omega) = \frac{\Omega}{v(\Omega)} = \frac{\Omega}{\dfrac{c}{n(\Omega)}} = \frac{\Omega}{c}n(\Omega) \tag{2.92}$$

since

$$c = v(\Omega)n(\Omega) \tag{2.93}$$

or,

$$v(\Omega) = \frac{c}{n(\Omega)} \tag{2.94}$$

Note that the refractive index $n_g(\omega)$ for near-IR waves differs from that for terahertz waves $n(\Omega)$. So the near-IR and terahertz waves travel at different velocities $v_g(\omega)$ and $v(\Omega)$ respectively.

Applying the formula of Taylor series expansion of a function $f(x)$ about a point a:

$$f(x) = f(a) + f'(a)(x-a) + \frac{f''(a)(x-a)^2}{2!} + \frac{f'''(a)(x-a)^3}{3!} \ldots . \tag{2.95}$$

we can express $k(\omega+\Omega)$ in equation (2.86) as a power series in Ω around ω obtaining

$$k(\omega) + \left|\frac{\partial k}{\partial \omega}\right|_{\omega}(\omega+\Omega-\omega) + \left|\frac{\partial^2 k}{\partial \omega^2}\right|_{\omega}(\omega+\Omega-\omega)^2 + \ldots - k(\omega) - k(\Omega) = 0 \tag{2.96}$$

or,

$$\left|\frac{\partial k}{\partial \omega}\right|_{\omega}(\omega+\Omega-\omega) - k(\Omega) = 0 \tag{2.97}$$

neglecting

$$\left|\frac{\partial^2 k}{\partial \omega^2}\right|_{\omega}(\omega+\Omega-\omega)^2 + \ldots \tag{2.98}$$

Equation (2.97) gives

$$\left|\frac{\partial k}{\partial \omega}\right|_{\omega}\Omega - k(\Omega) = 0 \tag{2.99}$$

or

$$\left|\frac{\partial k}{\partial \omega}\right|_{\omega} \Omega = k(\Omega) \tag{2.100}$$

or

$$\frac{\Omega}{v_{g}(\omega)} = k(\Omega) \tag{2.101}$$

because from the definition of group velocity

$$\left\{\left|\frac{\partial k}{\partial \omega}\right|_{\omega}\right\}^{-1} = v_{g}(\omega) \tag{2.102}$$

Since from equation (2.89)

$$v_{g}(\omega) = \frac{c}{n_{g}(\omega)} \tag{2.103}$$

Equation (2.101) becomes

$$\frac{\Omega}{\dfrac{c}{n_{g}(\omega)}} = k(\Omega) \tag{2.104}$$

or

$$\frac{n_{g}(\omega)}{c}\Omega = k(\Omega) \tag{2.105}$$

Inspection of equations (2.92) and (2.105) shows that

$$\frac{n_{g}(\omega)}{c}\Omega = \frac{\Omega}{c}n(\Omega) \tag{2.106}$$

or

$$n_{g}(\omega) = n(\Omega) \tag{2.107}$$

or

Group velocity refractive index of near-IR pulse at frequency ω
= Refractive index of terahertz pulse at frequency Ω. (2.108)

Thus terahertz waves can be generated via tuning the frequency ω of the near-IR laser to match the phases of near-IR and terahertz pulses. This changes the frequency difference Ω between the two input near-IR pulses; and the frequency difference Ω is the terahertz wave frequency. High bandwidths (0.3–7 THz) are obtained with GaP (Aoki *et al* 2017).

The efficiency of conversion of near-IR pulse energy to terahertz pulse energy is relatively low ~10^{-6}. Power of output phase-matched OR terahertz pulse sources typically ranges from nanowatts to submicrowatts for input near-IR pulse power of 1 W (Ferguson and Zhang 2002).

2.9 Detection of terahertz wave pulses

2.9.1 Electrooptic effects

In OR, the input signal to non-linear medium is a constant optical signal in the form of a strong laser beam at frequency ω. It produces a DC polarization but does not induce any emission of terahertz waves. In the specific case when this input signal is a laser pulse, it results in the emission of a terahertz wave pulse of frequency Ω as the output signal. For detection of this terahertz wave pulse, we must look for the inverse effect to OR. There are two such effects, one of which is the linear electrooptic effect or Pockel's effect, and the other is quadratic electrooptic effect or Kerr effect.

2.9.1.1 Linear electrooptic effect

In linear electrooptic effect, application of a DC electric field along with an incident terahertz wave pulse input of frequency Ω to a non-linear optical crystal causes a change in polarization as the output (figure 2.6). This change in polarization is observed as a refractive index variation. The modification of refractive index is proportional to the intensity of the electric field. Hence the phenomenon is said to be linear. The linear electrooptic effect is exhibited only by optical materials which are not symmetric to inversion, i.e., are non-centriosymmetric, e.g., lithium niobate ($LiNbO_3$), lithium tantalate ($LiTaO_3$), gallium arsenide (GaAs), gallium phosphide (GaP), potassium dihydrogen phosphate (KDP:KH_2PO_4), ammonium dihydrogen phosphate (ADP:$NH_4H_2PO_4$). Pockel coefficients typically lie between

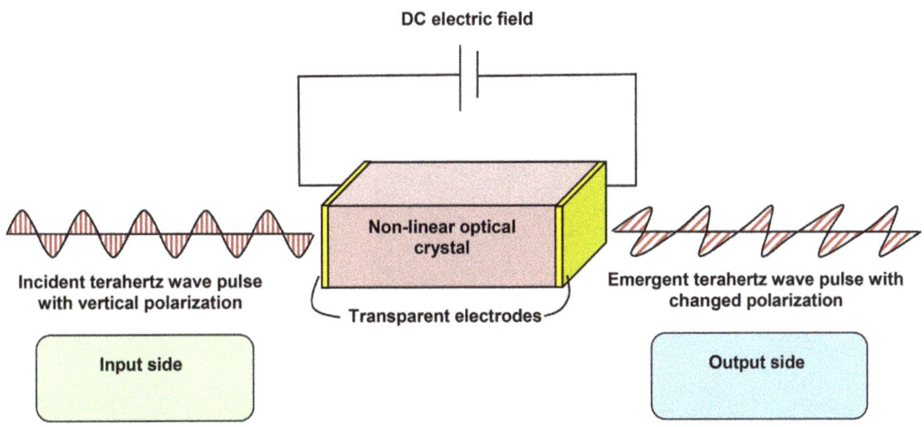

Figure 2.6. Linear electrooptic effect: change in polarization of a terahertz pulse after transmission through a non-linear crystal.

$(1\text{–}100) \times 10^{-12}$ m V^{-1} showing that the refractive index change for an electric field of 10^6 V m^{-1} is

$$\Delta n|_{\text{Linear}} = (1 - 100) \times 10^{-12} \times 10^6 = (1 - 100) \times 10^{-6} = 10^{-6} - 10^{-4} \quad (2.109)$$

2.9.1.2 Quadratic electrooptic effect

In the quadratic electrooptic effect, the refractive index change varies as the square of the electric field intensity. This property is displayed by all materials to varying extents. However, it is weaker than the linear electrooptic effect. For crystals, typical values of Kerr coefficients are $10^{-18}\text{–}10^{-14}$ m^2 V^{-2}, so that an electric field of strength 10^6 V m^{-1} is able to produce a change in refractive index

$$\begin{aligned}\Delta n|_{\text{Quadratic}} &= (10^{-18} - 10^{-14}) \times (10^6)^2 = (10^{-18} - 10^{-14}) \times 10^{12} \\ &= 10^{-6} - 10^{-2}\end{aligned} \quad (2.110)$$

Birefringence, also called birefraction or double refraction is the dependence of the refractive index of a material on polarization and direction of propagation of light. The name 'double refraction' shows that a ray of unpolarized light incident on such a material is split into two rays moving in different paths, an ordinary ray following the Snell's law of refraction and an extraordinary ray which does not follow this law (figure 2.7). These rays are polarized in orthogonal directions.

2.9.2 Electrooptic phase modulator

The principle of production of birefringence in an electrooptic crystal by an incident electromagnetic wave is utilized to build an electrooptic phase modulator for the wave (figure 2.8). The device used is called a Pockel's cell. The cell contains a

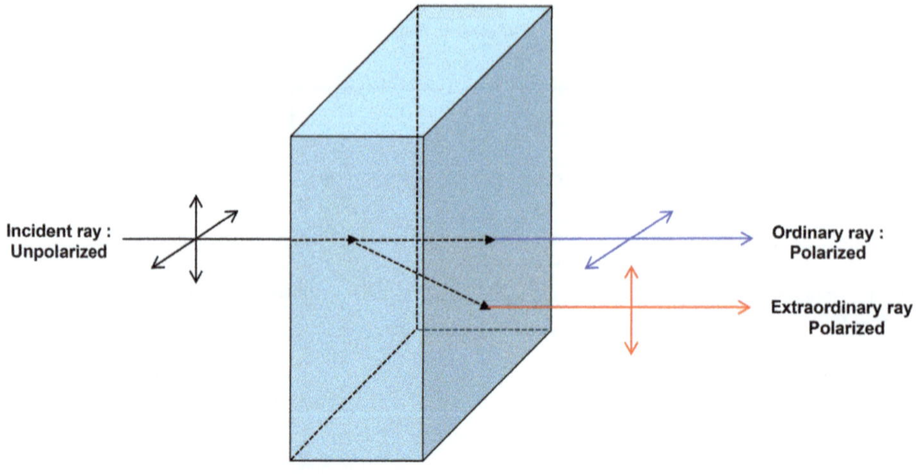

Figure 2.7. Birefringence or double refraction.

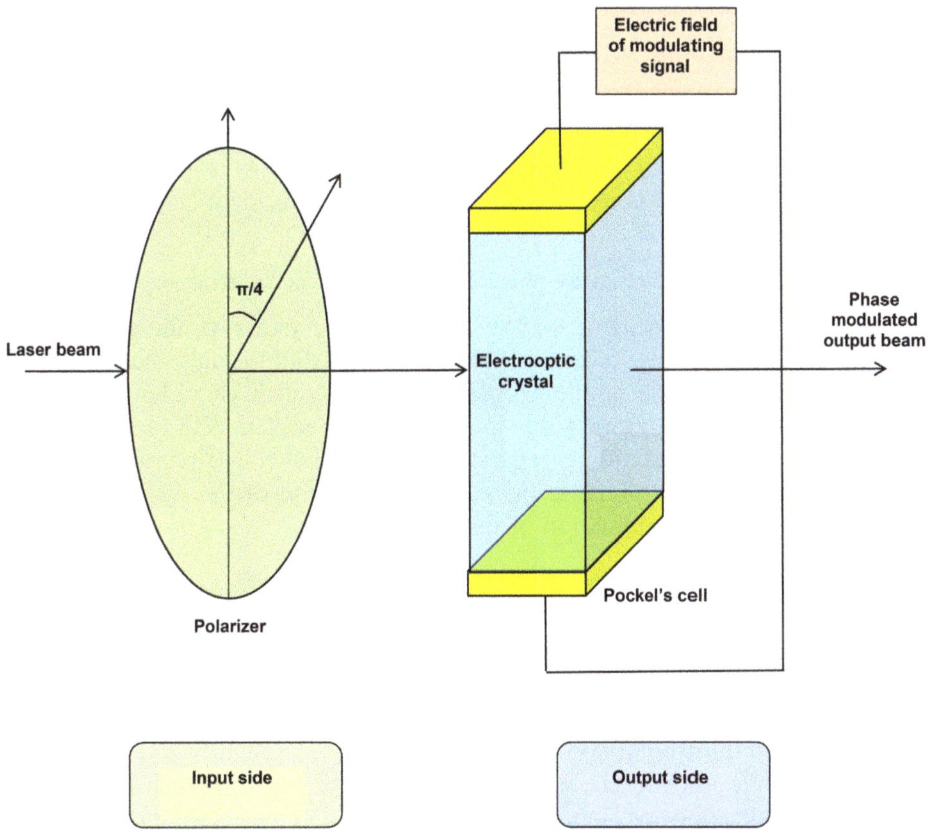

Figure 2.8. A transverse electrooptic phase modulator in which the modulating signal is applied as a voltage across the electrodes on the top and bottom faces of the electrooptic crystal and thereby the input laser beam is polarized in the direction of the electric field in the crystal.

uniaxial crystal. When an electric field is applied to it, it becomes a biaxial crystal. Suppose an electromagnetic wave such as a laser beam is incident on the crystal. Axes induced by the electric field applied to the crystal are oriented in defined directions. They subtend an angle of $\pi/4$ with respect to the plane of polarization of the laser beam. So the laser beam is resolved into two equal components which are polarized in mutually perpendicular directions. These rays travel along different directions. They also move with different velocities because of the induced changes in refractive index. The birefringence thus produced is given by the difference between refractive indices experienced by the extraordinary and ordinary rays; the former is called the extraordinary refractive index $n_{\text{Extraordinary ray}}$ and the latter ordinary refractive index $n_{\text{Ordinary ray}}$. The difference between refractive indices of the two rays is

$$\Delta n = (n_{\text{Extraordinary ray}} - n_{\text{Ordinary ray}}) \tag{2.111}$$

and varies with the electric field. If the length of the crystal is l, the phase retardation of the incident laser beam of wavelength λ is expressed as

$$\phi = \frac{2\pi}{\lambda}\Delta n l \tag{2.112}$$

This phase retardation is used for modulation of the incident laser beam.

2.9.3 Terahertz pulse detection by phase modulation of near infrared pulse

The terahertz pulse is incident on the electrooptic crystal. At the same time a femtosecond duration pulse of near-IR radiation is passed through the crystal. The terahertz pulse acts on the simultaneously propagating near-IR radiation pulse by linear electrooptic effect. During this interaction, the electric field of the terahertz pulse causes phase modulation of near-IR radiation pulse. No phase modulation occurs in the absence of terahertz pulse. Thus, detection of the terahertz pulse is enabled through phase modulation of the near-IR pulse.

2.10 Discussion and conclusions

Equation for propagation of electromagnetic waves in linear optical media is derived and modified by expressing the polarization in ascending powers of electric field. In addition to the first order or linear susceptibility, the polarization now contains second-order and third-order non-linear susceptibility terms, which give birth to new phenomena. While studying the second-order response of a non-linear medium to a single electromagnetic wave input, a DC term and a term containing double the input frequency are obtained in the polarization expression leading to the concepts of OR and second harmonic generation. The effect of OR is explained by shining a laser beam on a non-linear crystal sandwiched between parallel plates. In the second-order response of polarization of a non-linear medium to two-electromagnetic wave input, a term containing the sum of the two input frequencies and another term containing the difference between the two input frequencies are obtained. The former gives rise to sum-frequency generation and the latter to difference-frequency generation.

A very interesting observation is made when dealing with second-order response of a non-linear medium to an ultrashort pulse of single-electromagnetic wave input. This response causes emission of a pulse of terahertz radiation which is interpreted by considering that a large number of frequencies are produced within the bandwidth of the ultrashort laser pulse. Then difference-frequency mixing of these radiations generates the terahertz pulse. To ensure high efficiency of frequency mixing, the phase matching condition must be obeyed. This can be met only by a single frequency, and happens when the group velocity refractive index of the laser pulse equals the refractive index of the terahertz pulse.

The inverse effects to OR provide means for detection of terahertz radiation. Electrooptic effect is the change in polarization and hence refractive index of a medium on application of a DC electric field together with an incident terahertz wave. Accordingly as the change in refractive index is proportional to the first or second powers of the electric field, the electrooptic effects are referred to as linear and quadratic. An electrooptic modulator is built by utilizing the birefringence produced in an electrooptic crystal by an incident electromagnetic wave. The phase modulation of the laser pulse is applied for detecting the terahertz pulse. This is done by passing the femtosecond laser pulse concurrently with the incident terahertz pulse.

References

Aoki K, Savolainen J and Havenith M 2017 Broadband terahertz pulse generation by optical rectification in GaP crystals *Appl. Phys. Lett.* **110** 201103

Boyd R W 2020 *Non-Linear Optics* (London: Academic) pp 1–64

Casalbuoni S, Schlarb H, Schmidt B, Schmüser P, Steffen B and Winter A 2008 Numerical studies on the electro-optic detection of femtosecond electron bunches *Phys. Rev. Spec. Top. Accel. Beams* **11** 072802-1–072802-18

Ferguson B and Zhang X-C 2002 Materials for terahertz science and technology *Nat. Mater.* **1** 26–33

Laud B B 1992 *Lasers and Non-Linear Optics* (New York: Wiley) p 262

New G 2011 *Introduction to Nonlinear Optics* (Cambridge: Cambridge University Press) pp 7–8

Park T.-i., Gumbs G and Chen Y C 1999 Properties of the second-order nonlinear optical susceptibility $\chi^{(2)}$ in asymmetric undoped AlGaAs/InGaAs double quantum wells *J. Appl. Phys.* **86** 1467–171

Wilke I and Sengupta S 2008 Chapter 2: Nonlinear optical techniques for terahertz pulse generation and detection—optical rectification and electrooptic sampling *Terahertz Spectroscopy: Principles and Applications* ed S L Dexheimer (Boca Raton, FL: CRC Press) pp 41–72

IOP Publishing

Practical Terahertz Electronics: Devices and Applications, Volume 2
Optical devices and applications
Vinod Kumar Khanna

Chapter 3

Photoconductive antennas and photomixers

The principal components of a photoconductive antenna are described; namely the III–V semiconductor film, microstrip lines, dipole structure and silicon lens, and their functions are explained. The use of antenna for generation of broadband terahertz pulses and narrowband terahertz continuous waves is discussed. The operation of a photoconductive antenna time-domain spectroscopy system is outlined. After introducing the heterodyne principle underlying photomixing, the construction of a photomixer with laser diodes and photoconductive antenna is dwelled upon. Two types of photomixer setup for imaging or spectroscopy are made, one with cryogenic bolometers and the other by using coherent homodyne detection. Power output, radiation resistance and *R–C* roll-off of photoconductive antenna characterize its performance. An alternative approach to terahertz generation works with a large-area emitter without antenna. Instead of photoconductive detection, a semiconductor diode-based scheme serves as a valuable option. A unitravelling carrier photodiode (UTC-PD) can be designed to operate much faster than a PIN photodiode by employing an ultrathin absorber layer, making it very useful for terahertz generation. The working of a UTC-PD photomixer is briefly sketched.

In this chapter, the operation of a photoconductive antenna is described, the concept of photomixing is explained, and the application of photoconductive antenna in a photomixer is discussed. After highlighting the differences of a UTC-PD from a PIN photodiode, its use for photomixing is briefly presented.

3.1 Photoconductive antenna

A photoconductive antenna is used for generation and detection of:

(i) broadband pulsed terahertz radiation, or
(ii) narrowband continuous wave terahertz radiation (Burford and El-Shenawee 2017).

doi:10.1088/978-0-7503-4886-7ch3

It consists of:

(i) a semiconductor film,
(ii) microstrip lines, bias pads and dipole structure, and
(iii) a silicon lens.

Two types of photoconductive dipole antenna are illustrated in figure 3.1. Figure 3.2 shows the cross-sectional diagram of the antenna.

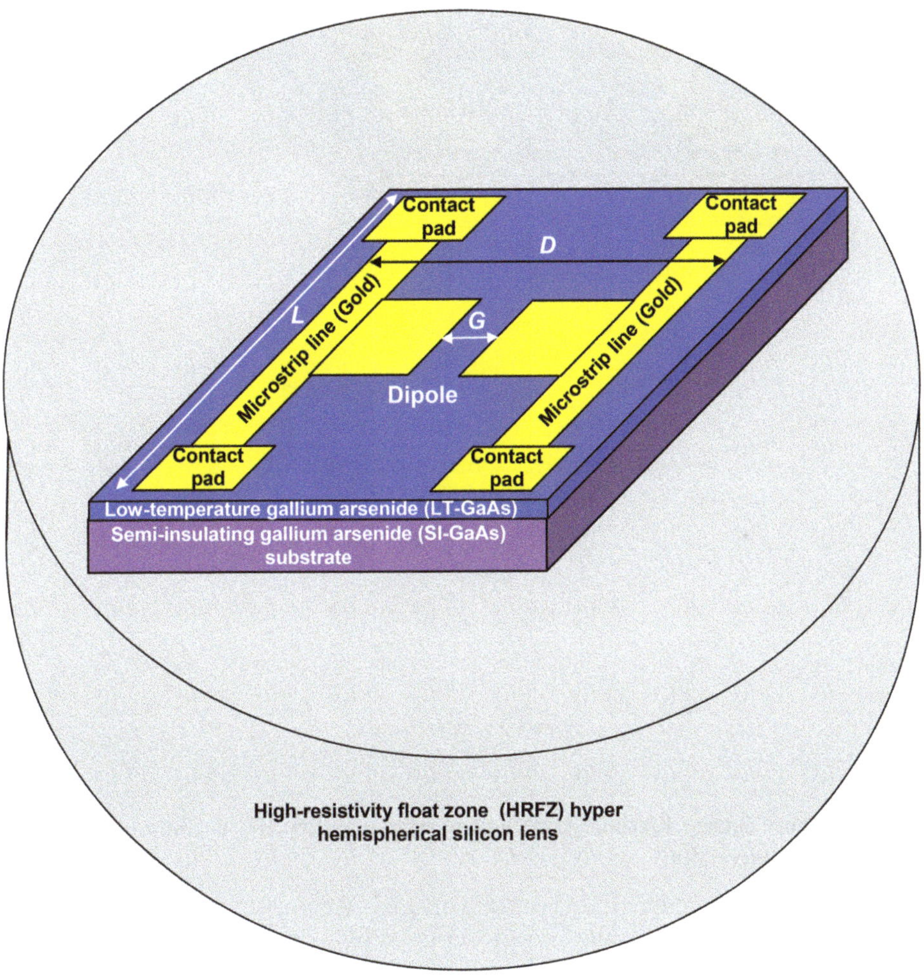

Figure 3.1. Photoconductive dipole antennas: (a) coplanar strip line and (b) bowtie.

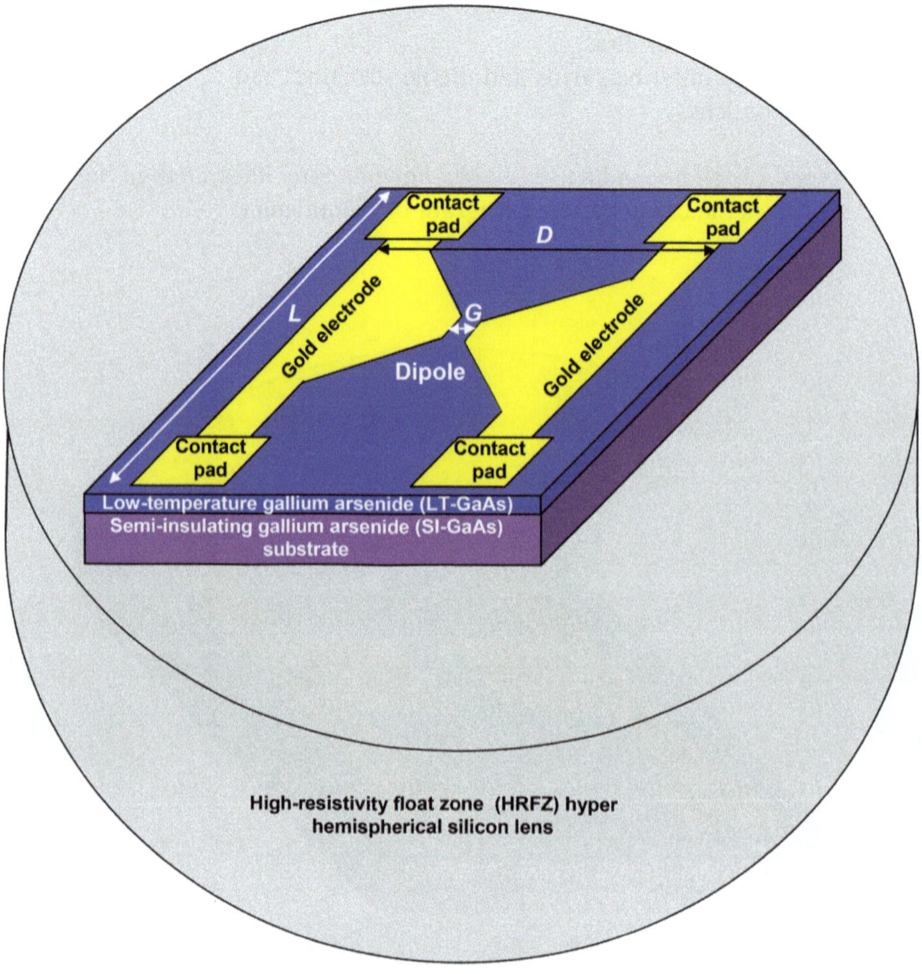

Figure 3.1. (Continued.)

3.1.1 The short carrier lifetime, sufficient-carrier-mobility III–V compound semiconductor film

A photoconductive antenna (PCA) is made on a III–V compound semiconductor such as a low-temperature epitaxially grown GaAs or InGaAs film on a semi-insulating GaAs substrate. The epitaxial film has an ultrashort carrier lifetime in the subpicosecond range but a high carrier mobility. To achieve ultrashort carrier lifetime, the film growth is performed by molecular beam epitaxy at a low temperature ~200 °C–300 °C with arsenic over-pressure. The intent is to incorporate crystal defects for lifetime

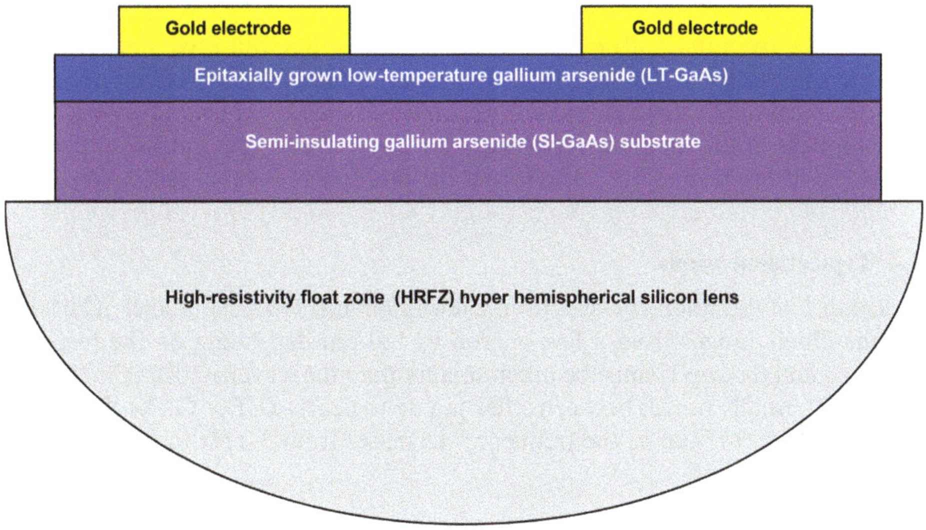

Figure 3.2. Cross-sectional diagram of photoconductive antenna.

killing. The film formed under these conditions contains 1% excess arsenic atoms residing in non-stoichiometric sites. These clusters of arsenic atoms introduce defect energy levels in the bandgap facilitating non-radiative carrier recombination and thereby providing a short carrier lifetime. Thus the non-stoichiometric composition provides a shorter lifetime. But it is also associated with an unwanted effect. It degrades the carrier mobility. For restoration of carrier mobility, the film is annealed at a temperature of 500 °C in arsenic ambient when arsenic atoms are rearranged in the gallium sites. After the thermal treatment, the carrier mobility improves to a value ~200 cm^2 V s^{-1}, but the carrier lifetime is still maintained at a low subpicosecond value. The film also shows a high breakdown field >4 × 10^5 V cm^{-1} (Brown *et al* 1993).

An alternative method to create crystal defects for carrier lifetime reduction employs the ion implantation technique involving bombarding of the lattice with a high-energy ion beam to dislodge atoms from lattice sites.

3.1.2 The microstrip lines, contact pads and dipole structure

On the substrate are patterned two parallel microstrip lines with large contact pads at the ends. At the center of the microstrip lines, there is a dipole structure. The contact pads provide large areas for bonding wires for electrical connection to an external circuit. A DC source is connected across the contact pads for electrical biasing so that one terminal is at a positive potential serving as the anode and the other is at a negative potential acting as the cathode.

3.1.3 The silicon lens

The substrate is mounted on a high-resistivity float zone silicon lens for coupling with the terahertz beam, shaping it and propagating it into free space.

The necessity of silicon lens arises from the fact that the terahertz waves tend to be reflected back from the walls of the GaAs substrate. They will leave the substrate if the receiving medium has the same refractive index, around 3.4. Silicon with a resistivity of 10 kOhm-cm fulfills this condition. A silicon lens is used for collimating the terahertz beam. In place of the usual hemispherical lens, a hyperhemispherical lens is preferred because the hemispherical lens suffers from the problem of internal reflections.

3.1.4 Typical dimensions

The length L of the antenna varies from a few millimeters to a centimeter. The length D of the dipole ranges from a few micron to 150 μm depending on the resonance frequency, and the gap G must be much smaller than the wavelength of the terahertz radiation. It usually ranges between a few μm up to nearly D. For GaAs, D increases from 14.7 μm to 147 μm as the frequency decreases from 3 THz to 0.3 THz.

Example 3.1

(a) A PCA is fabricated using GaAs thin film as the semiconductor. The refractive index of GaAs at terahertz frequencies is 3.4. The length of the dipole is chosen as 10 μm. Find the first resonance frequency and wavelength of the antenna.

(b) If the first resonance frequency is 1 THz, what is the length of the dipole?

Solution: (a) Resonance takes place in a linear conductor at a frequency for which the wavelength in free space is double the length of the conductor. This means that resonance occurs when the conductor has a length equal to half of the wavelength. A dipole antenna is usually used at this resonance frequency. For this reason, it is known as half-wave dipole antenna.

In a dipole antenna, the resonance is not limited to a single frequency but occurs at frequencies corresponding to integral multiples of half wavelength. Suppose the dipole antenna has a length D. If the dipole is in resonance with electromagnetic waves of wavelength λ_0, the condition for resonance in free space can be written as

$$D = n\frac{\lambda_0}{2} \tag{3.1}$$

where n can acquire integral values, 1,2,3..... But here free-space wavelength is not the issue of concern. Our interest lies in the wavelength in GaAs, which is a material of refractive index $\mu = 3.4$. To find this wavelength, equation (3.1) is modified as

$$D = n\frac{\lambda}{2\mu} \tag{3.2}$$

by making the substitution

$$\lambda_0 = \frac{\lambda}{\mu} \tag{3.3}$$

Since the wavelength λ and frequency ν are inter-related as

$$\lambda = \frac{c}{\nu} \tag{3.4}$$

Equation (3.2) reduces to

$$D = n\frac{c}{2\mu\nu} \tag{3.5}$$

giving

$$\nu = n\frac{c}{2\mu D} = \frac{nc}{2\mu D} \tag{3.6}$$

For first resonance, $n = 1$; hence from equation (3.6)

$$\begin{aligned} \nu &= \frac{c}{2\mu D} = \frac{2.9979 \times 10^8}{2 \times 3.4 \times 10 \times 10^{-6}} = \frac{2.9979 \times 10^{8+5}}{2 \times 3.4} \\ &= 0.440\,87 \times 10^{13} = 4.41 \times 10^{12}\ \text{Hz} = 4.41\ \text{THz} \end{aligned} \tag{3.7}$$

and from equation (3.4)

$$\begin{aligned} \lambda &= \frac{c}{\nu} = \frac{2.9979 \times 10^8}{4.41 \times 10^{12}} = \frac{2.9979 \times 10^{8-12}}{4.41} = 0.6798 \times 10^{-4}\ \text{m} \\ &= \frac{0.6798 \times 10^{-4}}{10^{-6}}\ \mu\text{m} \approx 0.6798 \times 100\ \mu\text{m} = 67.98\ \mu\text{m} \end{aligned} \tag{3.8}$$

(b) Applying equation (3.5) to first resonance frequency we have

$$\begin{aligned} D &= \frac{c}{2\mu\nu} = \frac{2.9979 \times 10^8}{2 \times 3.4 \times 1 \times 10^{12}} = \frac{2.9979 \times 10^{8-12}}{2 \times 3.4 \times 1} \\ &= 0.440\,87 \times 10^{-4}\ \text{m} = \frac{0.440\,87 \times 10^{-4}}{10^{-6}}\mu\text{m} \\ &= 0.440\,87 \times 10^{2}\ \mu\text{m} = 44.087\ \mu\text{m} \end{aligned} \tag{3.9}$$

Example 3.2

In a PCA fabricated with GaAs, calculate the lengths of the dipole for first resonance frequencies of 0.1 THz and 10 THz.

Solution: From equation (3.9),

$$\begin{aligned} D|_{0.1\ \text{THz}} &= \frac{c}{2\mu\nu|_{0.1\ \text{THz}}} = \frac{2.9979 \times 10^8}{2 \times 3.4 \ \times 0.1 \times 10^{12}} \\ &= 4.4087 \times 10^{8-12} = 4.4087 \times 10^{-4}\ \text{m} = \frac{4.4087 \times 10^{-4}}{10^{-6}}\ \mu\text{m} \\ &= 4.4087 \times 10^{-4+6}\ \mu\text{m} = 440.87\ \mu\text{m} \end{aligned} \tag{3.10}$$

Similarly from equation (3.9),

$$\begin{aligned} D|_{1\text{ THz}} &= \frac{c}{2\mu\nu|_{1\text{ THz}}} = \frac{2.9979 \times 10^{8}}{2 \times 3.4 \times 10 \times 10^{12}} = 0.440\,87 \times 10^{8-12-1} \\ &= 0.440\,87 \times 10^{-5}\text{ m} = \frac{0.440\,87 \times 10^{-5}}{10^{-6}}\ \mu\text{m} \\ &= 0.440\,87 \times 10^{-5+6}\ \mu\text{m} = 4.4087\ \mu\text{m} \end{aligned} \tag{3.11}$$

3.1.5 Generation of broadband pulses of terahertz wave

Looking at figure 3.3, the steps involved in the production of terahertz pulses are:

(i) An optical pulse of duration <1 ps from a femtosecond laser is focused at the spot of the antenna gap. The photons of the pulse have a frequency ν such that the photon energy

$$E = h\nu \tag{3.12}$$

is greater than the bandgap E_G of the semiconductor; h is Planck's constant. They collide with the lattice atoms removing in each collision an electron from its orbit and promoting it to the conduction band. At the same time, a hole is created in the valence band. As a result, an electron–hole pair is

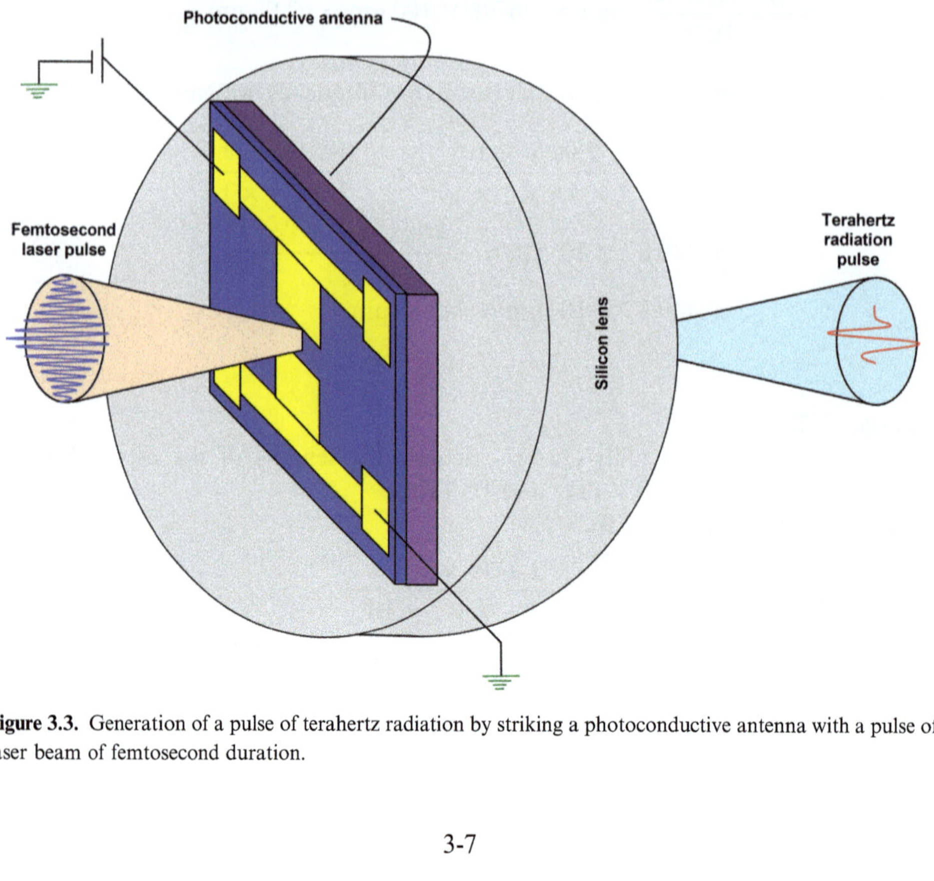

Figure 3.3. Generation of a pulse of terahertz radiation by striking a photoconductive antenna with a pulse of laser beam of femtosecond duration.

generated. Many electron–hole pairs are produced as a consequence of photons striking the semiconductor. These free carriers participate in electrical conduction until they are annihilated by recombination.

(ii) The generated free carriers are accelerated under the electric field created by the voltage applied to the contact pads. A transient photocurrent is produced with a rise time proportional to the rise time of the incident optical pulse. This photocurrent drives the dipole antenna, emitting a pulse of terahertz radiation.

(iii) The carrier lifetime in the semiconductor plays a crucial role in determining the time for which the photocurrent continues to flow after the full absorption of the optical pulse. Two temporal profiles need to be distinguished, one is the time profile of carrier generation which closely follows the time profile of optical pulse from the laser; the other is the time profile of the resulting photocurrent driving the antenna, whose decay after the carrier generation reaches peak depends on the rate of carrier annihilation by recombination and hence the carrier lifetime in the semiconductor. If the carrier lifetime is long, the carrier recombination will continue for sufficient time after cessation of optical pulse and peaking of the carrier generation profile. Consequently, the photocurrent pulse will be broader on the time scale. So, the output pulse of terahertz wave will also be broadened. In other words, the bandwidth of the output terahertz wave obtained will shrink. Therefore, for broadband terahertz generation, the carrier lifetime in the semiconductor must be adequately short. Then the photocurrent pulse will decay fast after the attainment of peak by the carrier generation pulse. In turn, the output terahertz pulse will also narrow down. A larger bandwidth of terahertz radiation will then be available as the output. So, an ultra-short carrier lifetime in the semiconductor is a mandatory precondition to achieve broadband terahertz generation.

For detection of terahertz waves, the terahertz antenna is used as shown in figure 3.4. Here the showering of the antenna with terahertz radiation produces a current across the antenna arms whose magnitude is an indicator of the incident radiation intensity.

3.1.6 Mechanisms of terahertz pulse generation

The production and emission of terahertz waves by the PCA is considered to be the combined result of several phenomena taking place in quick succession. These are:

(i) Transient acceleration of photocarriers in the semiconductor film: This happens when the electrons and holes are produced by the impinging laser beam on the semiconductor, and separate from each other in the presence of the bias applied to the contacts of the PCA.

(ii) Instantaneous fall in resistance of the semiconductor between the gap of the dipole: Oscillations are produced at terahertz frequency in the antenna metal viz., the poles/metallic arms/contact pads due to the response from

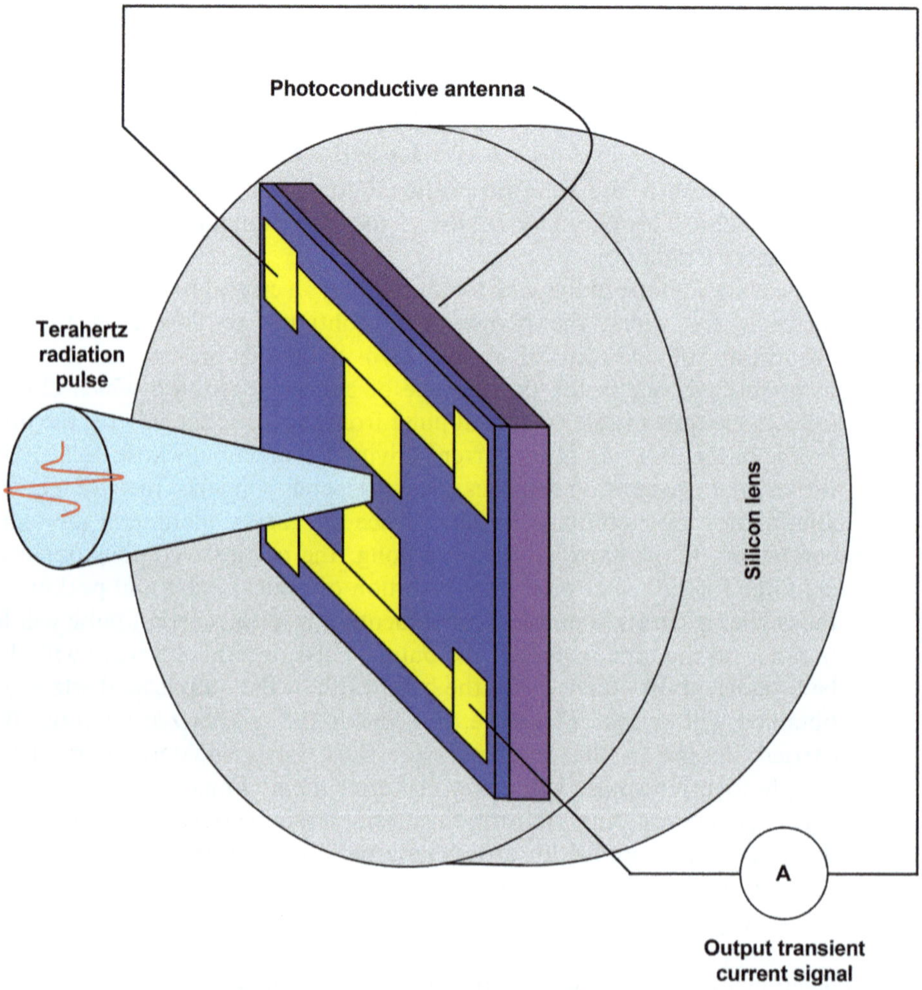

Figure 3.4. Detection of terahertz radiation by production of a transient current by striking a photoconductive antenna with a pulse of terahertz radiation.

the applied bias on the poles/metallic arms/contact pads to the rapid rise in free carrier concentration and decrease in resistance of the semiconductor in consequence to the optical excitation by laser pulse.

Note that before the pulse is applied, the antenna contact pads/arms are under the influence of an applied bias but the semiconductor film has a high resistance in absence of free carriers. So the parallel arms with the intervening high-resistance semiconducting film (dielectric) act as a capacitor storing electrostatic potential energy by virtue of the positive and negative charges build up on the two arms. There exists a distribution of electric field in the high-resistance semiconducting film and this distribution is disturbed on optical illumination to which a response in the form of high-frequency oscillation is anticipated from the pre-existing bias on the antenna contact pads/metallic arms. It is this response which has been mentioned above.

(iii) Direct collection of photocarrier current by the electrodes of the antenna: Carriers produced in vicinity to the antenna electrodes are immediately attracted by the electrodes. The resultant short-duration driving current due to carrier inflow induces oscillations in the antenna at terahertz frequencies

3.1.7 Generation of narrowband continuous wave terahertz radiation

Here the PCA is used as the heart of a photomixer device, which will be elaborated in section 3.3.

3.2 PCA time-domain spectroscopy system

So far we have learnt about the use of PCA as an emitter or detector of terahertz waves. For building a time-domain spectroscopy (TDS) system, the terahertz waves produced by the PCA illuminate the sample under test (figure 3.5). The beam emerging from the sample is detected by another PCA. Let us see how the PCA is

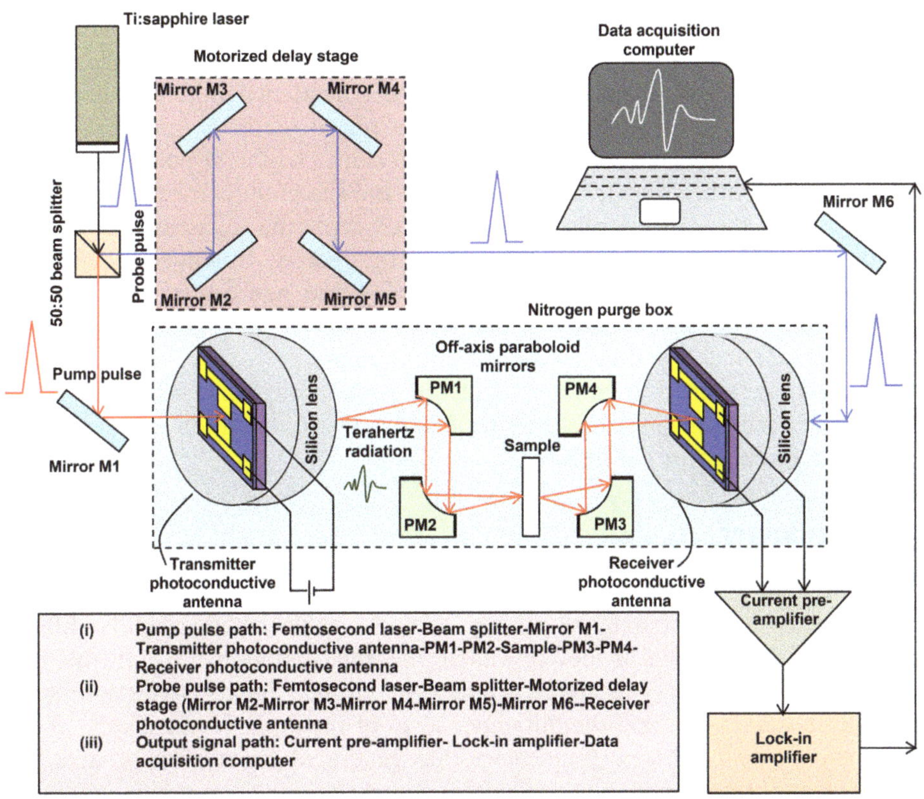

Figure 3.5. Scheme of electro-optic sampling of terahertz radiation pulses in a TDS system using photoconductive antennas for terahertz emission as well as detection.

used as the receiver. Unlike the transmitter antenna, no voltage source is connected across the receiver antenna. Instead a lock-in amplifier is connected across the contact pads of the receiver antenna.

The lock-in amplifier measures the transient voltage produced across the contacts of the receiver antenna by the terahertz beam striking the antenna on the semiconductor film at the spot between the contacts. The difference of a lock-in amplifier from an ordinary AC voltmeter must be clarified. For measurement, the lock-in amplifier must be supplied with a reference pulse in synchronization with the signal whose temporal variation is required. This pulse signal is used to identify the signal to be measured ignoring all signals which are not synchronized with the reference pulse. The synchronization procedure enables the lock-in amplifier to measure very weak signals in the nanovolt range while disregarding all other signals, however large. Thus it can extract the correct signal in a noisy environment. In distinction, the AC voltmeter measures the sum of all voltages appearing across the contacts.

To supply the necessary reference pulse for lock-in amplifier, a beam splitter is used to split a portion of the femtosecond pulse from the laser source illuminating the transmitter PCA. This optical pulse is passed through an adjustable optical delay line and is focused on the semiconductor film of the receiver PCA at the gap between the contacts. The optical pulse allows generation of an impulse of photocarriers at the gap region at a time determined by the delay line. Simultaneous arrival and overlapping of the optical pulse signal at the receiver and the transient electric field component of the incident terahertz pulse on the receiver antenna, results in the induction of a photocurrent proportional to the instantaneous electric field component of the incident terahertz pulse. The optical delay line is swept through various delay times and the photocarrier impulse signal produced from the optical pulse is convoluted with the transient electric field component of the incident terahertz pulse on the receiver antenna. By collection and correlation of the positions of optical delay and hence the induced photocurrent values, the time variation of the terahertz electric field is delineated. Thus the terahertz detection is done in a coherent manner.

The advantage gained by coherent detection is that a high signal-to-noise ratio is achieved. The method eliminates the effects of blackbody radiation and any sources of terahertz radiation at the receiver.

3.3 Photomixer

3.3.1 Heterodyne principle

It pertains to the combining or mixing of two signals having slightly different frequencies in a non-linear device called the mixer (figure 3.6). If the input frequencies are ν_1 and ν_2, the mixer supplies two output signals having frequencies equal to the sum $\nu_1 + \nu_2$ and the difference $\nu_1 - \nu_2$ of the input frequencies.

The two signals with new frequencies thus generated are referred to as heterodynes. Generally, the process is applied for frequency downconversion so that the lower of the two output frequencies $\nu_1 - \nu_2$ is only of interest. So, the higher frequency signal $\nu_1 + \nu_2$ is filtered out and the lower frequency signal $\nu_1 - \nu_2$ is retained for use.

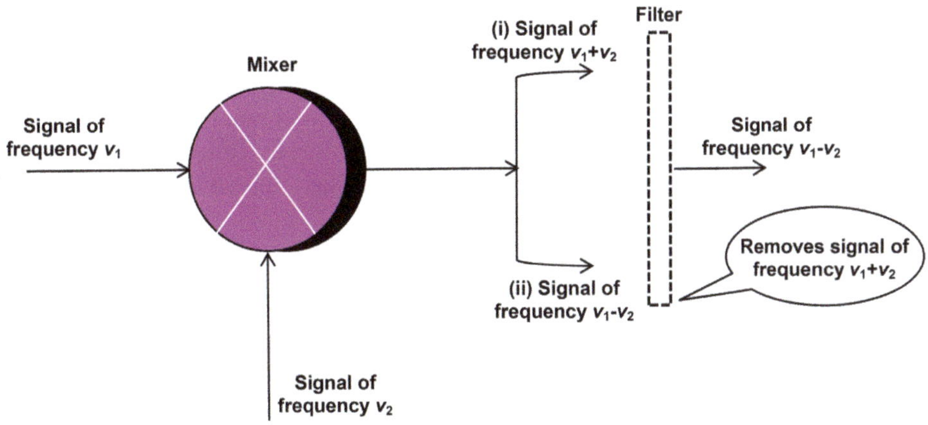

Figure 3.6. Principle of heterodyning.

Suppose two cosinusoidal waves of equal amplitude E_0 are mixed. The electric field-time (E-t) equations for the waves are:

$$E_1(t) = E_0 \cos(2\pi\nu_1 t) \tag{3.13}$$

and

$$E_2(t) = E_0 \cos(2\pi\nu_2 t) \tag{3.14}$$

When the two waves are superimposed on each other, the sum of electric fields is

$$\begin{aligned} E(t) &= E_1(t) + E_2(t) = A\cos(2\pi\nu_1 t) + A\cos(2\pi\nu_2 t) \\ &= 2A\left\{\cos 2\pi\left(\frac{\nu_1 + \nu_2}{2}\right)t + \cos 2\pi\left(\frac{\nu_1 - \nu_2}{2}\right)t\right\} \end{aligned} \tag{3.15}$$

by applying the cosine sum rule in trigonometry:

$$\cos\alpha + \cos\beta = 2\cos\left(\frac{\alpha + \beta}{2}\right)\cos\left(\frac{\alpha - \beta}{2}\right) \tag{3.16}$$

Thus superimposition of two waves produces components whose frequencies are equal to the sum and difference of the frequencies of the two waves mixed together.

From acoustics, we are aware of the phenomenon of beats produced by constructive and destructive interferences between two sound waves reaching the ear. The beating makes the sound alternate between zero and high or softness and loudness levels. The beats occur at a frequency equal to the magnitude of the difference between the frequencies of the two sound waves:

$$\nu_{\text{Beats}} = |\nu_1 - \nu_2| \tag{3.17}$$

Hence, heterodyning is analogous to beat production in acoustics.

The principle provides an elegant way for terahertz wave generation by applying the well-developed infrared laser technology, e.g., we can mix the signals from two infrared laser sources of slightly different frequencies, and choose the laser frequencies in such a manner that the frequency difference is in the range of terahertz frequencies, and of the desired terahertz frequency value. Then the output signal will be a terahertz source. The method has the advantage that it is continuously tunable in the terahertz band.

3.3.2 Construction of the photomixer

Figure 3.7 shows the photomixer whose principal components are the laser diodes and the PCA. A beam splitter and a mirror are also used.

3.3.2.1 Laser diodes

The photomixer consists of two tunable semiconductor diode lasers (Pliński 2010). These lasers operate independently. They are single-frequency lasers operating in a single resonator mode. Hence they are tunable single-mode lasers. Laser tuning is accomplished by one of the two ways: either by varying the temperature of the thermoelectric cooler of the laser, or by changing the laser driving current. The wavelength tuning ability is expressed in units of nm K^{-1} for temperature coefficient (e.g., 0.05 nm K^{-1}) and nm mA^{-1} for current coefficient (e.g., 0.002 nm mA^{-1}).

The frequencies of the lasers are often expressed in terms of a central frequency ν_0 as

$$\nu_1 = \nu_+ = \nu_0 + \frac{\Delta\nu_{\mathrm{THz}}}{2} \tag{3.18}$$

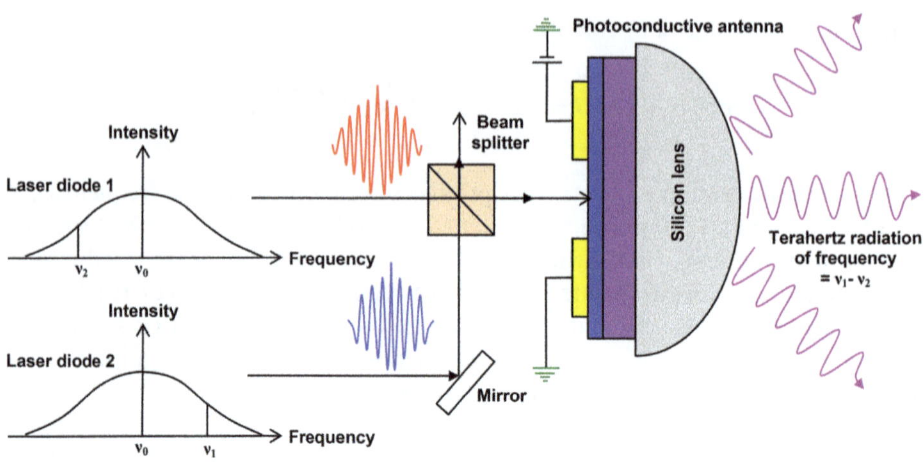

Figure 3.7. Photomixer in which overlapped laser beams of frequencies ν_1 and ν_2 are focused on a photoconductive antenna to produce terahertz waves with frequency equal to ν_1-ν_2.

and

$$\nu_2 = \nu_- = \nu_0 - \frac{\Delta\nu_{\text{THz}}}{2} \tag{3.19}$$

The equations for the two frequencies are sometimes written in a combined form as

$$\nu_{12} = \nu_0 \pm \frac{\Delta\nu_{\text{THz}}}{2} \tag{3.20}$$

or

$$\nu_\pm = \nu_0 \pm \frac{\Delta\nu_{\text{THz}}}{2} \tag{3.21}$$

Two types of single-frequency lasers are commonly used, namely, the distributed-feedback (DFB) laser and the distributed-Bragg-reflector (DBR) laser. In the DFB laser, the Bragg reflector is included within the structure of the laser diode itself, i.e., the Bragg reflector is distributed along the whole length of the active medium. In the DBR laser, the Bragg reflectors are located outside the active gain medium. They act as end mirrors.

The Bragg reflectors are periodic corrugated grating structures with distances in the range of wavelength of the radiation. They consist of a multilayer stack of thin films. These films are made of optical materials of high and low refractive indices. The high and low refractive index films are alternately placed. A quarter-wave mirror design is widely used. Here the thickness of each layer equals one quarter of the wavelength of radiation for which the Bragg reflector is made.

The multilayer structure comprising materials of different refractive indices ensures 100% reflectance over an extremely limited range of frequencies. The reflector is optimized to provide a high reflectivity at a particular frequency only. The Bragg reflector selectively reflects a single longitudinal mode as the lasing mode. Consequently, the laser functions on a single spatial and longitudinal mode providing single-frequency operation.

Example 3.3

An AlAs/GaAs Bragg mirror is fabricated for operation at an IR frequency of 282 THz with successively interposed films of AlAs (refractive index = 2.95) and GaAs (refractive index = 3.49). What thicknesses of the two films must be chosen for mirror realization?

Solution: For the IR frequency of 282 THz, the wavelength is

$$\begin{aligned}\lambda &= \frac{\text{Velocity of IR waves}}{\text{Frequency of waves}} = \frac{c}{\nu} = \frac{2.9979 \times 10^8}{282 \times 10^{12}} = 0.010\,63 \times 10^{8-12} \\ &= 0.010\,631 \times 10^{-4}\ \text{m} = 1.0631 \times 10^{-6}\ \text{m}\end{aligned} \tag{3.22}$$

In a Bragg mirror, all the films of high and low refractive materials (here GaAs and AlAs films) have thickness of one-quarter wavelength of terahertz radiation. Hence, thickness of GaAs film is

$$d_{GaAs} = \frac{\text{Wavelength of IR radiation}}{4 \times \text{Refractive index of the GaAs film}} = \frac{\lambda}{4\mu_{GaAl}} = \frac{1.0631 \times 10^{-6}}{4 \times 3.49}$$
$$= 0.07615 \times 10^{-6}\ \text{m} = \frac{0.07615 \times 10^{-6}}{10^{-9}}\text{m} \tag{3.23}$$
$$= 0.07615 \times 10^{-6+9}\ \text{nm} = 76.15\ \text{nm}$$

Similarly, thickness of AlAs film is

$$d_{AlAs} = \frac{\text{Wavelength of IR radiation}}{4 \times \text{Refractive index of AlAs film}} = \frac{\lambda}{4\mu_{AlAs}} = \frac{1.0631 \times 10^{-6}}{4 \times 2.95}$$
$$= 9.009 \times 10^{-8}\ \text{m} = \frac{9.009 \times 10^{-8}}{10^{-9}}\text{nm} \approx 90.09\ \text{nm} \tag{3.24}$$

3.3.2.2 Photoconductive antenna

The light beams from the laser sources are passed through a beam splitter and spatially overlapped. They are focused on the antenna gap. By superimposition of the laser beams, an interference pattern is created on the surface of the semiconductor. It consists of alternately dark and bright fringes running on the semiconductor. A DC voltage applied to the contact pads of the antenna leads to the flow of a photocurrent and launching of a terahertz wave from the antenna with frequency decided by the beat frequency ν_1-ν_2 of the laser beams, following the heterodyne principle. Due to the involvement of optical signals in photomixing, it is often referred to as optical heterodyning (Hindle *et al* 2008).

Example 3.4

The PCA in a photomixer is pumped with two laser sources of wavelengths 780 nm and 785 nm to produce a beat frequency in the terahertz region. Find the frequency of the wave obtained by photomixing.

Solution: Suppose λ_1, λ_2 are the wavelengths of the two laser sources. Let ν_1, ν_2 be their frequencies. The frequency of the output signal obtained is the difference between the frequencies of the two laser sources used. If c is the velocity of the waves, equation (3.17) can be rewritten as

$$\Delta\nu_{THz} = \nu_1 - \nu_2 = \frac{c}{\lambda_1} - \frac{c}{\lambda_2} = c\left(\frac{1}{\lambda_1} - \frac{1}{\lambda_2}\right) = \frac{c(\lambda_2 - \lambda_1)}{\lambda_1\lambda_2} \tag{3.25}$$

Putting the values of the wave velocity and wavelengths of the laser sources we have

$$\Delta\nu_{THz} = \frac{c(\lambda_2 - \lambda_1)}{\lambda_1\lambda_2} = \frac{2.9979 \times 10^8(785 - 780) \times 10^{-9}}{780 \times 10^{-9} \times 785 \times 10^{-9}} = \frac{2.9979 \times 5 \times 10^{8-9}}{780 \times 785 \times 10^{-9-9}}$$
$$= \frac{2.9979 \times 5 \times 10^{-1}}{780 \times 785 \times 10^{-18}} = 2.448\,06 \times 10^{-5} \times 10^{-1+18} \tag{3.26}$$
$$= 2.448\,06 \times 10^{-5-1+18} = 2.448\,06 \times 10^{12} = 2.45\ \text{THz}$$

3.3.3 Photomixer arrangement for imaging or spectroscopy

3.3.3.1 Imaging setup using cryogenic bolometer

The two laser sources of the photomixer feed light beams of the required frequency values ν_1, ν_2 to the PCA (figure 3.8). These frequencies are pre-selected to provide the terahertz waves of desired frequency difference ν_1-ν_2 as the output of the antenna. Optical isolators adjoining the laser sources prevent parasitic backscattering of the beams. During application of laser beams, square wave voltage pulses are applied to the contacts of the PCA. The frequency of pulses ranges from a few hertz to few kHz. So terahertz pulses at these frequencies are produced by the antenna. These pulses are collimated and focused by a pair of polyethylene lenses on the sample. After passing through and crossing the sample, the pulses are focused by another pair of polyethylene lenses on the cryogenic-bolometer, which provides a good signal-to-noise ratio. For convenience and portability, the cryo-bolometer is often substituted with a room-temperature detector albeit with low signal-to-noise ratio.

3.3.3.2 Imaging setup using coherent homodyne detection

An alternative option to the cryo-bolometer setup is an arrangement employing coherent homodyne detection. In this scheme (figure 3.9), a second PCA is used as the detector. Polyethylene lenses are used to focus the terahertz wave incident on the sample as well as the wave emerging out of the sample. One pair of lenses is used on the incidence side and another pair for the emerging wave. The radiation coming out from these lenses falls on the second PCA.

For synchronization of the measurement setup, the laser beam which was used for producing the terahertz radiation, i.e., the laser beam used for excitation of the wave in the first PCA is also used for exciting the second PCA at the receiver. A delay line placed in its route introduces the required time interval for coordinating the events to operate the system in harmony. Photocarriers thus generated produce a voltage signal at the receiving dipole antenna when the terahertz radiation transmitted through the sample strikes this antenna, thereby enabling detection.

3.4 Salient features of the PCA mixer

Let us look closely at the operation of the antenna, and bring out some noteworthy features.

3.4.1 Power limitation

We shall henceforth call the PCA mixer the antenna emitter (AE) for sake of brevity. In the AE, two laser beams strike a semiconductor in a small area between the dipole connected to microstrip lines and contact pads. At the same time, a DC potential difference is applied on the contact pads and the photocurrent produced is supplied to the antenna. The emitted terahertz waves pass through the semiconductor. This semiconductor has a large dielectric constant ε_r, which leads to a high value of refractive index. For GaAs, $\varepsilon_r = 12.9$; hence

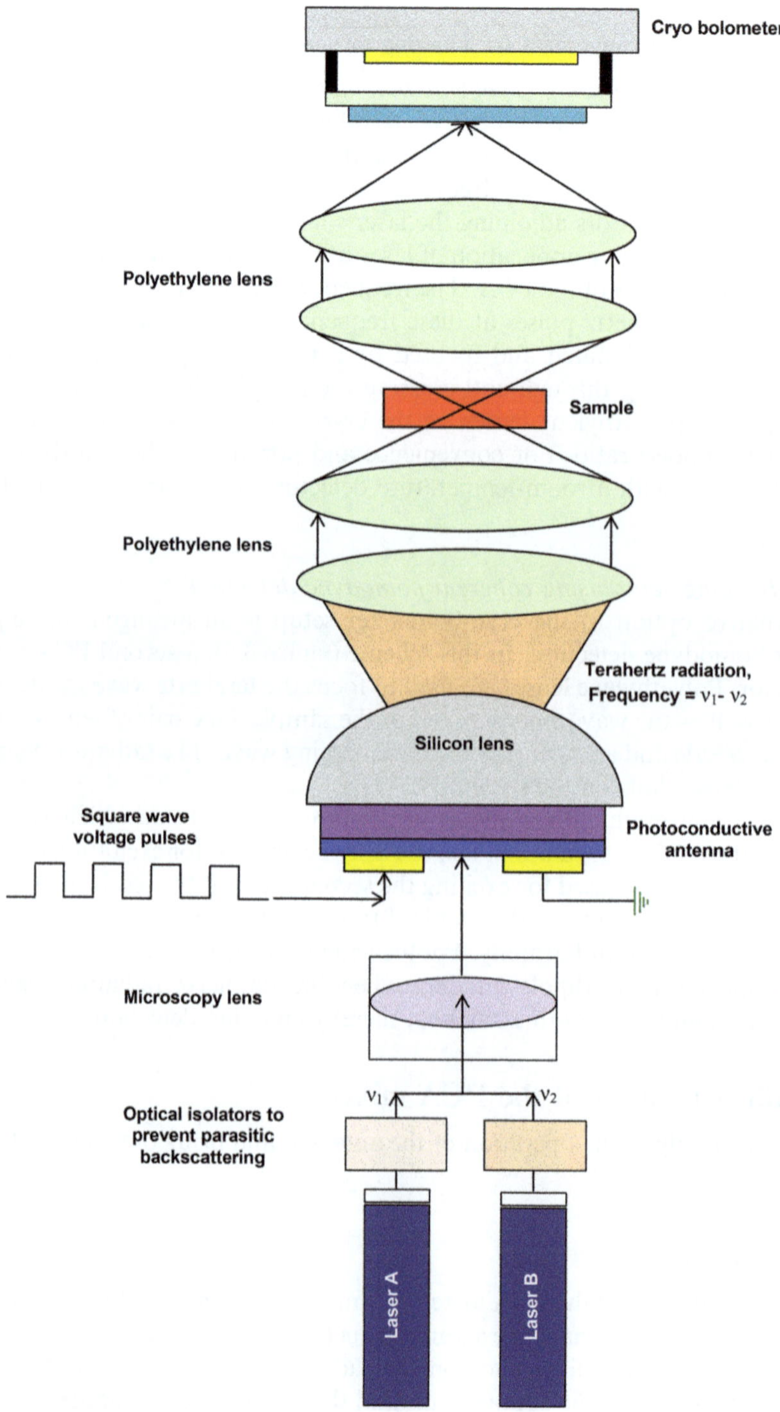

Figure 3.8. Cryo-bolometer-based imaging arrangement in which two lasers sources A and B of frequencies ν_1, ν_2 are used to produce a frequency ν_1-ν_2 with the help of a photoconductive antenna, and the terahertz radiation thus obtained by difference-frequency mixing impinges on the bolometer.

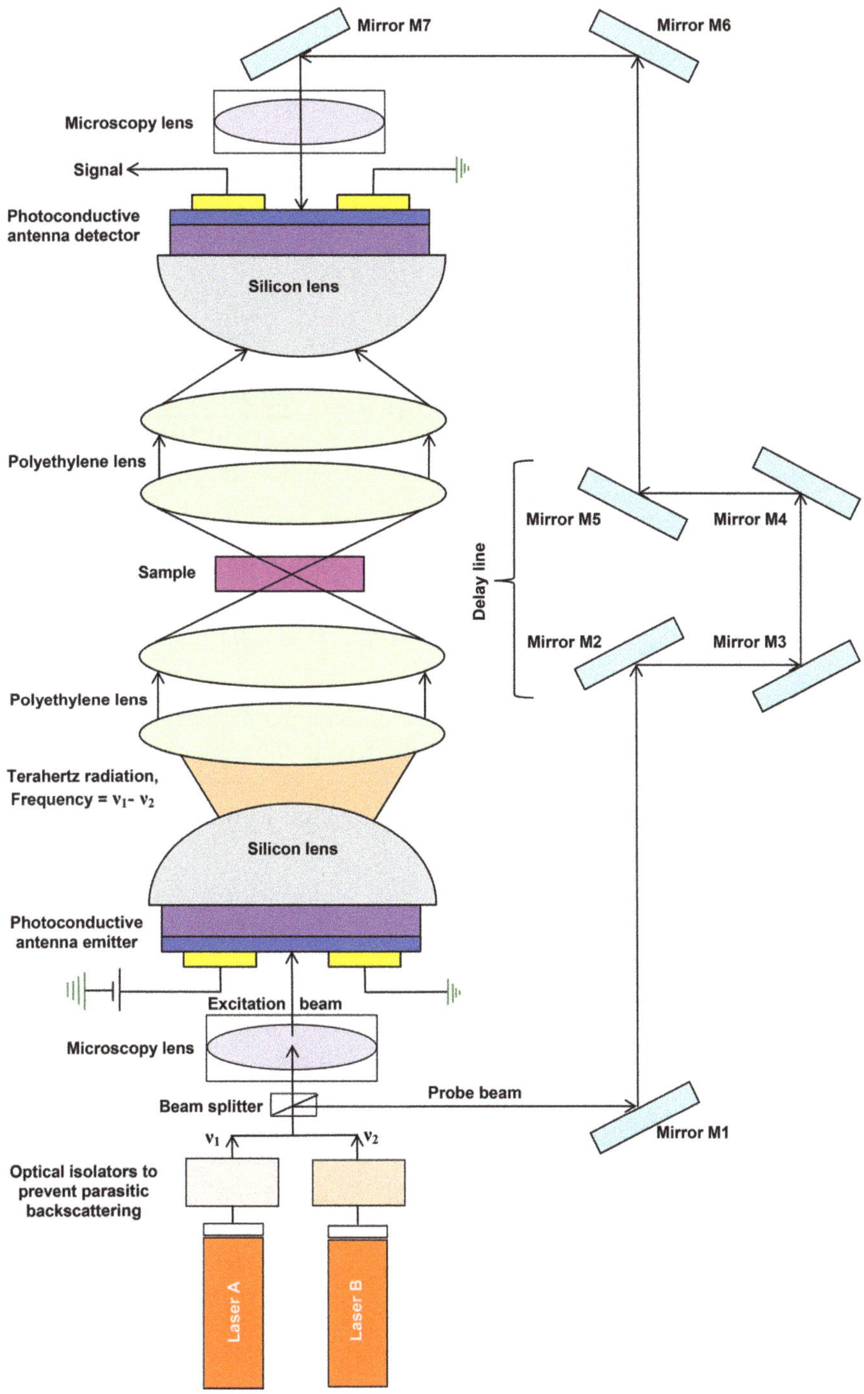

Figure 3.9. Coherent homodyne detection-based imaging arrangement using two photoconductive antennas with the second photoconductive antenna serving as the detector in place of the cryo-bolometer. The synchronization is achieved with a probe beam which is the same as the pump beam for the first photoconductive antenna with the necessary delay introduced by a delay line.

$$\mu = \sqrt{\varepsilon_r} = \sqrt{12.9} = 3.59 \tag{3.27}$$

Collimation of the terahertz beam is done by a hyperhemispherical lens made of silicon.

The important fact to be noted from these discussions is that the laser beams are focused into a spot of dimensions L_x and L_y. These dimensions are much smaller than the wavelength of terahertz waves, e.g., the wavelength of 1 THz terahertz waves is

$$\begin{aligned}\lambda_{1\ \text{THz}} &= \frac{c}{\nu} = \frac{2.9979 \times 10^8}{1 \times 10^{12}} \\ &= 2.9979 \times 10^{8-12} = 2.9979 \times 10^{-4}\ \text{m} = \frac{2.9979 \times 10^{-4}}{10^{-6}}\ \mu\text{m} \\ &= 2.9979 \times 10^{-4+6}\ \mu\text{m} = 2.9979 \times 10^{2}\ \mu\text{m} = 299.79\ \mu\text{m}\end{aligned} \tag{3.28}$$

So the area of the spot $L_x \times L_y$ must be much smaller than 300 μm.

For the PCA mixer, the terahertz power is given by (Preu *et al* 2011)

$$P_{\text{THz}}^{\text{AE}} = R_{\text{Radiation}} I_{\text{RMS}} = R_{\text{Radiation}} \left(\frac{1}{\sqrt{2}} I_{\text{THz}} \right)^2 = \frac{1}{2} R_{\text{Radiation}} I_{\text{THz}}^2 \tag{3.29}$$

where R_R stands for the radiation resistance of the antenna and I_{RMS} is the average or RMS component of the photocurrent produced in the semiconductor, I_{THz} is the peak terahertz component of this photocurrent. We defer discussion about radiation resistance to section 3.4.2. The current I_{THz} is related to the power of the incident laser beam P_{Laser}, as we see below:

$$\begin{aligned}P_{\text{Laser}} &= \frac{\text{Energy of terahertz radiation}}{\text{Time of illumination by laser}} \\ &= \frac{\text{Planck's constant} \times \text{Frequency of terahertz radiation emitted}}{\text{Time}} = \frac{h\nu}{t}\end{aligned} \tag{3.30}$$

or

$$t = \frac{h\nu}{P_{\text{Laser}}} \tag{3.31}$$

But

$$I_{\text{THz}} = \frac{\text{Electronic charge}}{\text{Time}} = \frac{e}{t} = \frac{e}{\dfrac{h\nu}{P_{\text{Laser}}}} = \frac{eP_{\text{Laser}}}{h\nu} \tag{3.32}$$

Substituting for I_{THZ} from equation (3.32) in equation (3.29) we get

$$P_{\text{THz}}^{\text{AE}} = \frac{1}{2} R_{\text{Radiation}} I_{\text{THz}}^2 = \frac{1}{2} R_{\text{Radiation}} \left(\frac{eP_{\text{Laser}}}{h\nu} \right)^2 \tag{3.33}$$

Equation (3.33) tells us

$$P_{\mathrm{THz}}^{\mathrm{AE}} \propto P_{\mathrm{Laser}}^{2} \tag{3.34}$$

because e, h are constants and ν is also constant for a given frequency. It is seen that the power of the generated terahertz radiation is proportional to the square of the laser beam power. At low laser beam powers, the small size of the spot illuminated by the laser beam does not matter because the heat can be easily dissipated but at high powers, the heat may become unmanageable. So, the PCA mixer suffers from the drawback of its unsuitability for high terahertz powers.

3.4.2 Radiation resistance of the antenna and *R–C* roll-off

Radiation resistance describes the recoil force exerted on the electron during photon emission. It decreases the momentum of the electron. The power consumed by the radiation resistance is transformed into electromagnetic waves. So the radiation resistance represents the desired function of the antenna. Besides the radiation resistance, there is a resistance R_{Losses} arising from ohmic losses in the antenna metallization, dielectric losses in insulating regions and ground losses due to current induction in the ground. Hence the total feedpoint resistance

$$R_{\mathrm{Feedpoint}} = R_{\mathrm{Radiation}} + R_{\mathrm{Losses}} \tag{3.35}$$

and the input power

$$P_{\mathrm{Feedpoint}} = \frac{1}{2} R_{\mathrm{Feedpoint}} I_{\mathrm{THz}}^{2} \tag{3.36}$$

From equations (3.29), (3.35) and (3.36), the antenna efficiency

$$\eta = \frac{P_{\mathrm{THz}}^{\mathrm{AE}}}{P_{\mathrm{Feedpoint}}} = \frac{\frac{1}{2} R_{\mathrm{Radiation}} I_{\mathrm{THz}}^{2}}{\frac{1}{2} R_{\mathrm{Feedpoint}} I_{\mathrm{THz}}^{2}} = \frac{R_{\mathrm{Radiation}}}{R_{\mathrm{Feedpoint}}} = \frac{R_{\mathrm{Radiation}}}{R_{\mathrm{Radiation}} + R_{\mathrm{Losses}}} \tag{3.37}$$

which implies that larger the radiation resistance, higher is the efficiency. The radiation resistance is proportional to the wave impedance

$$Z = \sqrt{\frac{\mu_0 \mu_{\mathrm{Effective}}}{\varepsilon_0 \varepsilon_{\mathrm{Effective}}}} \tag{3.38}$$

where μ_0, ε_0 are permeability and permittivity of free space respectively; and $\mu_{\mathrm{Effective}}$, $\varepsilon_{\mathrm{Effective}}$ are their effective values. Equation (3.38) can be expressed as

$$Z = \sqrt{\frac{\mu_0}{\varepsilon_0}} \sqrt{\frac{\mu_{\mathrm{Effective}}}{\varepsilon_{\mathrm{Effective}}}} = Z_0 \sqrt{\frac{\mu_{\mathrm{Effective}}}{\varepsilon_{\mathrm{Effective}}}} \tag{3.39}$$

where

$$Z_0 = \sqrt{\frac{\mu_0}{\varepsilon_0}} = \sqrt{\frac{1.2566 \times 10^{-6}\ \mathrm{H\ m^{-1}}}{8.854 \times 10^{-12}\ \mathrm{F\ m^{-1}}}} = \sqrt{\frac{1.2566 \times 10^{-6+12}}{8.854}} \tag{3.40}$$
$$= 0.3767 \times 10^3 \approx 377\ \Omega$$

For non-magnetic materials, $\mu_{\mathrm{Effective}} = 1$; hence from equations (3.39) and (3.40)

$$Z = \frac{377}{\sqrt{\varepsilon_{\mathrm{Effective}}}} \tag{3.41}$$

Here, $\varepsilon_{\mathrm{Effective}} = 12.9$ for GaAs; so equation (3.41) gives

$$Z = \frac{377}{\sqrt{12.9}} = \frac{377}{3.5917} = 104.96 \approx 105\ \Omega \tag{3.42}$$

The finite capacitance C of the PCA and the radiation resistance $R_{\mathrm{Radiation}}$ are connected in parallel to the photocurrent source. Consequently, the device exhibits a roll-off, i.e., a gradually reduced response with time constant τ_{RC} at high frequencies. Finite transport time of free carriers is responsible for this roll-off. The dependence of roll-off factor $\xi_{\mathrm{Roll\text{-}off}}$ on frequency ν is given by

$$\xi_{\mathrm{Roll-off}} = \frac{1}{1 + \left(\frac{\nu}{\nu_{\mathrm{RC}}}\right)^2} \tag{3.43}$$

where

$$\nu_{\mathrm{RC}} = 3 - \mathrm{dB\ cut} - \mathrm{off\ frequency} = \frac{1}{2\pi\tau_{\mathrm{RC}}} \quad \text{with} \quad \tau_{\mathrm{RC}} = R_{\mathrm{Radiation}}C \tag{3.44}$$

Example 3.5
The radiation resistance of a PCA is 70 Ω. A 3 dB cut-off frequency of 3 THz is required. Find the capacitance of the device.
Solution: From equation (3.44),

$$\nu_{\mathrm{RC}} = \frac{1}{2\pi\tau_{\mathrm{RC}}} = \frac{1}{2\pi R_{\mathrm{Radiation}}C} \tag{3.45}$$

or

$$C = \frac{1}{2\pi R_{\mathrm{Radiation}}\nu_{\mathrm{RC}}} = \frac{1}{2 \times 3.141\,59 \times 70 \times 3 \times 10^{12}} \tag{3.46}$$
$$= 7.579 \times 10^{-4} \times 10^{-12} = 7.579 \times 10^{-16}\ \mathrm{F}$$

3.5 Antenna-less large-area emitter approach of terahertz generation

So far the antenna-based approach to photomixing was outlined. But the question arises, can we generate terahertz waves without the antenna? If so, what will be the differences with respect to AE? Yes, it is possible using the large-area emitter (LAE) without the antenna. Figure 3.10 shows the directions of radiation emission for antenna-included and antenna-excluded methods.

We know that the emitted terahertz power varies quadratically with the incident laser power. In the LAE scheme, the laser beams are incident on a broader region whose dimensions are similar to or even larger than the wavelength of the terahertz radiation, hence providing efficient power dissipation. Thus the small area and hence low power capability problem encountered with PCA mixer is removed by using LAE.

The pattern of terahertz radiation emission from LAE differs from the AE. Here, unlike the terahertz propagation through the semiconductor, the terahertz beam follows two paths. One path is through the GaAs film below. The other path is in the direction of the reflected optical beam. It is found that the intensity of terahertz radiation emitted in all directions, which are different from these two directions, rapidly falls to zero as the dimensions of the area illuminated by the laser beams increase. This decay is due to destructive interference. Moreover, the terahertz beams emitted in the two aforesaid directions are of comparable diameters. Hence, by illuminating a larger area, the terahertz wave scattering to all directions except from that through the semiconductor and along the reflected optical beam can be minimized. Such simplification of optics for LAE in contrast to AE makes it very promising. Additionally, the fabrication procedure of LAE is much less demanding than that of AE. Together with the advantage of easier heat dissipation at high powers owing to larger illumination area, the LAE is very useful.

However, LAE also has shortcomings. There is no explicit antenna in LAE but it is implicitly present. The electron and hole pairs act as elementary Hertzian dipoles. Each incident photon in the laser beam produces a transient free electron–hole dipole. These dipoles are estimated to have small dipole lengths ~217 nm, and the dipole radiation resistance is extremely small $R_{\text{Radiation (dipole)}}$ ~1.1 mΩ (Preu *et al* 2011). The small radiation resistance of the elementary dipoles has two vital repercussions. Firstly, from equation (3.37) we find that the antenna efficiency is appreciably reduced. Secondly, from the point of view of *R–C* roll-off, the reduction in response is much less severe at high frequencies for LAE than AE ensuring better higher-frequency operation. This means that the 3 dB cut-off frequency can be shifted upwards deriving the benefit of smaller radiation resistance. Higher power outputs are achievable with LAE configuration for the resonances of the dipole above the 3 dB frequency of AE photomixer. So, the fall in efficiency is an adverse effect but the rise in frequency is beneficial.

3.6 Photomixer using UTC-PD

Unitravelling-carrier photodiode (UTC-PD) is another device that is used in terahertz emitters. The commonly used semiconductor GaAs in PCAs requires laser sources at frequencies above 3.4×10^{14} Hz (wavelengths below 0.87 μm) while the UTC-PD based terahertz emitters employ the compact laser sources developed for optical communication for frequencies of 1.99×10^{14} Hz (1.5 μm) (Deninger 2013).

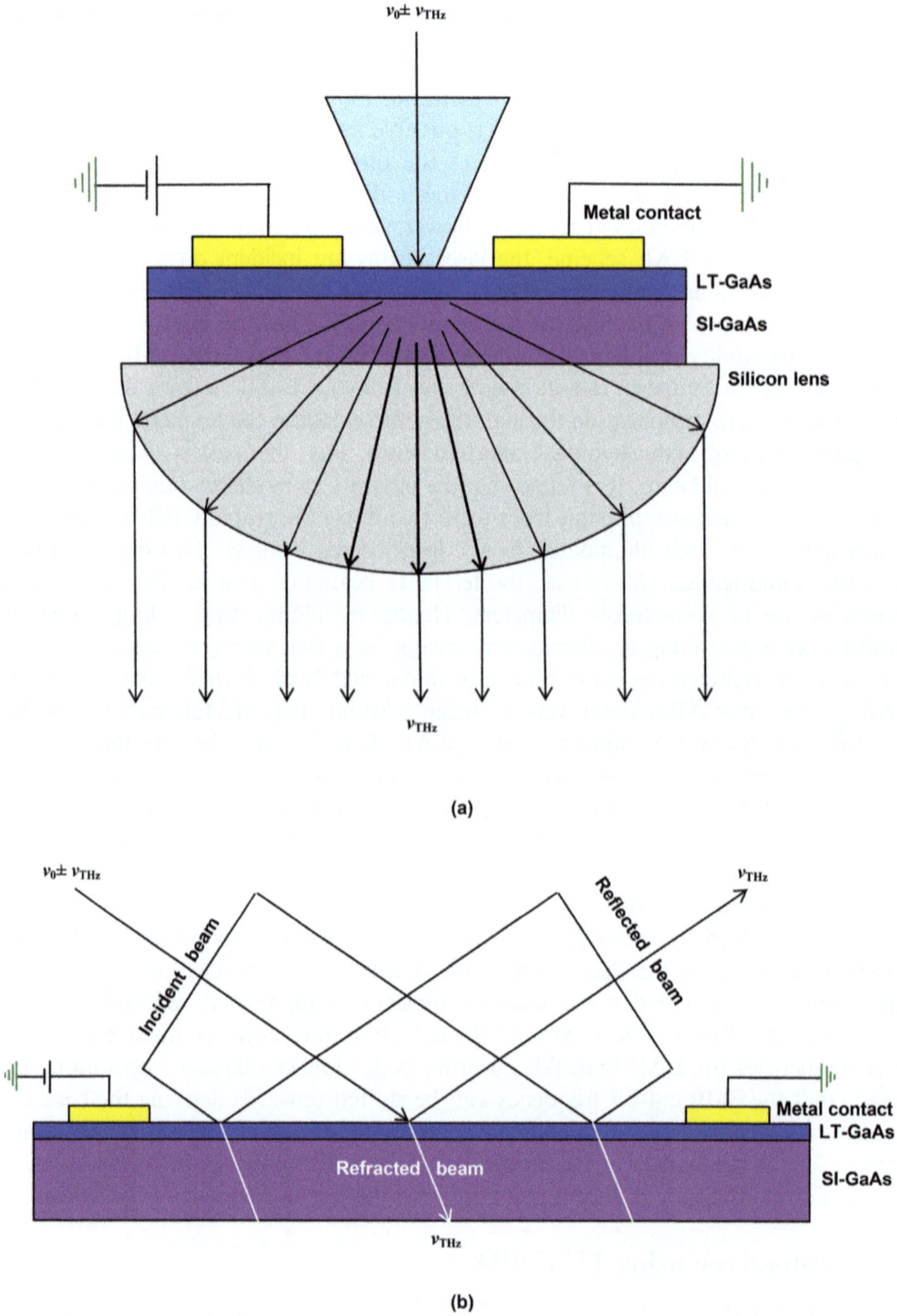

Figure 3.10. Radiation emission directions for the two approaches to terahertz generation by photoconductivity: (a) with antenna and (b) a large semiconductor area without antenna. The electrical connections for external bias are shown.

3.6.1 Unitravelling-carrier photodiode

As its name suggests, the UTC-PD is a photodiode which converts light or photons into an electric current. It is a modified form of PIN photodiode which being a slower device cannot cope with terahertz frequencies. Let us know about the PIN photodiode, the reason due to which it fails to work at terahertz frequencies, and the necessary improvements made in a UTC-PD for fast operation.

3.6.1.1 PIN photodiode

The PIN photodiode consists of an intrinsic (I) region interposed between a P-type region on one side and an N-type region on the other side (figure 3.11). The P-type region is called the anode and N-type region is known as the cathode. The P- and N-regions are heavily doped and the intrinsic region is generally wide. This wide intrinsic or undoped region acts as the absorber layer for photons. Due to the larger

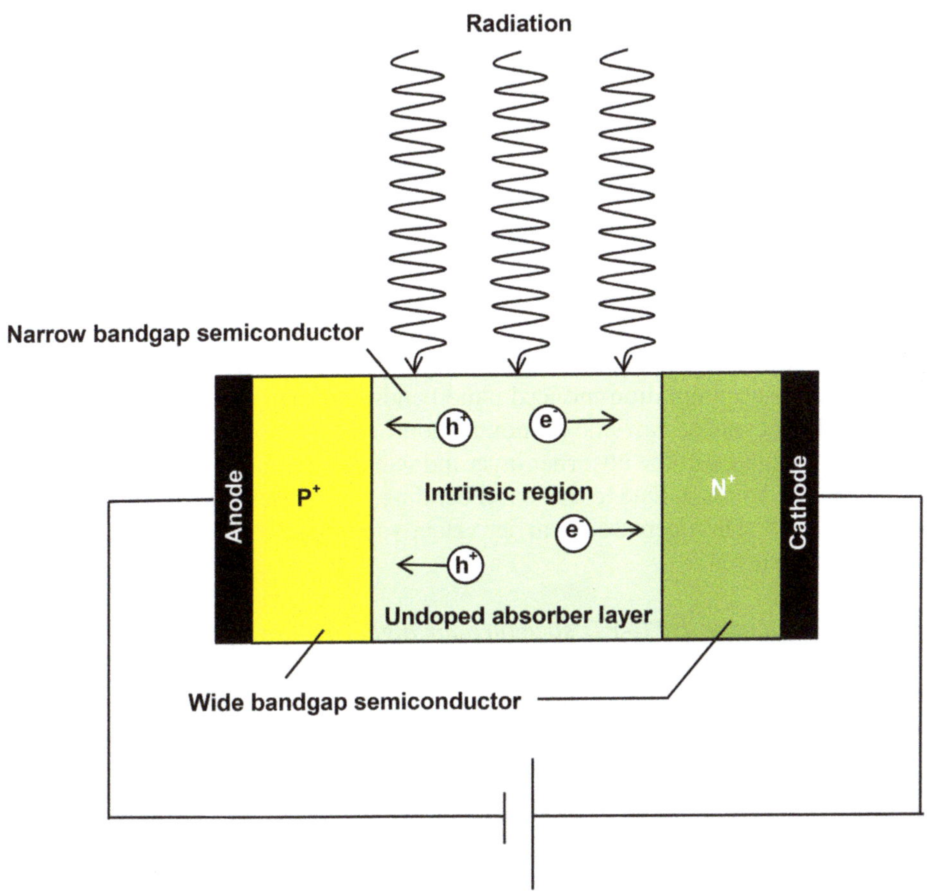

Figure 3.11. Schematic diagram of a heterojunction PIN diode showing the heavily-doped P and N end regions made of a wide bandgap semiconductor separated by a broad intrinsic middle region made of a narrow bandgap semiconductor and acting as the absorber layer. The electrical drifting of radiation-created electrons and holes towards opposite electrodes in reverse-biased configuration is shown.

volume provided by the wide absorber layer to incident radiation than a normal P–N junction diode, its sensitivity is much higher than that of a normal diode.

The P- and N-regions are made of a wide bandgap semiconductor e.g., AlGaAs while the intrinsic region is a low bandgap semiconductor such as GaAs resulting in a heterojunction diode. The difference in bandgap enables the creation of a barrier height which facilitates the injection of holes from the P^+ anode and impedes the back flow of electrons from the I-region to P^+ anode. The injected carriers are confined within the intrinsic region thereby lowering the forward voltage drop across the PIN diode.

For using the PIN photodiode as a photodetector, it is operated in the reverse-biased mode. A depletion region is formed across the junction. An incident photon striking an electron in the undoped absorber layer dislodges it from the chemical bond resulting in the creation of an electron vacancy or hole. Thus an electron–hole pair is produced. The electrons and holes are swept to the respective contacts by the electric field across the junction. As a consequence, a reverse bias current is recorded. The greater the intensity of incident light, the larger is the number of electron–hole pairs and the higher is the reverse bias current of the diode. So the reverse bias current gives an indication of the light intensity.

3.6.1.2 Deficiencies of PIN photodiode structure

The PIN photodiode suffers from two major shortcomings which incapacitate it for high-frequency operation:

(i) Bipolar conduction: The current is carried by two charge carriers, viz., electrons and holes. Electrons have a smaller effective mass and so are more agile. On the other hand, the holes have a larger effective mass making them slower than electrons. As the net current is the sum of electron and hole currents, it is decreased because of the lower hole velocity.

(ii) Carrier accumulation-induced band bending: When the intensity of incident light increases, carriers cannot respond to the field fast enough and are accumulated in the absorber layer inducing a bending of the bands by space charge buildup. Due to this band bending, the carrier velocity is appreciably reduced. This decrease of carrier velocity manifests itself as saturation of the diode current.

3.6.1.3 Provisions in UTC-PD structure for higher photoresponse

For sake of clarity, let us look at an InGaAs/InP UTC-PD diode (figure 3.12). Like the PIN photodiode, it is a heterojunction diode. Here InGaAs is the narrow bandgap semiconductor. If the alloy composition is taken as $In_{0.53}Ga_{0.47}As$, the energy bandgap is 0.75 eV. InP is a wide bandgap semiconductor with energy gap of 1.344 eV.

The main layers in the InGaAs/InP structure are: (i) P^{++} InGaAs contact layer, (ii) P^{++} InGaAsP blocking layer, (iii) P^+ InGaAs absorber layer, (iv) I-InGaAs spacer layer, (v) I-InGaAsP spacer layer, (vi) I-InP spacer, (vii) N^+ InP cliff layer, (viii) N-InP collector layer, (ix) N^+ InP subcollector layer and (x) N^+ InGaAs contact layer (Banik *et al* 2008).

Figure 3.13 shows the energy-band diagrams of the PIN diode and UTC-PD under reverse bias.

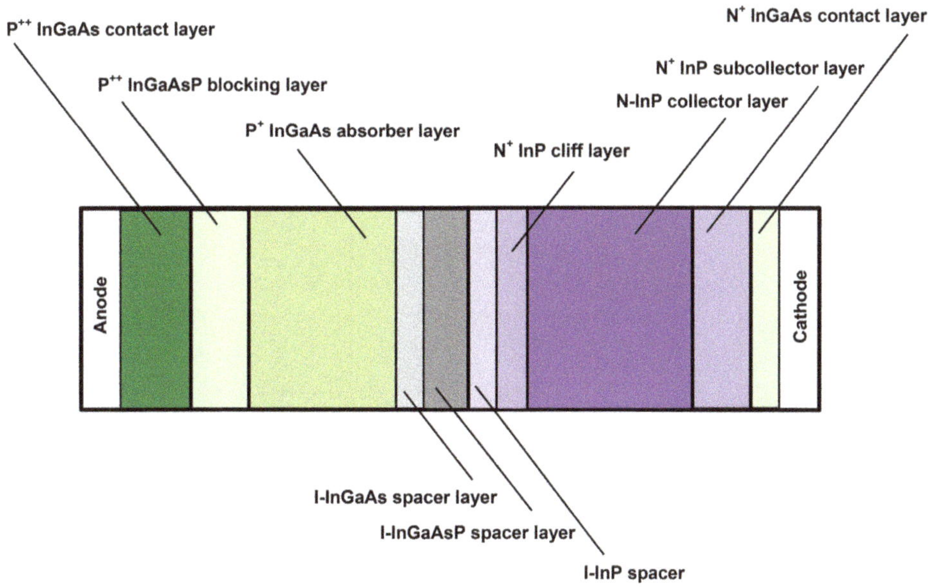

Figure 3.12. The ten constituent layers in the InGaAs/InP UTC-PD structure: three P-type layers, three intrinsic spacer layers and four N-type layers.

Example 3.6

In a PCA, the semiconductor film is made of GaAs while in a UTC-PD, the P^+ InGaAs absorber layer is used. The energy gap of GaAs is 1.424 eV while that of InGaAs is 0.75 eV. Find the minimum frequencies of laser sources required for producing electron–hole pairs in the two semiconductors.

Solution: Since the energy E of a photon is related to its frequency ν through the Planck's constant h as

$$E = h\nu \tag{3.47}$$

we have

$$\nu = \frac{E}{h} = \frac{\text{Energy band gap}}{h} = \frac{\text{Energy band gap}}{4.13567 \times 10^{-15}\,\text{eV. s}} \tag{3.48}$$

Hence,

$$\nu_{\text{GaAs}} = \frac{\text{Energy band gap of GaAs}}{4.13567 \times 10^{-15}\,\text{eV. s}} = \frac{1.424\ \text{eV}}{4.13567 \times 10^{-15}\,\text{eV. s}} = 3.443 \times 10^{14}\ \text{Hz} \tag{3.49}$$

Similarly,

$$\nu_{\text{InGaAs}} = \frac{\text{Energy band gap of InGaAs}}{4.13567 \times 10^{-15}\,\text{eV. s}} = \frac{0.75\ \text{eV}}{4.13567 \times 10^{-15}\,\text{eV. s}} = 1.8135 \times 10^{14}\ \text{Hz} \tag{3.50}$$

Comment: A UTC-PD requires a lower frequency laser source than a GaAs PCA.

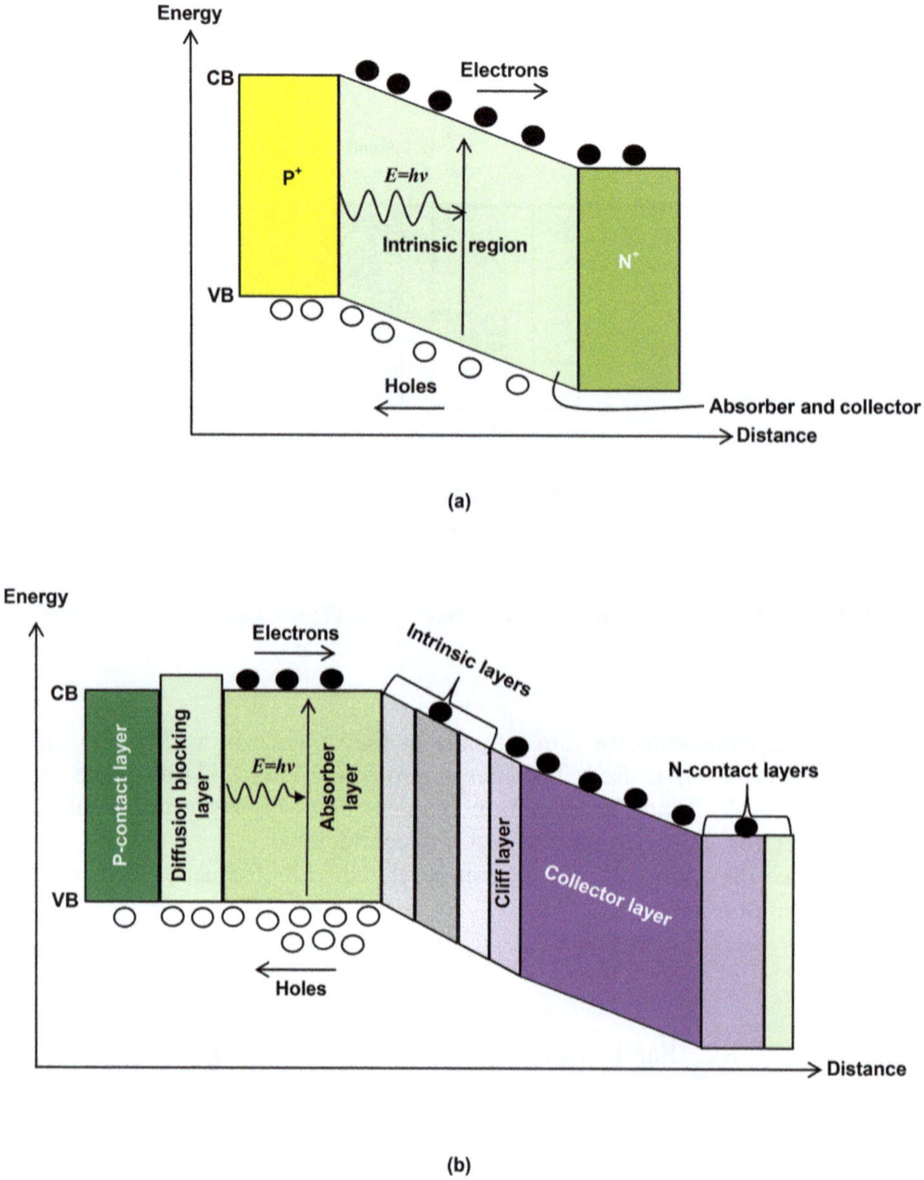

Figure 3.13. Reverse-bias energy-band diagrams of: (a) PIN diode and (b) UTC photodiode.

If we look closely at the structural layouts of PIN-photodiode (figure 3.11) and UTC-PD (figure 3.12), we notice that in the PIN-photodiode, the intrinsic region performs dual functions acting as the absorber region as well as the carrier collection region. In a UTC-PD, the absorber region is P^+ InGaAs layer while the collector region is the N-InP collector layer. So in UTC-PD, the absorber and collector regions are two distinct and separate regions. Their thicknesses can be independently decided and controlled. The P^+ InGaAs absorber layer can be made very thin. Then the holes produced in this layer have to travel a very short distance to reach the contact. As a consequence, the holes quickly move to the contact. Any minority carrier electrons generated by photon impact in the absorber layer will be stopped by the diffusion blocking layer. Hence, the space charge produced comprises only one type of carrier, namely the electrons. Since the UTC-PD utilizes only electrons as active charge carriers, it is referred to as a unicarrier device.

The minority-carrier electrons generated in the P^+ InGaAs absorber layer diffuse to the collector and are accelerated by the applied electric field towards the collector. In the depleted collector, the electron transport is quasi-ballistic. The ballistically high velocity makes the time delay in electron transport almost insignificant.

Further, the electron velocity being higher than hole velocity, current saturation is delayed to higher levels of optical intensity. The high electron velocity in depletion region is supported even at a lower electric field or even with built-in electric field of the junction. Thus UTC-PD can operate at higher power levels and faster speeds at lower applied voltages than PIN-photodiode. It shows both high speed and excellent output characteristics.

3.6.2 UTC photomixer

For generation of electromagnetic waves at terahertz frequencies, photomixing is carried out by employing a UTC-PD (figure 3.14), and the principle used is optical-to-electrical (O/E) conversion (Wakatsuki *et al* 2012).

As already described in the case of PCA, two higher frequency signals of frequencies ν_1 and ν_2 are combined together to yield the difference-frequency signal at frequency ν given by

$$\nu = \nu_1 - \nu_2 \tag{3.51}$$

where $\nu_1 > \nu_2$. This difference-frequency signal given by equation (3.51) is called the optical beat signal. It is supplied to a UTC-PD using a setup consisting of two parts: an optical part and an electrical part. In the optical part, an optical fiber is used along with lenses for collimation of the beat signal. The UTC-PD chip is mounted on a metallic block connected to a waveguide. This waveguide serves as an antenna. The connection of metallic block with the waveguide is done through a quartz coupler. The coupler is a three-component arrangement. These components are: (i) an impedance converter, (ii) a strip line and (iii) a radiator. A butterfly package is used for assembly of all the components.

The frequency of the optical beat signal is varied from 0.2 THz to 0.65 THz. The output power changes with frequency. It attains a maximum value of 0.5 mW at 0.35 THz. The 3 dB bandwidth is 0.12 THz while the 10 dB bandwidth is 0.26 THz.

Using UTC-PD photomixing technique, continuous wave terahertz radiation at frequencies >5 THz has been generated (Ishibashi and Ito 2020).

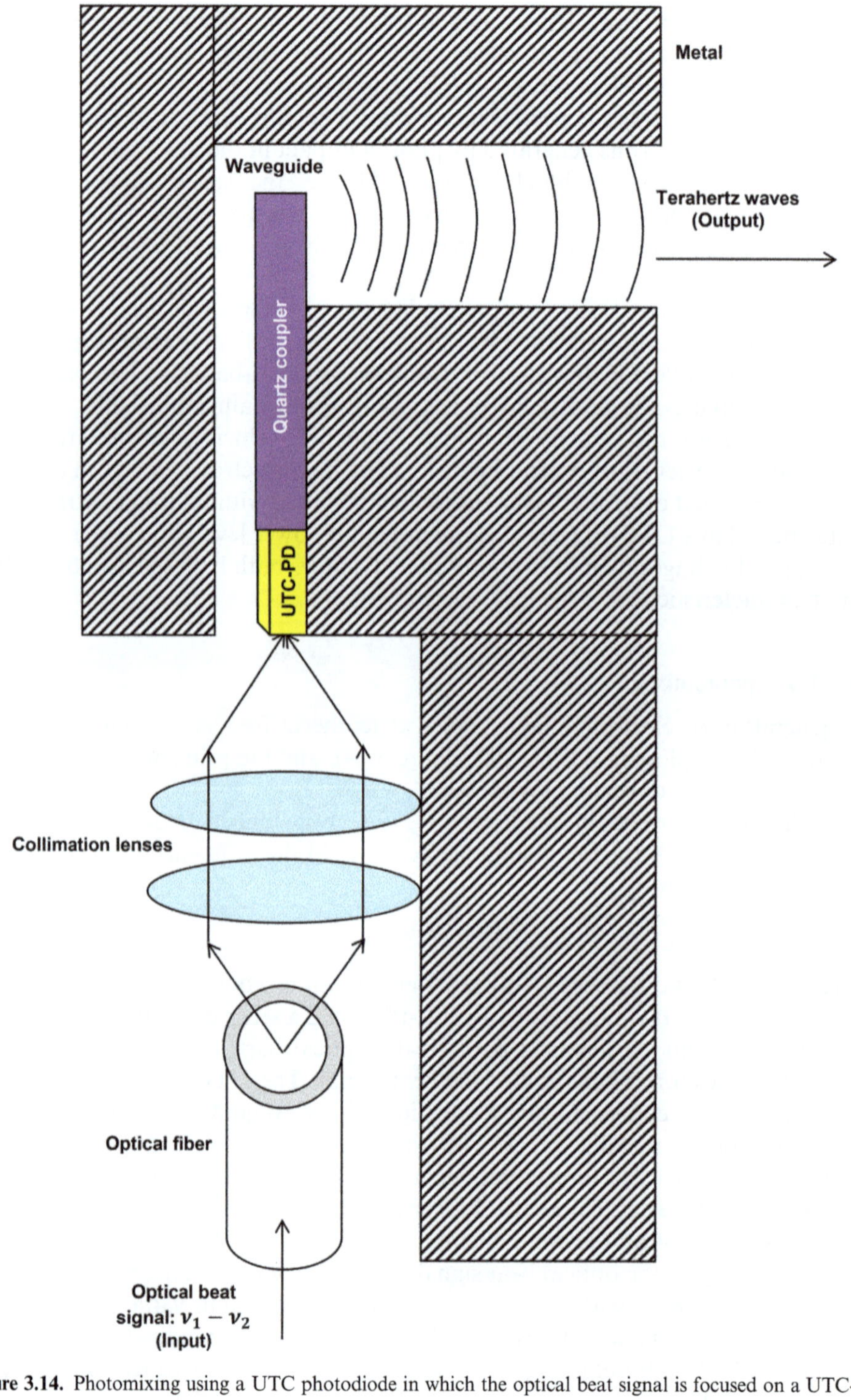

Figure 3.14. Photomixing using a UTC photodiode in which the optical beat signal is focused on a UTC-PD chip to generate terahertz waves. The waves produced are taken out via a quartz coupler and launched by a waveguide antenna.

3.7 Discussion and conclusions

PCAs can produce broadband pulses of terahertz radiation as well as narrow bandwidth continuous wave terahertz waves. Their room-temperature operation provides a convenient working environment. Photomixers working on optical heterodyning principle combine together the merits of frequency tuning and power capability together with room-temperature operation to generate and detect terahertz radiation enabling several applications in communication, astronomy and various fields. The UTC-PD photomixers require laser sources of lower frequencies as compared to PCAs (Deninger 2013).

References

Banik B, Vukusic J, Hjelmgren H and Stake J 2008 Optimization of the UTC-PD epitaxy for photomixing at 340 GHz *Int. J. Infrared Millimeter Waves* **29** 914–23

Brown E R, McIntosh K A, Smith F W, Manfra M J and Dennis C L 1993 Measurements of optical-heterodyne conversion in low-temperature-grown GaAs *Appl. Phys. Lett.* **62** 1206–8

Burford N M and El-Shenawee M O 2017 Review of terahertz photoconductive antenna technology *Opt. Eng.* **56** 010901-1–20

Deninger A 2013 11-State-of-the-art in terahertz continuous-wave photomixer systems *Handbook of Terahertz Technology for Imaging, Sensing and Communications* Series in Electronic and Optical Materials (Cambridge: Woodhead Publishing) pp 327–73

Hindle F, Cuisset A, Bocquet R and Mouret G 2008 Continuous-wave terahertz by photomixing: applications to gas phase pollutant detection and quantification *C.R. Phys.* **9** 262–75

Ishibashi T and Ito H 2020 Uni-traveling-carrier photodiodes *J. Appl. Phys.* **127** 031101-1–031101-10

Pliński E F 2010 Terahertz photomixer *Bull. Pol. Acad. Sci. Tech. Sci.* **58** 463–70

Preu S, Döhler G, Malzer S, Wang L J and Gossard A C 2011 Tunable, continuous-wave terahertz photomixer sources and applications *J. Appl. Phys.* **109** 061301-1–56

Wakatsuki A, Muramoto Y and Ishibashi T 2012 Development of terahertz-wave photomixer module using a uni-traveling-carrier photodiode *NTT Tech. Rev.* **10** 1–7

Chapter 4

Quantum cascade and optically-pumped gas lasers

Due to the non-capability of semiconductor diode lasers to provide terahertz emission, the obvious recourse is taken to lasers working on new principles. One category of these lasers is the quantum cascade laser (QCL) which works through intersubband transitions and therefore is liberated from the constraint of dependence on semiconductor energy gap. A major technological breakthrough in these lasers is the realization of room temperature operation by frequency mixing in a monolithic structure providing high second order nonlinear susceptibility. The other class is the optically pumped gas laser which uses a carbon dioxide laser to excite transitions in methanol gas. Through a scheme of four energy levels, the methanol laser provides terahertz emission. A milestone achievement in these lasers was the discovery that the methanol laser could be made with smaller cavities using higher pressures. Details of operational mechanisms and updated research status of the QCL and optically pumped gas laser are presented.

Two cornerstone developments in terahertz laser technology, the QCL and the optically pumped molecular gas laser constitute the focal theme of this chapter. These lasers work on fundamentally different principles. They are distinguished from the bulk semiconductor diode laser in table 4.1.

4.1 Features of semiconductor diode lasers and their failure to produce terahertz waves

In a semiconductor diode laser, conduction band electrons and valence band holes are injected into the active region by forward biasing the diode. In the active region, the electrons in the conduction band recombine with holes in the valence band across the semiconductor bandgap (Sirtori and Teissier 2006). Hence, this laser works by

doi:10.1088/978-0-7503-4886-7ch4

Table 4.1. Bulk semiconductor diode, quantum cascade and molecular gas lasers.

Property	Bulk semiconductor diode laser	Quantum cascade laser	Molecular gas laser
Lasing medium	Bulk semiconductor	Semiconductor heterostructures	Molecular gases (e.g., CO_2, N_2)
Type of energy transitions	Electronic transitions between conduction and valence bands	Inter-subband transitions in quantum wells	Molecular transitions involving vibrational and rotational energy levels of molecules
Polarity	Bipolar involving electrons and holes	Unipolar involving electrons only	
Pumping	Electric current	Electric current	Electrical gas discharge pumping, optical pumping
Terahertz frequency operation	No	Yes	Yes

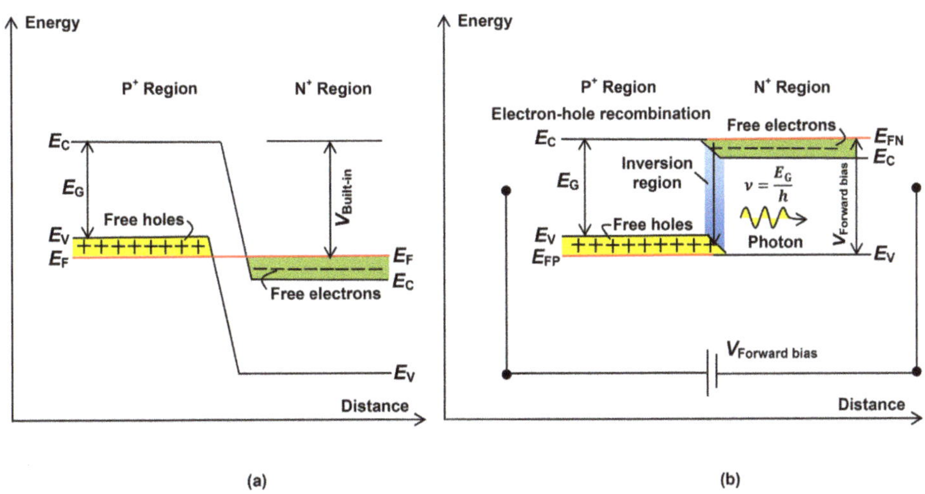

Figure 4.1. Energy-band diagram of the conventional interband semiconductor laser diode: (a) without bias and (b) under applied bias.

band-to-band or interband transitions through bipolar processes in which two types of carriers, electrons and holes participate (figure 4.1). Moreover, an electron annihilates a hole, and a photon is ejected. The lost electron is not utilized any further. So, two characteristics are noticeable, the operation via interband transitions and non-availability of an electron in future processes consequent upon its recombination.

The bandgap of the semiconductor determines the frequency and hence energy of released photons. Non-existence of materials with suitable bandgaps has rendered the frequency range 0.5–5 THz inaccessible to semiconductor bulk lasers. Alloys having different bandgaps are formed through various combinations of materials. In this way, photons of required frequencies can be produced. Nonetheless, there is a restriction to bandgap adjustment by alloying. Therefore, tailoring bandgaps in this manner for lasing at desired frequencies offers limited possibilities.

4.2 Quantum cascade laser

4.2.1 QCL materials

A commonly used material system for terahertz frequency range is GaAs/$Al_xGa_{1-x}As$/GaAs. At 300 K, the dependence of direct bandgap of $Al_xGa_{1-x}As$ on aluminum fraction x is given by the equation

$$E_{G,\,\mathrm{Direct}}(x) = 1.424 + 1.247x \text{ eV for } x < 0.45 \tag{4.1}$$

varying from 1.424 eV at $x = 0$ to $1.424 + 1.247 \times 0.44 = 1.97$ eV for $x = 0.44$. The conduction band offset for GaAs/AlGaAs is 390 meV. The material system for mid-IR range is InGaAs/AlInAs/InP. The conduction band offset for InGaAs/AlInAs is 520 meV.

The GaAs/AlGaAs and InGaAs/AlInAs lasers are illustrated in figure 4.2.

4.2.2 Principle of a quantum cascade laser

4.2.2.1 Quantum confinement

As its name suggests, the QCL is a quantum mechanical device using semiconductor heterostructures (figure 4.3). It has a nanostructured optical gain medium comprising specially designed quantum wells. These quantum wells are made by stacking alternating layers of semiconductor material, and are enclosed in a waveguide forming the optical cavity. In these quantum wells, the electrons are restricted to discrete energy states called subbands. The phenomenon is called quantum confinement.

4.2.2.2 Injector and active regions

The QCL consists of a large number of stages or periods (figure 4.4), each consisting of an injector region, which is doped, and an active region in which radiative transition takes place from an upper energy level E_3 of the quantum well to a lower level E_2, followed by non-radiative transition to the ground level E_1, after which the electron tunnels to the next period; the emptying of the ground level is known as depopulation. 8–10 quantum wells and barriers may be fabricated per period.

The lower laser level and depopulation level are separated from each other by the longitudinal optical (LO) phonon energy, which is ~30 meV for InGaAs. This separation makes the depopulation of lower laser level efficient. Also, the injector level is located at the correct position between two contiguous periods to ensure proper electron injection into the upper laser level of the next period (Jirauschek and Kubis 2014).

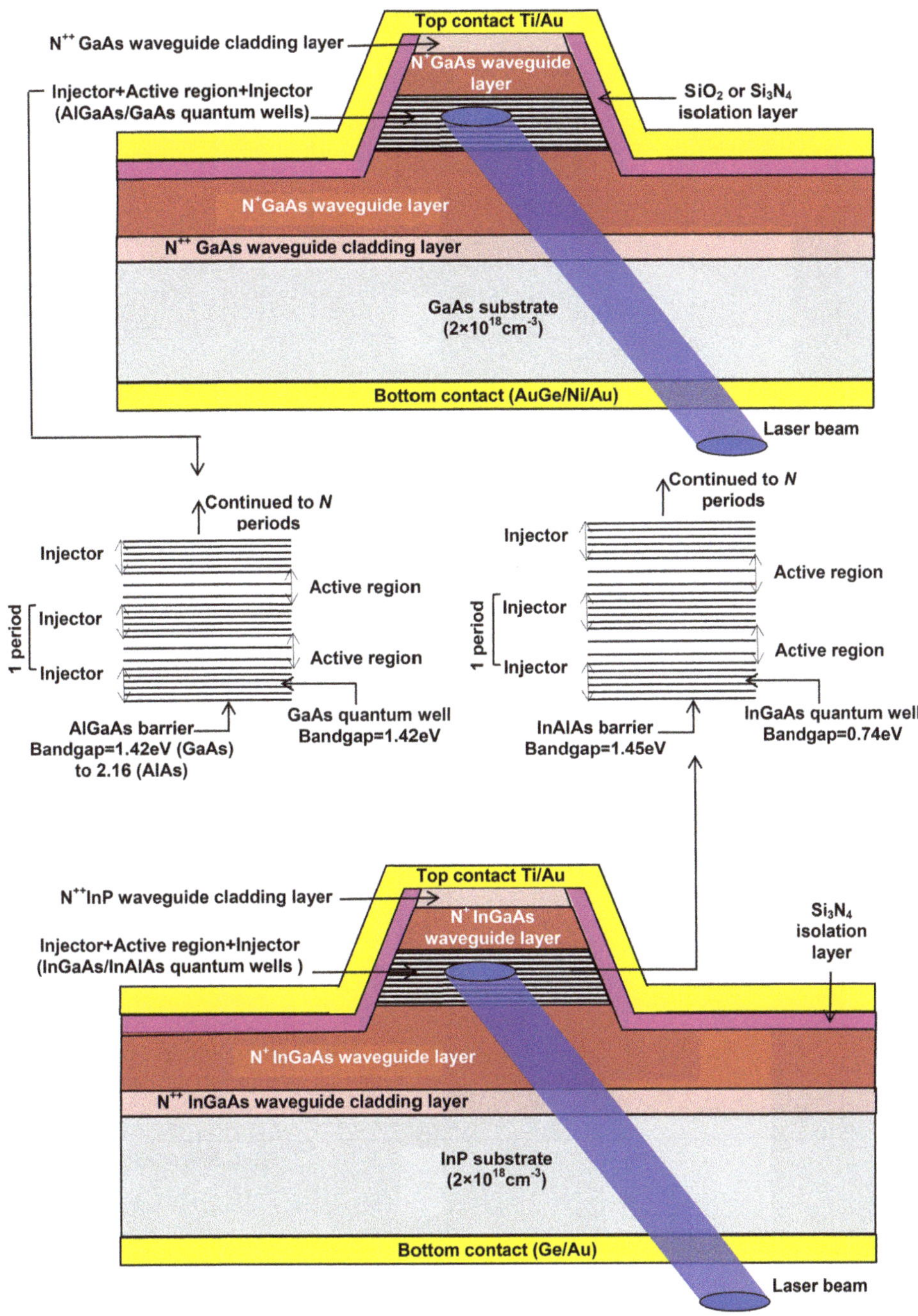

Figure 4.2. Cross-sectional diagrams of AlGaAs/GaAs and InGaAs/InAlAs quantum cascade laser ridges.

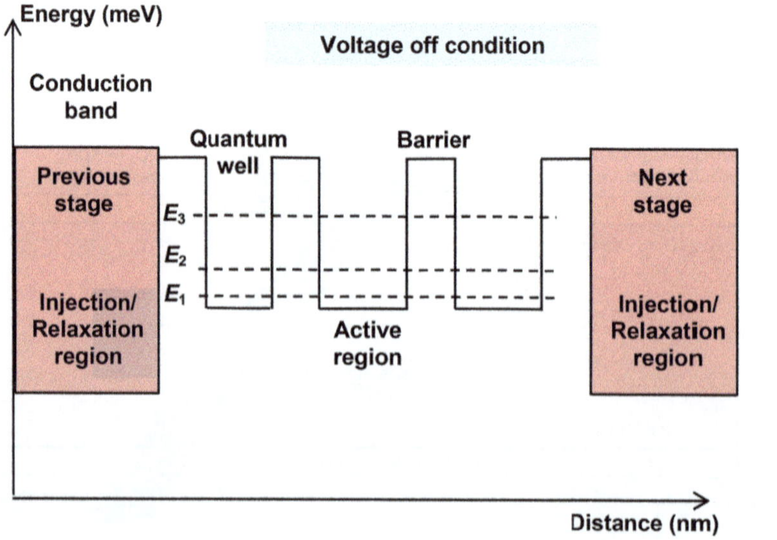

(a)

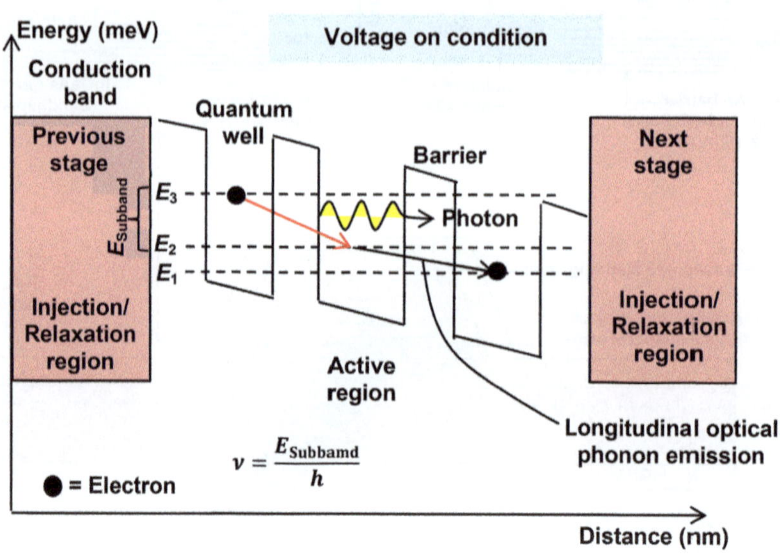

(b)

Figure 4.3. Elementary diagram showing the construction (injection/relaxation regions and active region) and energy levels of the quantum cascade laser: (a) without applied bias and (b) with bias.

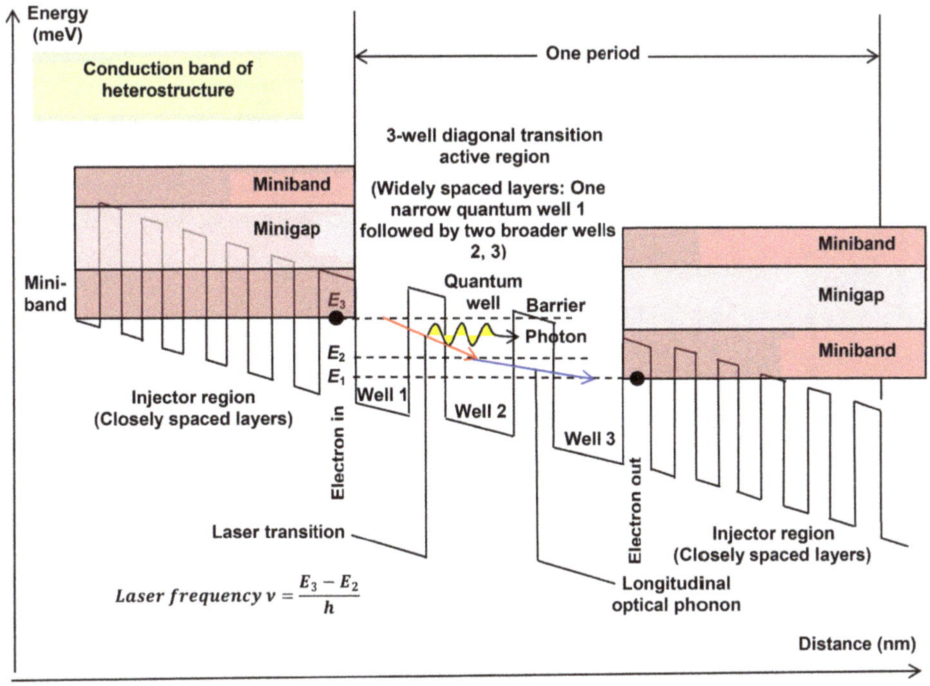

Figure 4.4. Injection and active regions of quantum cascaded laser.

4.2.2.3 Population inversion

Under the influence of voltage applied across the laser, the electron moves through the lasing medium from one period to a succeeding period. Moving down the different periods, the electron enters the active regions of successive periods on its path, where it moves from an excited energy state to a lower energy state in the conduction band. In every active region, a photon is liberated. The electron continues to move. There can be as many as 100 active regions. Thus a population inversion is obtained between two subbands, an essential condition for laser emission.

4.2.2.4 Intersubband mechanism

As the QCL operation entails electron transitions between subbands, it is essentially an inter-subband laser as opposed to the inter-band diode laser. In a QCL, transitions occur between subbands in the conduction band, not conduction-to-valence band transitions. Also, in distinction to a diode laser where an electron is used only once, here the same electron is used several times, thereby making the quantum efficiency > 1 and providing higher output power. The QCL greatly outperforms the diode laser in terms of output power. The 'single electron, multiple photons approach' brings in the idea of cascading to which the laser owes its name (figure 4.5). Moreover, the electronic transitions in a QCL being low energy jumps result in production of photons from mid to far infrared or terahertz frequencies.

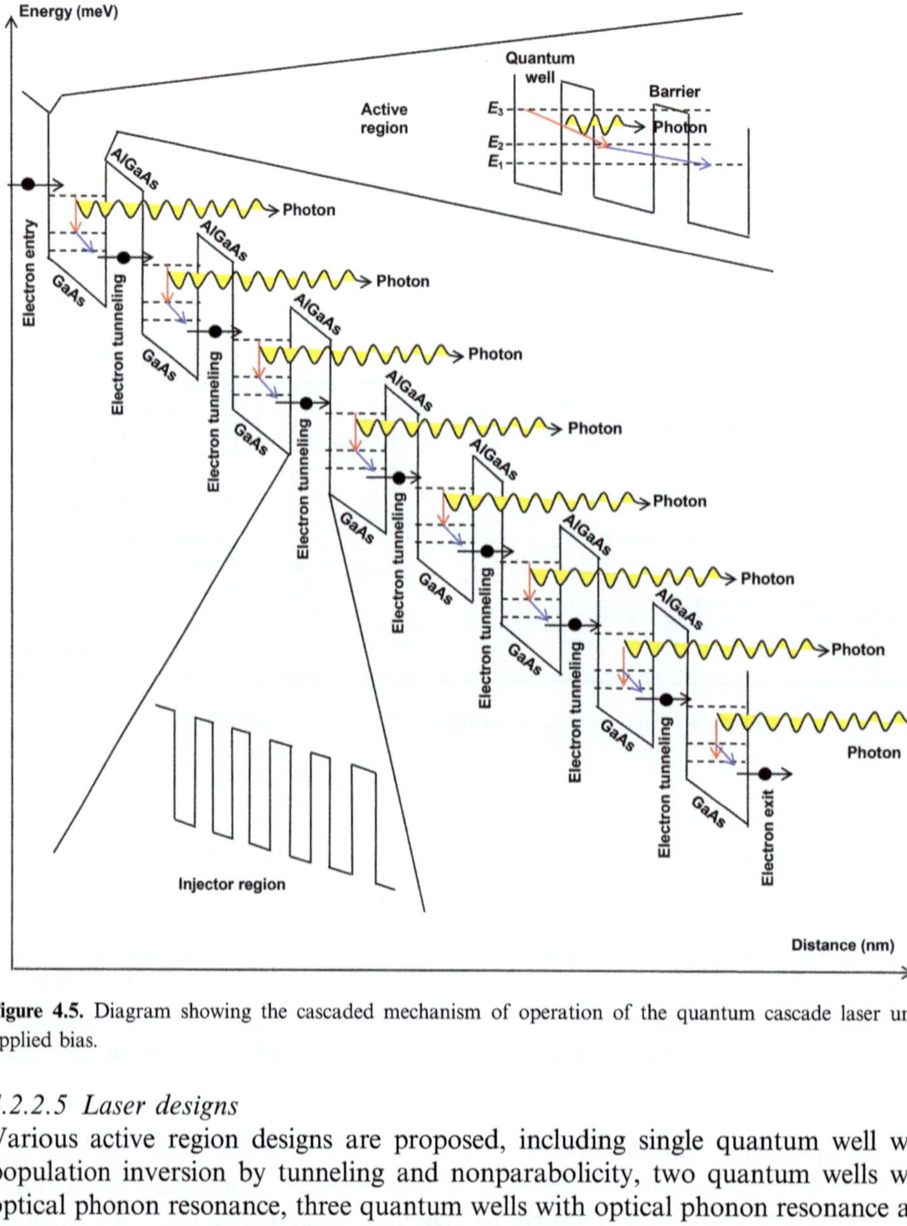

Figure 4.5. Diagram showing the cascaded mechanism of operation of the quantum cascade laser under applied bias.

4.2.2.5 Laser designs

Various active region designs are proposed, including single quantum well with population inversion by tunneling and nonparabolicity, two quantum wells with optical phonon resonance, three quantum wells with optical phonon resonance and tunneling, superlattice relying on difference in phase space, and chirped lattice working on phase space argument (Kaur *et al* 2020).

In a three-quantum well active region design, the first narrow width quantum well preceding the later two larger width wells ensures that electron leakage does not take place directly from the injector to the E_2 or E_1 energy levels instead of E_3 energy level. This happens because the energy levels are higher in a thinner well as compared to a thicker well, which increases the probability of an electron from injector to tunnel to the higher energy E_3 level. The three-well design can work on both vertical electron transitions and diagonal electron transitions. The diagonal transition is shown in the diagrams.

4.2.3 Frequency tunability of QCL

A unique attribute of QCL is that the frequency of emission is not an intrinsic property of the semiconductor but a consequence of the design of constituent layers. Population inversion is not due to an inherent physical property of the system. The wavefunctions need to be appropriately engineered for this purpose.

The frequency of a QCL depends on the thicknesses of the ultrathin layers constituting the quantum well structures, viz., the widths of wells and barriers, which are adjustable design parameters. It is independent of the bandgaps of constituent semiconductor materials. So, it is not necessary to change the material every time lasing is necessary at a new frequency.

Since the output frequency is controlled by the structure of the layers, and not by the semiconductor material used, any desired frequency can be achieved by proper band structural manipulation involving layer thicknesses in a fundamentally different way than implementable with a diode laser. Stated explicitly, variations in the same material system allow frequency changes in place of using a material with a different bandgap value, as needed for a diode laser. The heterostructure design determines the electrical and optical properties of a QCL.

The highest frequency is decided by the conduction band offset between the barrier and quantum well layers. There is no frequency limitation on the lower side except for the reststrahlen (residual radiation) band (Wang *et al* 2010).

4.2.4 Unipolarity of QCL

Unlike the diode laser which is bipolar in action because of involvement of two types of charge carriers, electrons and holes, the QCL is a unipolar semiconductor laser whose operation relies on only one type of charge carrier. Electrons only (and no holes), are responsible for releasing energy as photons.

4.2.5 Cryogenically-cooled THz QCLs

The QCL is one example of a solid-state laser source which can provide emission over a broad span of infrared and terahertz spectra. Albeit terahertz emission is presently available at low temperatures.

Köhler *et al* (2002) demonstrated a terahertz injection laser using GaAs/AlGaAs/GaAs heterostructure. The laser emits a single mode at 4.4 THz with output power of 2 mW up to 50 K.

Laser emission was observed at 3 THz up to 178 K in a three-quantum well active region with Cu cladding in metal–metal waveguides in place of gold (Belkin *et al* 2008a).

Lasing at 1 THz up to 215 K and at 3 THz up to 225 K was shown by applying strong magnetic fields > 16 T to suppress inter-Landau-level non-radiative scattering (Wade *et al* 2009).

Laser operation at 199.5 K was shown at 3.22 THz by Fathololoumi *et al* (2012) by optimization of oscillator strength of resonant phonon-based three-well design.

Performance matrices of QCLs have been progressively improving with emission frequencies 1.2–5.4 THz, pulse mode power output ~1 W and continuous wave output > 100 mW, but cooling requirements continue (Fujita *et al* 2018). QCLs working at room temperature or at thermoelectric cooler range ~240 K are needed.

4.2.6 Obstacles to room-temperature terahertz production by QCL

Unlike infrared QCLs, THz QCLs suffer from the drawback of the requirement of cryogenic operating temperatures. The difficulty arises from the inability to inject electrons selectively in the upper energy state for laser transition. The energy spread of the injected electrons is comparable to the energy difference between the laser levels. In consequence to the poor injection efficiency from the injector into higher energy laser state, population inversion is not easily attainable to support laser action.

4.2.7 Circumventing hindrances to room-temperature QCL operation

The problem is addressed by an indirect route based on frequency mixing concept. Terahertz radiation is generated by using two mid-IR pump waves in a non-linear optical crystal (Belkin *et al* 2008b). The technique is known as difference-frequency generation (DFG). The DFG can occur in crystalline materials that lack inversion symmetry.

As already explained in detail in chapter 2, sections 2.3 and 2.4, in linear optics, the polarization $P(t)$ or dipole moment per unit volume induced in a material is taken to be proportional to the electric field $E(t)$ as given by equation (2.41). The constant of proportionality $\chi^{(1)}$ is called the linear susceptibility:

$$P(t) = \varepsilon_0\chi^{(1)}E(t) \tag{4.2}$$

At high light intensities such as those of laser beams, the optical response of the medium is no longer linear and can be generalized by expressing polarization $P(t)$ as a power series in electric field $E(t)$ as

$$P(t) = \varepsilon_0\chi^{(1)}E(t) + \varepsilon_0\chi^{(2)}E^2(t) + \varepsilon_0\chi^{(3)}E^3(t) + \dots \tag{4.3}$$

where $\chi^{(2)}$ and $\chi^{(3)}$ are referred to as second- and third-order optical susceptibilities; see equation (2.42).

In DFG, two mid-IR laser beams having closely spaced angular frequencies ω_1, ω_2 interact in a medium with second order optical susceptibility $\chi^{(2)}$. This interaction results in the production of terahertz waves of angular frequency

$$\omega = \omega_1 - \omega_2 \tag{4.4}$$

or, in terms of linear frequency

$$\nu = \nu_1 - \nu_2 = \frac{c}{\lambda_1} - \frac{c}{\lambda_2} = c\left(\frac{1}{\lambda_1} - \frac{1}{\lambda_2}\right). \tag{4.5}$$

For λ_1=7.6 µm, and λ_2=8.7 µm (Belkin *et al* 2007),

$$\begin{aligned}\nu &= 2.9979 \times 10^8\left(\frac{1}{7.6 \times 10^{-6}} - \frac{1}{8.7 \times 10^{-6}}\right)\\ &= 2.9979 \times 10^8 \times 10^6\left(\frac{1}{7.6} - \frac{1}{8.7}\right)\\ &= 2.9979 \times 10^{14}(0.131\,578\,9 - 0.114\,942\,5)\\ &= 2.9979 \times 10^{14} \times 0.016\,636\,4 = 0.049\,874 \times 10^{14}\ \text{Hz} = 4.987\ \text{THz}\end{aligned} \tag{4.6}$$

with

$$\begin{aligned}\lambda &= \frac{c}{\nu} = \frac{2.9979 \times 10^8}{4.987 \times 10^{12}} = 0.601\,14 \times 10^{8-12} = 0.601\,14 \times 10^{-4}\ \text{m}\\ &= \frac{0.601\,14 \times 10^{-4}}{10^{-6}}\,\mu\text{m} = 60.11\ \mu\text{m} \approx 60\ \mu\text{m}\end{aligned} \tag{4.7}$$

To obtain giant second-order optical susceptibility $\chi^{(2)}$, a monolithically integrated, dual-wavelength QCL is designed for intracavity difference-frequency generation, i.e., DFG occurring within a cavity. In this laser, high $\chi^{(2)}$ is attained by engineering the active region of the laser. This susceptibility is associated with intersubband transitions in quantum wells coupled with each other. InGaAs/AlInAs heterostructures are used (figure 4.6). These structures are grown by molecular beam epitaxy on an InP substrate.

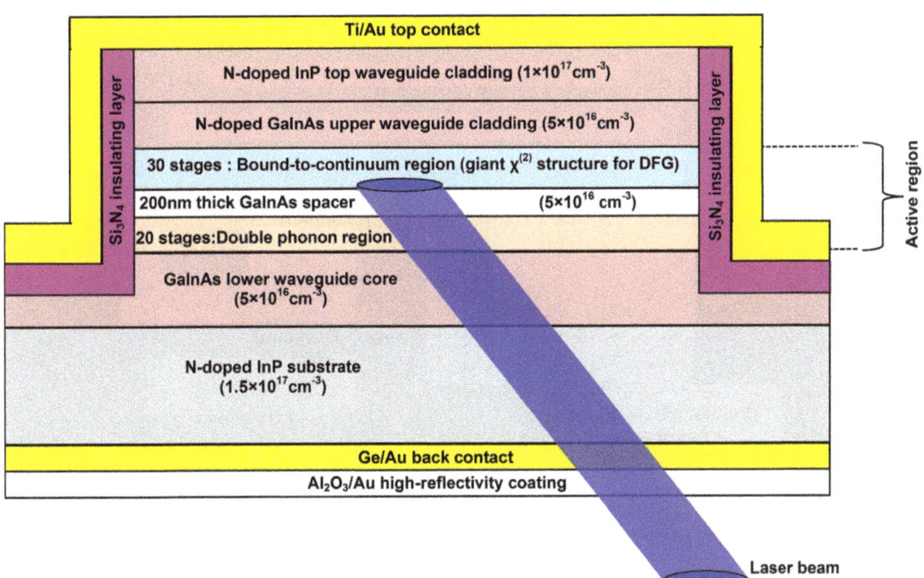

Figure 4.6. Layers constituting the structure of the InGaAs/AlInAs heterostructure dual-wavelength quantum cascade laser ridge for intracavity difference frequency generation.

Two substacks for the two wavelengths constitute the active region of the QCL:

(i) A substack for the emission wavelength 8.7 μm: It contains 30 stages of a 'bound-to-continuum' structure. This structure possesses giant $\chi^{(2)}$ for DFG.

(ii) A substack for emission wavelength 7.6 μm: It contains 20 stages of a two-phonon structure. It has a comparatively low $\chi^{(2)}$ for DFG.

As the name indicates, the 'bound-to-continuum' and two-phonon structures support different electron injection and extraction processes to and from the active region of the QCL (figure 4.7). In the 'bound-to-continuum' structure, the radiative transition occurs between a bound isolated state and a miniband. In the two-phonon structure, the transition takes place along a ladder of three states. Each of these three states is separated by an optical phonon (Faist *et al* 2002).

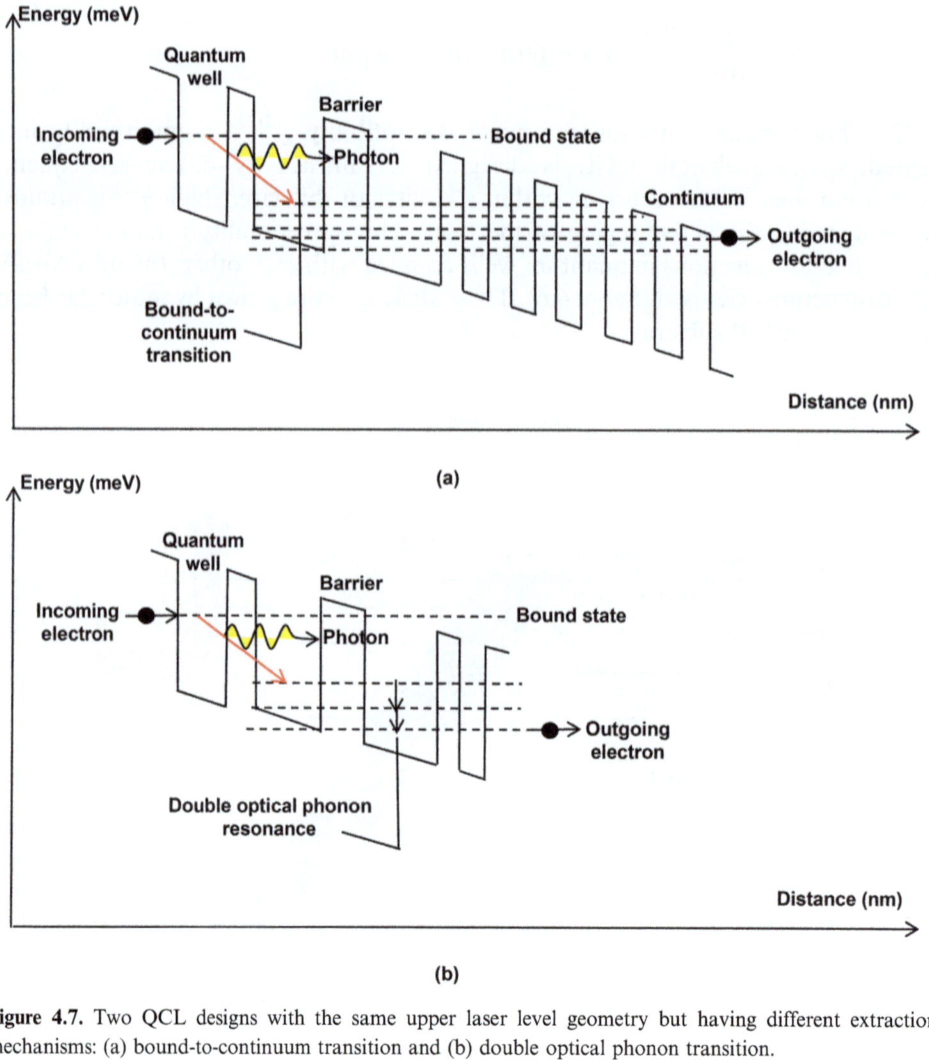

Figure 4.7. Two QCL designs with the same upper laser level geometry but having different extraction mechanisms: (a) bound-to-continuum transition and (b) double optical phonon transition.

For fabrication of the laser, an N-doped GaInAs layer is grown on an N-doped InP substrate to act as the core of lower waveguide followed by the growth of 20-stage stack (ii) plus 30-stage stack (i) with the two stacks separated by an N-doped GaInAs spacer layer. Stack (i) is covered with N-doped GaInAs layer acting as the upper waveguide cladding. Further ahead, an N-doped InP layer is overgrown for providing top waveguide cladding. After processing into deep edged ridge waveguides, an Si_3N_4 insulating film is formed on the lateral walls of the ridge. The ensuing steps include the deposition of a Ti/Au top contact, a non-alloyed Ge/Au back contact and a high-reflectivity film on the back facet.

The output power achieved with this QCL is 7 μW at 80 K. It decreases to 1 μW at 250 K and falls to 300 nW at room temperature (300 K) (Belkin *et al* 2008b).

Example 4.1

The energy gap of $In_xGa_{1-x}N$ varies from 0.69 eV for InN to 3.4 eV for GaN for values of x from 1 to 0. What is the frequency of laser emission from this material for these two limits of energy gap? In which segments of electromagnetic spectrum are these frequencies located?

Solution:

$$\text{Energy of photon emitted } (E) = \text{Energy gap of the semiconductor used } \left(E_g\right) \tag{4.8}$$

But

$$E = h\nu \tag{4.9}$$

where h is Planck's constant and ν is the frequency of emitted photon.

From equations (4.8) and (4.9),

$$h\nu = E_g \tag{4.10}$$

For InN, we can write from equation (4.10)

$$6.626 \times 10^{-34}\nu = 0.69 \times 1.602 \times 10^{-19} \tag{4.11}$$

or,

$$\nu = \frac{0.69 \times 1.602 \times 10^{-19}}{6.626 \times 10^{-34}} = \frac{0.69 \times 1.602 \times 10^{-19+34}}{6.626} = 0.1668 \times 10^{15} \tag{4.12}$$
$$= 1.668 \times 10^{14} \text{ Hz}$$

which lies in the infrared range of frequencies: 3×10^{11} Hz to 4.30×10^{14} Hz.

For GaN, we have from equation (4.10)

$$6.626 \times 10^{-34}\nu = 3.4 \times 1.6 \times 10^{-19} \tag{4.13}$$

or,

$$\nu = \frac{3.4 \times 1.602 \times 10^{-19}}{6.626 \times 10^{-34}} = \frac{3.4 \times 1.602 \times 10^{-19+34}}{6.626} = 0.822 \times 10^{15} \tag{4.14}$$
$$= 8.22 \times 10^{14} \text{ Hz}$$

which lies in the ultraviolet range of frequencies: 7.5×10^{14} Hz to 3×10^{16} Hz.

Thus as the band gap of $In_xGa_{1-x}N$ changes from 0.69 eV for InN to 3.4 eV for GaN for values of x decreasing from 1 to 0, the frequency of laser emission increases from infrared to ultraviolet part of the electromagnetic spectrum.

Example 4.2

Wavelength of semiconductor diode lasers is limited to 2.5 μm? Find the band gap of the semiconductor to which this wavelength relates. What is the frequency limit? Does it fall in the terahertz range?

Solution: Energy E of photon, its wavelength λ and frequency ν are interrelated through Planck's constant h as

$$E = h\nu = h\frac{c}{\lambda} \tag{4.15}$$

from which

$$\begin{aligned} E_g = E = h\frac{c}{\lambda} &= 6.626 \times 10^{-34}\frac{2.9979 \times 10^8}{2.5 \times 10^{-6}}\text{J} \\ &= 6.626 \times 10^{-34} \times \frac{2.9979 \times 10^8}{2.5 \times 10^{-6}} \times \frac{1}{1.602 \times 10^{-19}}\text{eV} \\ &= \frac{6.626 \times 2.9979 \times 10^{-34+8+6+19}}{2.5 \times 1.602} = 4.96 \times 10^{-1}\,\text{eV} = 0.496\,\text{eV} \\ &\approx 0.5\,\text{eV} \end{aligned} \tag{4.16}$$

Frequency of emitted photon is

$$\begin{aligned} \nu = \frac{c}{\lambda} = \frac{2.9979 \times 10^8}{2.5 \times 10^{-6}} &= \frac{2.9979}{2.5} \times 10^{8+6} \\ &= 1.199 \times 10^{14}\,\text{Hz} \approx 1.2 \times 10^{14}\,\text{Hz} \end{aligned} \tag{4.17}$$

It is outside the terahertz range. It falls in the infrared range of frequencies.

Example 4.3

Calculate the difference of energy levels between which electronic transitions will produce waves of frequencies from 0.1 THz to 10 THz.

Solution: Let the energy levels be E_1, E_2. For 0.1 THz waves

$$\begin{aligned} E_2 - E_1 &= h\nu = 6.626 \times 10^{-34} \times 0.1 \times 10^{12}\,\text{J} = \frac{6.626 \times 10^{-34} \times 0.1 \times 10^{12}}{1.602 \times 10^{-19}}\text{eV} \\ &= \frac{6.626 \times 0.1}{1.602} \times 10^{-34+12+19}\,\text{eV} = 0.4136 \times 10^{-3}\,\text{eV} \approx 0.41\,\text{meV} \end{aligned} \tag{4.18}$$

For 10 THz waves

$$\begin{aligned} E_2 - E_1 &= h\nu = 6.626 \times 10^{-34} \times 10 \times 10^{12}\,\text{J} = \frac{6.626 \times 10^{-34} \times 10 \times 10^{12}}{1.602 \times 10^{-19}}\text{eV} \\ &= 41.36 \times 10^{-3}\,\text{eV} \approx 41.4\,\text{meV} \end{aligned} \tag{4.19}$$

Example 4.4
Find the frequency of terahertz waves generated by electronic transitions between energy levels separated by: (a) 10 meV and (b) 20 meV.
Solution: (a) We know that

$$h\nu = \Delta E \tag{4.20}$$

or,

$$\nu = \frac{\Delta E}{h} = \frac{10 \times 10^{-3} \times 1.602 \times 10^{-19}}{6.626 \times 10^{-34}} = \frac{10 \times 1.602}{6.626} \times 10^{-3-19+34} \tag{4.21}$$
$$= 2.4177 \times 10^{12} \approx 2.42 \text{ THz}$$

(b) From equation (4.21)

$$\nu = \frac{\Delta E}{h} = \frac{20 \times 10^{-3} \times 1.602 \times 10^{-19}}{6.626 \times 10^{-34}} = 4.83 \text{ THz} \tag{4.22}$$

4.3 Optically-pumped methanol gas laser

4.3.1 Optical pumping

Here a strong source of light is used to excite the active medium to higher energy levels for creating population inversion. The optical sources used for pumping include gas discharge lamps, laser diodes or other kinds of lasers.

Gas lasers are usually excited by electrical discharges. This is because optical pumping is an inefficient technique for gases. For efficient pumping, it is desired that the absorption spectrum of the gas laser should match with the emission spectrum of the pumping source. The absorption spectrum of gases is narrow and the emission spectrum of the pumping sources is broad so that a large proportion of light is wasted. So optical pumping is generally not applied to gas lasers. However, a few situations are exceptions to this practice. In the terahertz portion of the spectrum, a comparatively larger number of discrete laser lines have been observed using optical pumping than by electrical or any other excitation method (Mukhopadhyayay and Singh 1998). Hence optical pumping is an enabling technological tool for coherent radiation emission in the terahertz portion of the electromagnetic spectrum. It is used in terahertz gas lasers.

A high-power carbon dioxide laser is commonly used as the pump laser for the terahertz laser. It is an infrared laser with wavelengths 9–11 μm (frequencies: 33.31–27.25 THz). Examples of gases used in the terahertz laser as active medium are CH_3OH (methanol), CH_3OD (methanol-OD or mono-Deuteromethanol), CH_3F (fluoromethane), CF_4 (carbon tetrafluoride), etc. We proceed below by first describing the operation of CO_2 IR laser. This will be followed by pumping of methanol molecules for terahertz laser action.

4.3.2 CO_2 gas IR laser as optical pumping source

The CO_2 gas laser uses a gas discharge tube. The discharge tube is filled with a mixture of gases consisting of carbon dioxide (10%–20%), nitrogen (10%–20%), hydrogen (a few percent), some water vapor, xenon and helium. One version of the gas discharge tube is a sealed tube. Another type of discharge tube is a flow-through tube. In this tube, carbon dioxide and nitrogen are continuously pumped into the tube to produce a plasma discharge. At the same time, the hot gases are pumped out of the tube. Thus the flow-through tube works by pumping in gases and by exhaustion.

Two types of flow-through discharge tube-based carbon dioxide lasers are illustrated in figure 4.8. At one end, the input side, the gas discharge tube is closed by a reflector. At the opposite end, the output side, the tube is closed by a partially reflecting mirror. The laser beam can pass out through this end; hence it works as an output coupler.

The gas discharge is maintained either with a DC source or with an AC source of frequency 20–50 kHz. So the pumping method for this laser is electrical gas discharge. To remove the heat, the low-power tubes are air-cooled while high-power tubes are water-cooled.

During the gas discharge, electrons impacting nitrogen molecules excite them to metastable vibrational levels. By metastability we mean that these energy levels are theoretically unstable but practically relatively long-lived. These nitrogen molecules are de-excited by collision with carbon dioxide molecules. During collision, the nitrogen molecules transfer their vibrational level energy to carbon dioxide molecules. As a result, the carbon dioxide molecules are transferred to higher vibrational mode quantum states. When these carbon dioxide molecules fall to one particular vibrational mode, they radially emit at 9.6 μm. Carbon dioxide molecules dropping to another vibrational mode lead to emission at 10.6 μm. These are the lasing transitions responsible for operation of the laser. It is obvious that the photons released due to transitions between these quantum states have lower energies than visible and near-infrared regions because the excitation energies of vibrational modes are low.

After undergoing the emitting transition, the carbon dioxide molecules move to their ground states by colliding with cold helium atoms. The resulting hot helium atoms must be cooled for continuance of operation. Thus helium helps in depopulation of laser level and also aids in heat removal. Hydrogen and water vapor serve to re-oxidize carbon monoxide formed during discharge to carbon dioxide.

High-quality infrared optics is essential for CO_2 lasers. The mirrors are silvered. Ge or ZnSe is used for windows/lenses. Average power outputs vary from tens of Watts to several kilowatts with conversion efficiency more than 10%. At high powers, Au mirrors are used. The windows/lenses are made of ZnSe or diamond. Hardness and high thermal conductivity of diamond provides ruggedness.

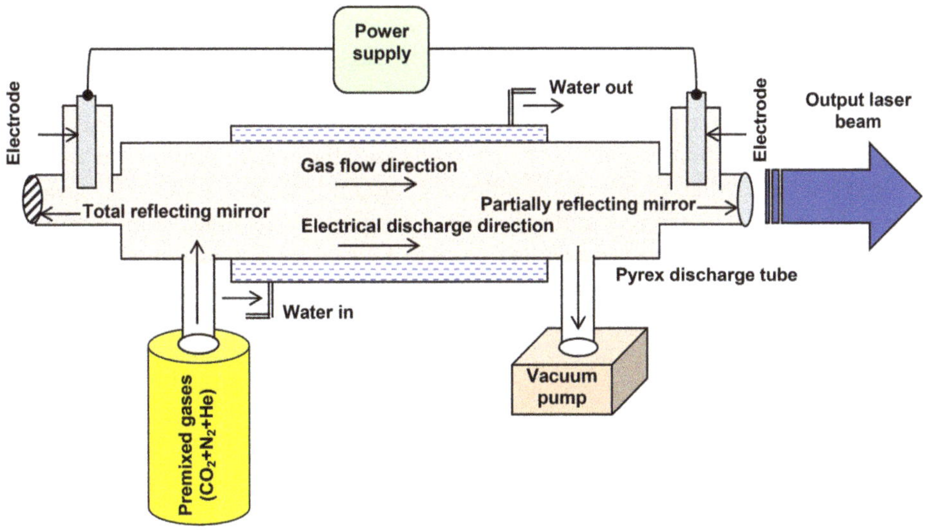

(a)

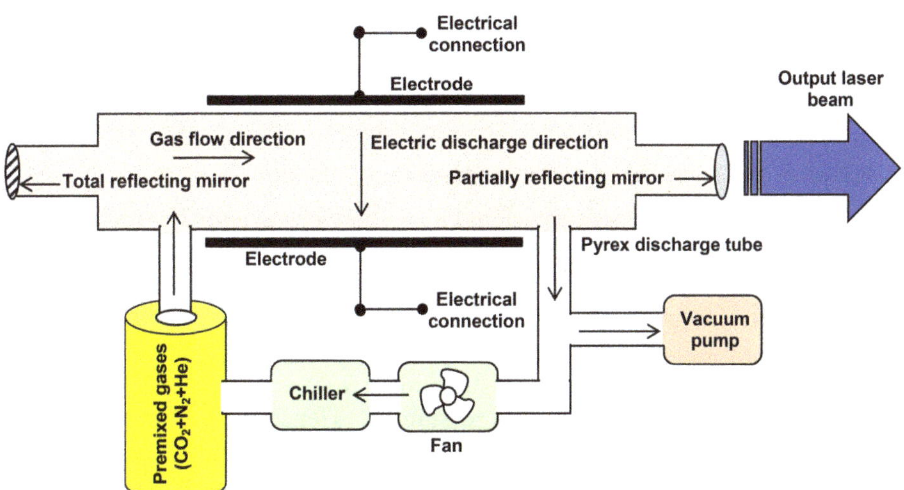

(b)

Figure 4.8. Two types of carbon dioxide gas laser: (a) water-cooled longitudinally excited laser and (b) convectively-cooled transversely excited atmosphere (TEA) laser.

4.3.3 Methanol as active molecular medium being pumped

Two desirable properties expected in the active medium used for a terahertz laser are:

(i) it should have a high absorption coefficient for IR radiation which will be used for pumping,

(ii) the molecules of the medium should possess a large permanent dipole moment > 0.1 Debye.

These properties are found in light polyatomic molecules such as methanol, which has a dipole moment of 0.9 Debye parallel to the axis of near symmetry and a dipole moment of 1.4 Debye perpendicular to it.

Molecules with large permanent dipole moments show a rich spectrum of rotational transitions in the terahertz region. For such molecules, the excited vibrational states are coupled to several other vibrational rotational states and all these states relax speedily. Such molecules can be pumped optically into a rotational level in a vacant vibrational state at low pressure. Then laser transitions can take place in the excited vibrational level. If the frequency of the pumping radiation is higher than the frequency of emitted laser radiation then the optical pumping and laser action can occur in the same vibrational band. Generally, the frequency of the pumping radiation matches with the IR absorption transition in the medium being pumped.

Reasons favoring choice of methanol molecules as the lasing medium are:

(i) the availability of suitable molecular energy levels capable of providing a large number of emission lines in the terahertz region;

(ii) the high density of IR absorption spectra of the C–O stretch band in methanol owing to which there is great likelihood of its coincidence with discrete lines of carbon dioxide laser; and

(iii) the excellent IR absorption strength of methanol at the 10 μm emission wavelength of CO_2 laser, thereby making pumping with CO_2 laser very efficient.

4.3.4 Energy-level scheme of methanol laser

Let us first explain three-energy level and four-energy level laser systems (figure 4.9).

In a three-energy level laser system (figure 4.9(a)), a molecule in the ground state E_0 is pumped to the higher energy level E_1 from which it undergoes a fast non-radiant transition to lower energy level E_2. Then from energy level E_2, the molecule experiences a slow lasing transition to ground state E_0, emitting a photon. For population inversion, the lifetime for transition from state E_2 to state E_0 must be much longer than that from state E_1 to state E_2. Hence the molecular population builds up in the state E_2.

In a four-level system (figure 4.9(b)), the gas molecule is pumped from the ground state E_0 to an intermediate upper state E_1. From the intermediate upper state, the molecule undergoes rapid decay to upper lasing level E_2. From upper lasing level, the molecule is stimulated to the lower lasing level E_3. Finally, the molecule is de-excited to the ground state E_0. For creation of population inversion and sustaining of laser action, the necessary condition is that the transition time from state E_2 to state E_3 must be much greater than the transition time from state E_3 to state E_0 and this time from state E_3 to state E_0 is nearly equal to the time for transition from state E_1 to state E_2. This means that the lifetime of state E_2 must be appreciably long. Then state E_2 becomes a bottleneck to deactivation of energy creating a traffic jam. So population of molecules can be accumulated in state E_2.

Let us now move to the CO_2 infrared pumped methanol laser (figure 4.10). The energy-level diagram for terahertz emission is a three-level system (Xu *et al* 1996),

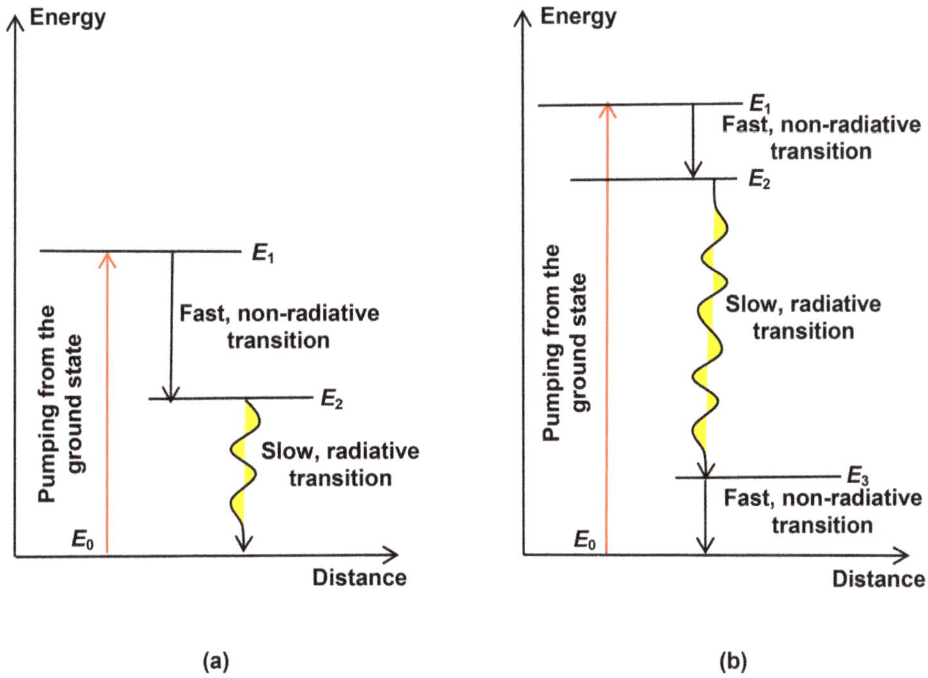

Figure 4.9. Energy level diagrams of (a) three-energy level laser system and (b) four-energy level laser system.

as shown in figure 4.10(a). The IR CO_2 laser pumps the methanol molecules from the ground vibrational state E_0 to an excited vibrational state E_1. Stimulated emission occurs between energy levels E_1 and E_2. This transition is purely rotational or rotational torsional transition in a particular vibrational state. There is a strong probability of transition between closely spaced vibrational states. Finally, the molecule is de-excited to the ground state E_0. The transition from level E_1 to level E_2 is the lasing transition. The transition from level E_3 to level E_0 is the refilling transition.

A four-energy level transition is speculated by Xu *et al* (1996). Here (figure 4.10(b)) starting from the energy state E_1, the methanol molecules move to the higher energy level E_2 induced by IR pumping. From energy level E_2, the energy and population flow out along an interaction channel coupling the levels over to the energy level E_3 from which terahertz emission lasing takes place at a later time to the lower energy level E_4.

As we see, the population inversion is obtained by optical pumping between two rotational states in an excited vibrational level of methanol molecule. This inversion provides the necessary gain or amplification. The frequency of transition between the two rotational states decides the frequency of emitted terahertz radiation.

4.3.5 Organization of optically-pumped methanol laser

The IR CO_2 pump laser is a tunable laser with power in the range 10–100 W. The resonator of methanol laser is much like a conventional gas resonator with the

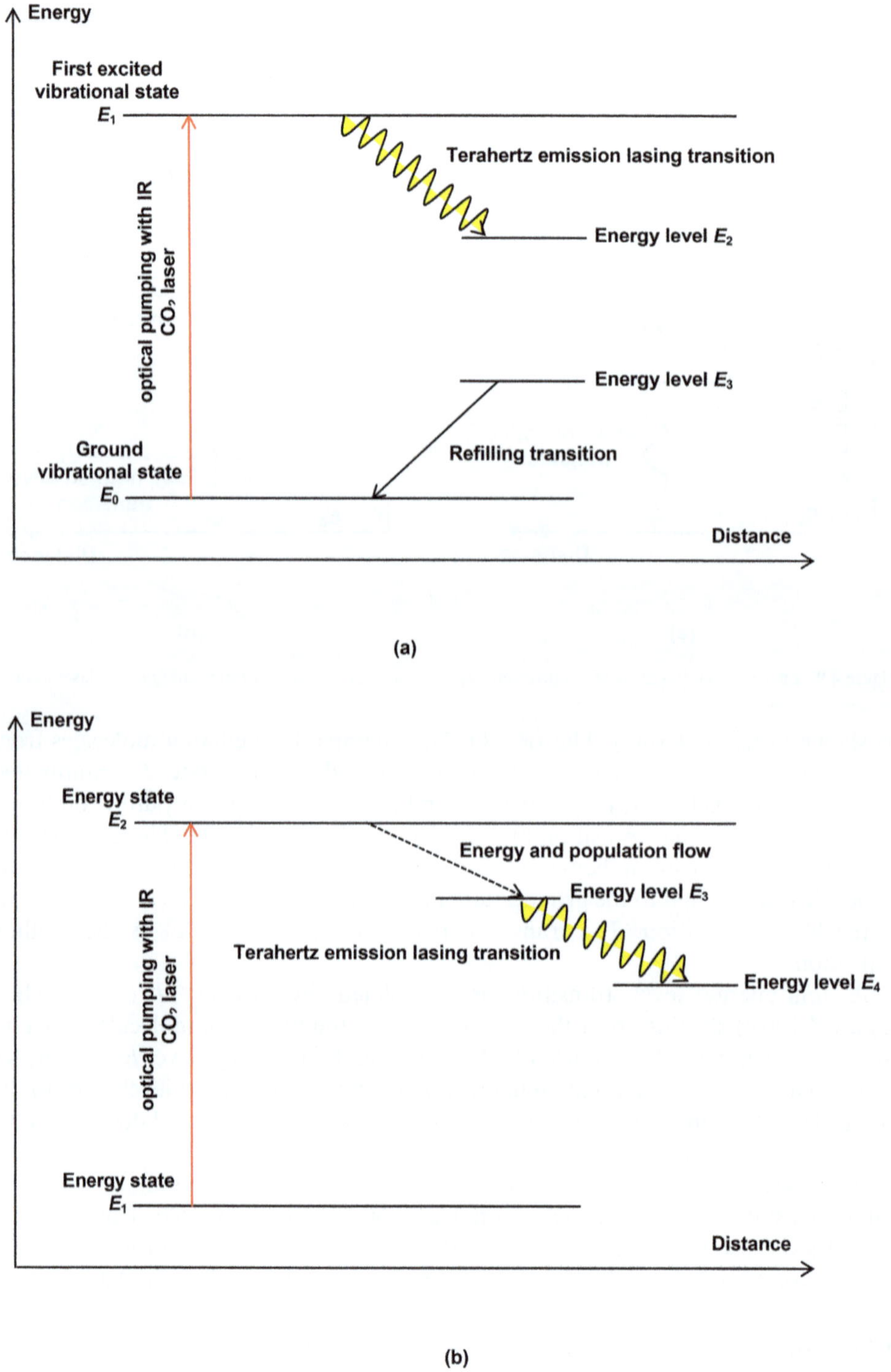

Figure 4.10. CO_2 infrared laser pumped methanol laser: (a) Three-energy level laser diagram, (b) Four-energy level laser diagram and (c) organization of the laser.

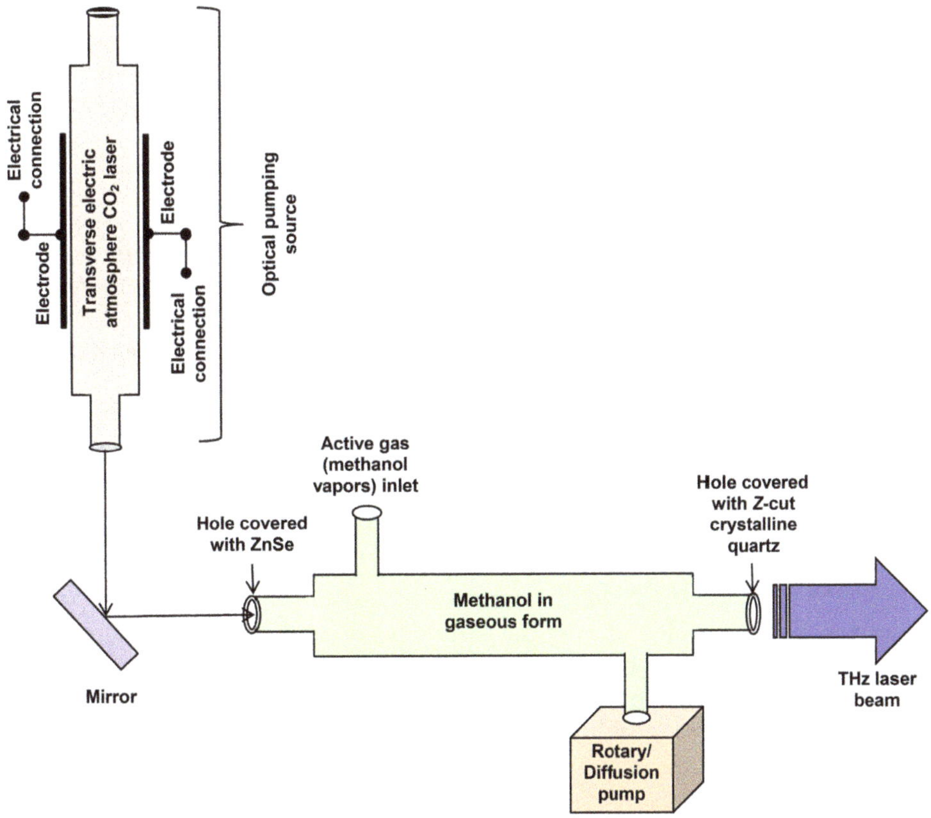

(c)

Figure 4.10. (Continued.)

electrical discharge portion absent (figure 4.10(c)). It is a vacuum-tight enclosure built with proper high-vaccum components. It contains methanol vapors. The 10 cm diameter cavity is around a meter long. It is maintained at a low pressure < 0.1 Torr (Wang *et al* 2018). The constructional materials of the resonator must be compatible to IR and terahertz radiations. The resonator is a waveguide made of dielectric or metallic materials. Two flat reflectors close the ends of the resonator. The reflectors for the two ends are different because one is an input coupler for feeding the IR CO_2 pump laser beam and the other is output coupler through which the emitted radiation is taken out. Accordingly, the input coupler is transparent to IR radiation but opaque to terahertz radiation. Similarly, the output coupler is semi-transparent to terahertz radiation but opaque to IR radiation. For the input coupler, a hole is drilled in the resonator mirror. The hole is covered with ZnSe, which passes infrared rays but stops terahertz waves. Similarly, the hole in the resonator mirror of the output coupler is covered with Z-cut crystalline quartz, which partially allows passage of terahertz waves and blocks infrared rays. In both cases, the size of the hole determines the degree of coupling.

The design of the resonator cavity controls the feedback mechanism for laser action. The phase shift during one round trip can be altered by adjusting the length of the resonator. In this way optimal output power is attained.

4.3.6 Advantages and disadvantages of methanol laser

The methanol laser supplies terahertz radiation of high spectral brightness with narrow emission lines $\leqslant$ 1 MHz. With 10.6 μm IR pump beam, the methanol laser supplies terahertz radiation at 118.83 μm, 2.523 THz with IR-to-THz conversion factors~$10^{-2} - 10^{-3}$, and power levels of several tens of milliwatts; some continuous wave lasers provide 100 mW (Shumyatsky and Alfano 2011).

A disadvantage of methanol laser is its small spectral tunability ~2 cm^{-1} = (2/cm) × 29.979 cm/ns = 59.958 GHz ~ 60 GHz. This happens because the laser operation is related to discrete frequencies involving molecular transitions.

4.3.7 Optically-pumped gas laser as a high-power, compact and continuous-wave source of terahertz radiation

A common belief prevalent during the early days of terahertz lasers was that they were bulky and unwieldy tubes. It was thought that the resonant cavity must be large in size and evacuated to a low pressure. Wang *et al* (2018) showed the inefficiency of these large size cavities working at low pressures. The main problem with these cavities was the creation of a vibrational bottleneck. This bottleneck is due to a relaxation mechanism restricted by diffusion arising from collisions of molecules with walls of the tube. By this mechanism, any population inversion exceeding a peak pressure was nullified. Therefore, an increase in intensity of pumping beyond a certain limit did not yield any success. Subsequently, it was noticed that terahertz laser operation was possible at higher pressures and in smaller cavities. This happened by virtue of collisional processes that excited vibrational levels of higher energies. The bottleneck was thus mitigated with inclusion of higher energy vibrational levels.

Wang *et al* (2018) further showed that small cavities operated at high pressures were needed for attainment of high powers. A model was developed encompassing the physical phenomena contributing to the lasing process. With this model, accurate predictions at high pressures could be made, as confirmed by experimental verification. The model gave the correct regime of operation of terahertz lasers. They redesigned the cavity and fabricated a terahertz laser having cavity 1000 times smaller in size. The efficiency of this laser was 10 times greater than previous lasers, thus establishing the optically pumped gas laser as a small-size, powerful terahertz source.

The optical pumping laser is a CO_2 laser with wavelength of 10.7 μm. The resonant cavity is a copper tube of length 14.3 cm with radius 0.5 mm (figure 4.11). The length of cavity is optimized to be equal to ½ the decay length of the CO_2 laser beam used for pumping. The gas used by them is fluoromethane-13C ($^{13}CH_3F$).

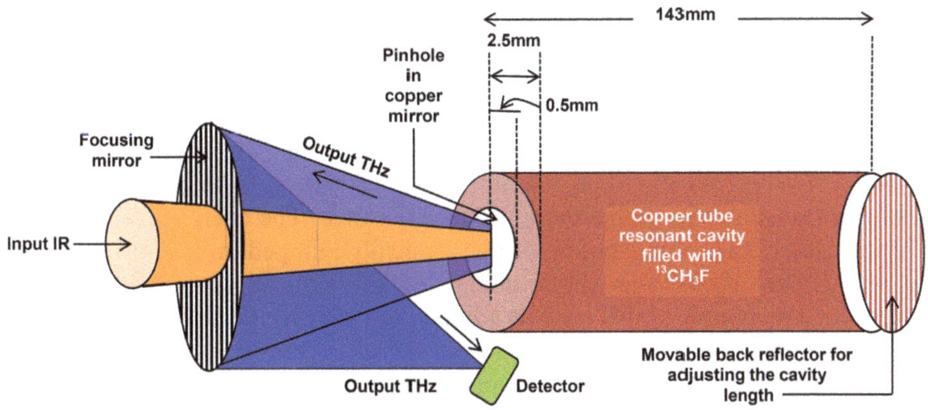

Figure 4.11. Compact optically pumped terahertz CH_3F gas laser.

Coincidence of a spectral line in CO_2 laser with rotational-vibrational transition in CH_3F results in two rotational inversions of population, the direct one with frequency 0.245 351 THz and a refilling rotational transition with frequency 0.248 559 THz.

Input power is 10 W. The pumping beam enters the cavity through a pinhole in the copper mirror at the front end of the tube. A copper-plunger mirror in the rear end of the tube is used for adjusting the length of the cavity for resonance of the lasing transition with the cavity. The terahertz radiation comes from the pinhole in the copper mirror located at the front end.

It was found that the lasing was impeded at high pressures when the length of the cavity was larger than necessary. Moreover, cavities of diameter 15 mm and length < 20 cm provided high power generation at pressures > 400 mTorr over a broad pressure range. The flat and broad gain profile allows tuning over a wide frequency range of ± 10 MHz.

A heterodyne receiver is used to validate laser operation at 0.25 THz. The lasing transitions and their intensities are measured for different pressures and pumping powers.

4.4 Discussion and conclusions

The reason for inability of a semiconductor diode laser to function at terahertz frequencies was explained followed by the elaboration of the principle of operation of a QCL through inter-subband transitions. Frequency tuning of the QCL can be implemented by varying the thicknesses of quantum wells and barrier layers, dispelling the need for using materials with different bandgaps. The unipolar nature of this laser was emphasized. Reported cryogenically cooled QCLs using GaAs/AlGaAs/GaAs heterostructure were described and the hindrances to their room temperature operation were pinpointed. A monolithic QCL was designed on the concept of intracavity difference frequency generation to work at room temperature by engineering the active region to obtain high second order nonlinear susceptibility.

Internal difference-frequency generation enabled room temperature operating QCL signifies a turning point in development of these devices.

Necessity of optical pumping for lasing action in terahertz region was emphasized and the justification for favoring the CO_2 laser was presented. Methanol was favored as the active medium because of its high absorption coefficient for pumping radiation. The permanent dipole moments of its molecules facilitate numerous rotational transitions in the terahertz frequency range. The energy level scheme of methanol laser and organization of the optically-pumped methanol laser were explained. The merits and shortcomings of the optically-pumped laser were mentioned. The notion of this laser as a bulky and inconvenient source was broken by the finding that laser operation is feasible at higher pressures with smaller cavity sizes, establishing it as a high-power, compact emitter of terahertz radiation. Non-requirement of low pressures and large cavities was a landmark event for these lasers.

References

Belkin M A, Capasso F, Belyanin A, Sivco D L, Cho A Y, Oakley D C, Vineis C J and Turner G W 2007 Terahertz quantum-cascade-laser source based on intracavity difference-frequency generation *Nat. Photonics* **1** 288–92

Belkin M A, Fan J A, Hormoz S, Capasso F, Khanna S P, Lachab M, Davies A G and Linfield E H 2008a Terahertz quantum cascade lasers with copper metal-metal waveguides operating up to 178 K *Opt. Express* **16** 3242–48

Belkin M A, Capasso F, Xie F, Belyanin A, Fischer M, Wittmann A and Faist J 2008b Room temperature terahertz quantum cascade laser source based on intracavity difference-frequency generation *Appl. Phys. Lett.* **92** 201101-1–3

Faist J, Hofstetter D, Beck M, Aellen T, Rochat M and Blaser S 2002 Bound-to-continuum and two-phonon resonance quantum-cascade lasers for high duty cycle, high-temperature operation *IEEE J. Quantum Electron.* **38** 533–45

Fathololoumi S, Dupont E, Chan C W I, Wasilewski Z R, Laframboise S R, Ban D, Mátyás A, Jirauschek C, Hu Q and Liu H C 2012 Terahertz quantum cascade lasers operating up to 200 K with optimized oscillator strength and improved injection tunneling *Opt. Express* **20** 3866–76

Fujita K, Jung S, Jiang Y, Kim J H, Nakanishi A, Ito A, Hitaka M, Edamura T and Belkin M A 2018 Recent progress in terahertz difference-frequency quantum cascade laser sources *Nanophotonics* **7** 1795–817

Jirauschek C and Kubis T 2014 Modeling techniques for quantum cascade lasers *Appl. Phys. Rev.* **1** 011307-1–51

Kaur H, Singh M, Chauhan A and Sharma R 2020 Active region designs and mounting schemes of THz quantum-cascade laser *Mater. Today Proc.* **26** 3458–61

Köhler R, Tredicucci A, Beltram F, Beere H E, Linfield E H, Davies A G, Ritchie D A, Iotti R C and Rossi F 2002 Terahertz semiconductor-heterostructure laser *Nature* **417** 156–9

Mukhopadhyay I and Singh S 1998 Optically pumped far infrared molecular lasers: spectroscopic and application aspects *Spectrochim. Acta Part A* **54** 395–410

Sirtori C and Teissier R 2006 Chapter 1: Quantum cascade lasers: overview of basic principles of operation and state of the art *Intersubband Transitions in Quantum Structures* ed R Paiella (New York: McGraw-Hill) pp 1–44

Shumyatsky P and Alfano R R 2011 Terahertz sources *J. Biomed. Opt.* **16** 033001-1–9

Wade A, Fedorov G, Smirnov D, Kumar S, Williams B S, Hu Q and Reno J L 2009 Magnetic-field-assisted terahertz quantum cascade laser operating up to 225 K *Nat. Photonics* **3** 41–5

Wang C A, Goyal A, Huang R, Donnelly J, Calawa D, Turner G, Sanchez-Rubio A, Hsu A, Hu Q and Williams B 2010 Strain-compensated GaInAs/AlInAs/InP quantum cascade laser materials *J. Cryst. Growth* **312** 1157–64

Wang F, Lee J, Phillips D J, Holliday S G, Chua S-L, Bravo-Abad J, Joannopoulos J D, Soljačić M, Johnson S G and Everitt H O 2018 A high-efficiency regime for gas-phase terahertz lasers *PNAS* **115** 6614–9

Xu L-H, Lees R M, Vasconcellos E C C, Zerbetto S C, Zink L R and Evenson K M 1996 Methanol and the optically pumped far-infrared laser *IEEE J. Quantum Electron.* **32** 392–99

Part II

Applications

Chapter 5

Communication

Significance and relevance of terahertz waves as a vehicle for communication is carved out by bringing out its merits and demerits relative to remaining components of the electromagnetic spectrum, including the familiar and widely-practiced radio frequency and microwave communication, the restrictedly used visible/infrared communication, and the hostile gamma ray/x-ray communication systems. Establishment of reliable THz communication links calls for clear perception of absorption losses during propagation through the atmosphere besides free space path loss (FSPL). Careful selection of transmission windows in the terahertz spectrum must be emphasized. Transmission capabilities of present networks can be drastically improved by relieving the load at chosen points by replacement with terahertz wave-based networks. Judicious utilization of the terahertz spectrum to complement the pre-existing telecommunication facilities will lead to networks that can convey vast amounts of data at faster speeds ushering a revolution in wireless and wired networks. Terahertz wave modulation techniques are described.

After familiarization with the vast gamut of electronic, photothermal and opto-electronic devices in parts I–III of the book, this concluding part IV divulges the expansive applications in which these devices are being engaged. Communications, spectroscopy and imaging are three prominent areas in which terahertz electronics is already making inroads, and is likely to make far-reaching impact in the near future. In this chapter we take up the topic of terahertz communications. The success stories of spectroscopy and imaging will be covered in forthcoming chapters.

It is expedient to begin with basics of communication and move gradually to terahertz frequencies in order that the significance of communication in this portion of the electromagnetic spectrum is appreciated relative to the spectrum in its entirety.

doi:10.1088/978-0-7503-4886-7ch5

5.1 Recapitulation of common communication terms and laws

Some network communication terms look so identical causing misconceptions. To prevent confusion, we must clarify the subtle differences between them at the very outset. We must also restate some basic formulae and laws of communication systems, which will be helpful in discussions.

5.1.1 Data rate or data transfer rate

This is the speed of data transmission over a channel measured in bits/s or bytes/s; 1 byte = 8 bits.

5.1.2 Bandwidth

This is the amount of data that can be transferred over a channel in a given interval of time, reckoned in bits/s or bytes/s. Note carefully the difference between bandwidth given here for data communications and the bandwidth in the context of signal processing where it represents a range of frequencies between upper and lower cut-off frequencies; hence measured in Hertz units. Always use the relevant definition applicable to the situation being dealt with.

5.1.3 Bandwidth and speed

Bandwidth is a representation of the amount of data that can be received/sent per second while data rate represents the speed at which that data is received/sent, fast or slow, e.g., bandwidth gives how many bits one can download from or upload to a network channel in a second and data rate measures how fast or slow those bits can be downloaded/uploaded.

Speed can be less than bandwidth due to network congestion during traffic. The familiar analogy of water flowing through a pipe helps in easy comprehension of the difference between bandwidth and speed. Bandwidth is the metaphor for diameter of the pipe in cms, whereas speed is similar to the volume rate of flow of water in liters per second. A larger diameter pipe is equivalent to a broader bandwidth but if the speed is less than the specified amount, the pipe (bandwidth) is not fully filled so that the rated capability of the broad bandwidth is not tapped.

5.1.4 Throughput

This is the actual amount of data transmitted through a channel in a given interval of time taking losses and other factors into consideration. It is measured in bits/s or bytes/s. Throughput differs from bandwidth because bandwidth represents the data transference capability of the channel while throughput indicates how much data is able to reach the destination successfully.

5.1.5 Network capacity

This is the theoretical maximum amount of data that can be transferred over the network per unit time = theoretical maximum data rate. Its unit is bits/s or bytes/s.

5.1.6 Nyquist bit rate formula for a noiseless channel

Even for a perfect noiseless channel, there exists an upper bound to the data rate r_{max} related by Nyquist bit rate formula with bandwidth B and the number N of quantized levels used for data representation as (Nyquist, 1928, Forouzan and Fegan 2007)

$$r_{max} = 2B\log_2 N \tag{5.1}$$

The rate r_{max} gives the channel capacity. Bandwidth for a channel is a fixed number, not variable. To increase the data rate, the number of levels must be raised. But when a large number of levels is used, the spacing intervening any two levels decreases. So, two proximate levels may not be easily discernible. Hence probability of error increases impacting the reliability of the system unfavorably.

5.1.7 Noise in channels

Practically, channels are prone to various types of noise such as thermal noise, impulse noise, cross-talks, etc. Random signals produced due to vibrations of atoms in the receiver act as a major noise source in wireless systems. Because the received signal $r(t)$ is the sum of transmitted signal $s(t)$ and noise signal $n(t)$, as expressed by the equation

$$r(t) = s(t) + n(t) \tag{5.2}$$

the noise signal $n(t)$ is said to be additive. In addition, it is referred to as 'white' implying uniformity of power across the complete frequency band similar to the white color composed of all the colors in visible spectrum. Hence, it is called additive white Gaussian noise (AWGN).

5.1.8 Shannon's law for a noisy channel

Shannon–Hartley capacity theorem or Shannon's law is a statement connecting the capacity of a channel with its bandwidth, and signal and noise powers. For a noisy channel, the theoretical maximum amount of error-free data rate is known as the Shannon information capacity C. Its dependence on the channel bandwidth B and the root-mean-square signal-to-noise ratio ($RMS\ SNR$) is expressed as (Hartley 1928, Shannon 1948, 1949, Shannon and Weaver 1949, Stallings 2007, Rioul and Magossi 2014)

$$C = B\log_2(1 + RMS\ SNR) \tag{5.3}$$

To calculate logarithm with respect to base 10, equation (5.3) is rewritten as

$$C = B \times \frac{1}{\log_{10}2}\log_{10}(1 + RMS\ SNR) \approx 3.32B\log_{10}(1 + RMS\ SNR) \tag{5.4}$$

using the change-of-base formula

$$\log_b x = \frac{\log_a x}{\log_a b} \tag{5.5}$$

SNR in decibels is

$$SNR(dB) = 10\log_{10} RMS\ SNR \tag{5.6}$$

or

$$RMS \quad SNR = 10^{\frac{SNR(dB)}{10}} \tag{5.7}$$

5.1.9 Bandwidth and capacity

In equation (5.3), bandwidth is a fixed quantity. So bandwidth remaining constant, the capacity becomes proportional to SNR, and therefore, for a given noise level, it is proportional to signal power. Thus signal power is the parameter that must be increased to obtain a larger capacity.

Briefly, bandwidth is always less than capacity. It is a subset of capacity. In a noise-free channel, it is less than capacity by a factor which depends on the number of levels in signal processing, as determined by the Nyquist bit rate, equation (5.1). In a noisy channel, it is less than capacity by a factor which varies with SNR, equation (5.3).

Thus we can arrange the following in descending order:

$$\text{Capacity} \rightarrow \text{Bandwidth} \rightarrow \text{Throughput} \tag{5.8}$$

5.1.10 Latency (s)

The time delay incurred in a network is spoken of as its latency. Latency is generally measured as a round trip delay after the instruction for data transfer has been given, viz., the time taken by data to travel from source to destination over the channel and back to source again.

Example 5.1 A noiseless channel has a bandwidth of 0.7 Tbps. What is the maximum data rate achieved while transmitting a signal with eight discrete levels?

Solution: Here $N = 8$.

By Nyquist bit rate formula (equation (5.1)),

$$\begin{aligned} r_{\max} &= 2B\log_2 N = 2 \times 0.7\text{Tbps} \times \log_2(8) = 1.4\text{Tbps} \times \frac{\log_{10}(8)}{\log_{10}(2)} \\ &= 1.4\text{Tbps} \times \frac{0.903}{0.301} = 4.2\text{Tbps} \end{aligned} \tag{5.9}$$

Example 5.2

(a) If the SNR of a data network is 25 dB and peak data rate is 1 Tbps, what is the bandwidth? (b) When the SNR is 40 dB, by what factor does the peak data rate rise for the same bandwidth? (c) At an SNR of 0 dB for the 'no loss-no gain' case, how much does the peak data rate fall keeping bandwidth fixed?

Solution: (a) Applying equation (5.7),

$$RMS\ SNR = 10^{\frac{SNR(dB)}{10}} = 10^{\frac{25}{10}} = 10^{2.5} = 316.228 \tag{5.10}$$

Then equation (5.4) gives

$$\begin{aligned}\text{Peak data rate} &= C = 3.32B\log_{10}(1 + RMS\ SNR) = 3.32B\log_{10}(1 + 316.228)\\ &= 3.32B\log_{10}(317.228) = 3.32B \times 2.5014 = 8.305B\end{aligned} \tag{5.11}$$

Since the peak data rate is 1 Tbps, we can write

$$1\ \text{Tbps} = 8.305B \tag{5.12}$$

from which

$$B = \frac{1\ \text{Tbps}}{8.305} = 0.12\ \text{Tbps} \tag{5.13}$$

(a) At SNR = 40 dB, from equation (5.7)

$$RMS\ SNR = 10^{\frac{SNR(dB)}{10}} = 10^{\frac{40}{10}} = 10^4 \tag{5.14}$$

Then from equations (5.4) and (5.13)

$$\begin{aligned}C &= 3.32B\log_{10}(1 + 10^4) = 3.32 \times 0.12\ \text{Tbps}\log_{10}(10\,001)\\ &= 0.3984 \times 4 = 1.5936\ \text{Tbps} \approx 1.6\ \text{Tbps}\end{aligned} \tag{5.15}$$

From equation (5.5) and given that the peak data rate for SNR = 25 dB is 1 Tbps, we have

$$\frac{C_{40\ \text{dB}}}{C_{25\ \text{dB}}} = \frac{1.6\ \text{Tbps}}{1\ \text{Tbps}} = 1.6 \tag{5.16}$$

(b) For SNR = 0 dB, from equation (5.7)

$$RMS\ SNR = 10^{\frac{SNR(\text{dB})}{10}} = 10^{\frac{0}{10}} = 10^0 = 1 \tag{5.17}$$

Then from equations (5.4) and (5.13)

$$\begin{aligned} C &= 3.32B\log_{10}(1 + RMS\ SNR) = 3.32 \times 0.12\ \text{Tbps}\log_{10}(1+1) \\ &= 0.3984\log_{10}(2) = 0.3984 \times 0.3010 = 0.1199\ \text{Tbps} \approx 0.12\ \text{Tbps} \end{aligned} \tag{5.18}$$

Hence, from equation (5.18) and given that the peak data rate for SNR = 25 dB is 1 Tbps, we have

$$\frac{C_{0\,\text{dB}}}{C_{25\,\text{dB}}} = \frac{0.12\ \text{Tbps}}{1\ \text{Tbps}} = 0.12 \tag{5.19}$$

Example 5.3: Raw and effective bandwidths

Find the effective bandwidth of a network of raw bandwidth 0.85 Tbps if the sender overhead is 0.2 μs, receiver overhead is 0.3 μs, size of the message is 10000 bytes and transmission distance is 50 m? What is the ratio of effective bandwidth to raw bandwidth for this network?

Solution: Raw bandwidth excludes the consideration of sender and receiver overheads as well as the message flight and transmission times. Effective bandwidth is a more realistic representation of signal transmission and reception because of inclusion of delays incurred in headers and trailers.

$$\text{Effective bandwidth} = B_{\text{Effective}} = \frac{\text{Message size}}{\text{Latency}} \tag{5.20}$$

where

$$\begin{aligned} \text{Latency} &= \text{Sender overhead} + \text{Time of flight} + \text{Transmission time} \\ &+ \text{Receiver overhead} \end{aligned} \tag{5.21}$$

Here

$$\begin{aligned} \text{Time of flight} &= \frac{2 \times \text{Distance}}{\text{velocity}} = \frac{2 \times 50\ \text{m}}{2.9979 \times 10^{8}\ \text{m s}^{-1}} \\ &= \frac{100}{2.9979} \times 10^{-8} = 0.3336 \times 10^{-6}\ \text{s} \end{aligned} \tag{5.22}$$

$$\begin{aligned} \text{Message size} &= 10\,000 \times 8 = 80\,000\ \text{bits} = \frac{80000}{10^{12}}\ \text{Tb} \\ &= 8 \times 10^{-8}\ \text{Tb} \end{aligned} \tag{5.23}$$

because 10000 bytes = 8 × 10 000 bits. Further, since 0.85 Tb are transmitted in 1 s

$\therefore$ 1 Tb is transmitted in (1/0.85) s

$\therefore$ 8×10^{-8} Tb is transmitted in $(8 \times 10^{-8}/0.85)$ s = 9.41×10^{-8} s.

Hence,

$$\text{Transmission time} = 9.41 \times 10^{-8}\ \text{s} \tag{5.24}$$

Putting the values from problem statement and equations (5.21)–(5.24) in equation (5.20), we get

$$\begin{aligned}
B_{\text{Effective}} &= \frac{\text{Message size}}{\text{Latency}} \\
&= \frac{\text{Message size}}{\text{Sender overhead} + \text{Time of flight} + \text{Transmission time} + \text{Receiver overhead}} \\
&= \frac{8 \times 10^{-8}\,\text{Tb}}{0.2 \times 10^{-6} + 0.3336 \times 10^{-6}\text{s} + 9.41 \times 10^{-8} + 0.3 \times 10^{-6}} \\
&= \frac{8 \times 10^{-8}\,\text{Tb}}{2 \times 10^{-7} + 3.336 \times 10^{-7}\text{s} + 0.941 \times 10^{-7} + 3 \times 10^{-7}} \\
&= \frac{8 \times 10^{-8}\,\text{Tb}}{9.277 \times 10^{-7}\,\text{s}} = 0.08623\text{Tbps}
\end{aligned} \tag{5.25}$$

Since the raw bandwidth is given as 0.85 Tbps, from equation (5.25)

$$\frac{B_{\text{Effective}}}{B_{\text{Raw}}} = \frac{0.08623\text{Tbps}}{0.85\text{Tbps}} = 0.1014 \approx 0.1 \tag{5.26}$$

The effective bandwidth is 1/10 the raw bandwidth.

5.2 Effects of frequency of electromagnetic waves on their propagation characteristics, antenna dimensions and information-carrying capacity

While initiating discussions on terahertz communication, the thought that immediately strikes the mind is that we are already accustomed to communication at radio and microwave frequencies. Yes, of course! So the new event that is happening is that the frequency of electromagnetic waves is being raised further. Note that we are speaking about terrestrial communication, i.e., between two places on the globe.

5.2.1 The central role of frequency of communication signal

Frequency and equivalently wavelength of electromagnetic waves is a key parameter which decides how these waves will be used for communication, whether they will be made to travel along or close to the Earth's surface or ground, whether they will be projected into the sky to be reflected back to Earth from a layer in the atmosphere or whether they will be hurled as directional beams across line of sight (LOS) distances in the atmosphere from one sky-embracing tower to another or into outer space to artificial satellites (uplink) for reflection back to Earth (downlink) for international broadcasts. Additionally, frequency or wavelength determines the size of the antenna. For resonance, the size of the antenna must match the wavelength.

5.2.2 Pertinent questions

So the obvious questions that arise are: How can the terahertz portion of the electromagnetic spectrum be gainfully exploited? What significant achievements are expected from terahertz frequencies? What are the main obstacles and challenges faced? How can and to what extent can these be overcome?

To respond to these questions, we must look back and examine what has already happened in the frequency arena in bygone years, and where do we stand today. Then only can we plan where do we want to go? For that reason, we shall climb up the frequency ladder step by step moving from low frequency upwards, pause at every step, contemplate and try to explore: What has happened? What is happening? How is it significant in the context of terahertz communication?

The radio spectrum covers the frequency range from 30 Hz to 3 GHz, and the microwave region spans over 3 GHz to 300 GHz. The effects of frequency are well enshrined in the categorization of radio waves into ground, sky and free space waves (figure 5.1).

5.2.3 Ground waves

5.2.3.1 The frequency range

The frequency range is 500 kHz to 2 MHz, and with rapidly falling off energy up to 3 MHz. So communication with ground waves is not possible at frequencies above 3 MHz due to the signal becoming fainter.

5.2.3.2 Propagation along the surface of the Earth

The ground waves are sent by vertical antennas and in the state of vertical polarization along the Earth's surface; hence also called surface waves. The statement regarding polarization means that the electric field is in the vertical plane with magnetic field oriented horizontally. The vertical polarization is necessary because the electric field in the horizontal direction will be electrically short-circuited by the conducting Earth. So the waves will become feeble and decay within a short distance. The waves follow the curvature of the Earth and are easily diffracted around mountains, and bend around skycrapers in cities or other obstructions, to reach the receiver.

5.2.3.3 Distance covered by the wave

Transmission distance range of ground waves falls as the frequency increases. This can be understood from the fact that in a high-frequency wave, e.g., a sine wave, the troughs touch or come closer to the Earth a greater number of times over a shorter distance by virtue of the smaller wavelength of this wave. So by absorption during every touching event or by induction of opposite charges in the Earth through electrostatic induction, which partially neutralizes the electric field of the wave, the wave is weakened considerably after travelling a few kms. But a low frequency wave has a longer wavelength. So, the number of times the troughs touch the Earth or come near its surface over a given distance is much less. As a result, a low-frequency wave can cover a longer distance.

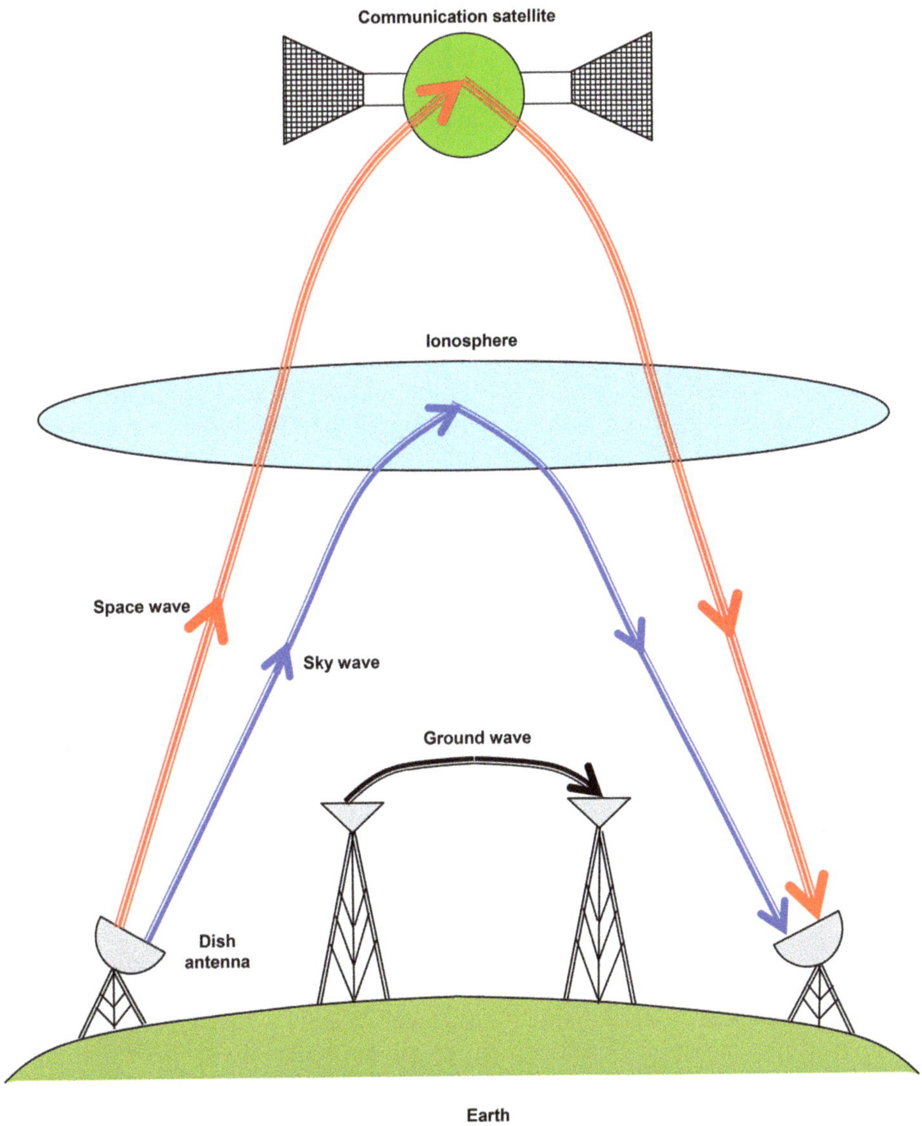

Figure 5.1. Ground, sky and space waves for electronic communication.

5.2.3.4 Losses incurred on the path

The low frequency ground waves are low energy photons by virtue of their low frequencies because the photon energy is obtained by multiplying the frequency by Planck's constant. Hence, these waves cannot excite atomic or molecular collisions, and therefore experience fewer losses when moving through the atmosphere. The atmospheric attenuation becomes a dominant loss mechanism at high frequencies, as we shall see for microwaves and terahertz waves.

5.2.3.5 Size of antenna

Concerning antenna requirements, the lower the frequency the longer the wavelength and therefore the larger the antenna. So, big antennas are needed, both for transmission and reception of ground waves.

5.2.3.6 Bandwidth of signal

Frequency is the parameter on which data carrying capacity of a wave depends, as indicated by bandwidth of the signal. A kHz frequency signal will provide at the most a bandwidth in kbps, definitely not in Mbps. Similarly, a MHz frequency signal allows a maximum of Mbps scale bandwidth, certainly not in Gbps. This can be appreciated by considering a sinusoidal signal and associating a zero or one each time the signal crosses the x-axis. Clearly, the number of zero crossings in a fixed interval of time is larger for a high frequency signal than for a low frequency signal. Hence more data can be assigned to a high-frequency signal. Consequently, a high frequency signal can carry more data than one of lower frequency. The reader must have perceived that the high data carrying capability in a high-frequency signal comes from its proficiency of changing a greater number of times in a second.

In this way, a high frequency signal provides a larger bandwidth. From these arguments we at once surmise that the ground waves suffer from the shortcoming of very small bandwidth so that large volumes of data cannot be transmitted through these waves.

5.2.4 Sky waves

5.2.4.1 The frequency range

The frequency range used for sky wave propagation is 3 MHz to 30 MHz. As the name implies, these waves are launched into the sky. But they are not lost in space.

5.2.4.2 The ionosphere

In the atmosphere there is a layer of ionized air called the ionosphere which stretches from 60 to 300 km from the Earth's surface. The ions of the atmosphere are produced by the incessant bombardment of air molecules by sunlight and cosmic rays from deep space. Neither the upper atmospheric layers above 300 km altitude nor the lower layers below 60 km are sufficiently ionized. The reason is that the atmosphere above 300 km is rarefied so that even if it is fully ionized, because of fewer air molecules present, this portion of the atmosphere does not become adequately charged to reflect back the sky waves. The atmospheric layers below 60 km height are also not much ionized because the energy of Sun rays and cosmic radiation is exhausted by collisions with air molecules during 300 – 60 = 240 km journey through air. Hence this 240 km thick ionized layer keeps hanging in the air at 60 km height above Earth's surface.

5.2.4.3 Successive reflections between the ionosphere and Earth

In a single reflection of sky waves transmitted from Earth and reflected back to Earth from the ionosphere, radio communication over a distance of 4000 km is accomplished. By successive reflections of sky waves between the Earth's surface and ionosphere, it is possible for sky waves to cover intercontinental distances reaching distant places across the globe. By this method, communication across the world is achieved.

5.2.4.4 Energy losses

The sky waves incur energy losses during propagation through the atmosphere. These air losses are larger than for ground waves.

5.2.4.5 Antenna size

Regarding antennas, the shorter wavelengths of sky waves than ground waves result in the requirement of proportionately smaller sizes of transmission and reception antennas for sky waves than the ground waves.

5.2.4.6 Bandwidth of signal

It is easy to guess from reasons elucidated previously for ground waves that the use of higher frequencies than ground waves makes broader bandwidths available for sky waves.

5.2.5 Space waves and satellite links

Sky waves of frequencies > 30 MHz can penetrate the ionosphere. So, the signal is not received back on Earth by reflection. Hence this method is limited to 30 MHz frequency. Then what? Recourse is taken to space wave propagation.

5.2.5.1 The frequency range

Space waves propagation employs frequencies from 30 MHz to 300 MHz and upwards with microwaves (300 MHz to 300 GHz).

5.2.5.2 Terrestrial communication through the atmosphere

For terrestrial links through the atmosphere, the space waves move from one antenna to another either directly or by reflection from the Earth's surface. This movement takes place through the troposphere, a layer of the atmosphere extending between 8 and 14 km from the ground, having the minimum thickness at the poles and maximum thickness at the equatorial regions.

5.2.5.3 Communication via satellites

Space waves move along LOS straight line paths. An Earth-based transmitting station and a satellite in space can see each other. For linkages via satellites, the LOS space waves are launched from the transmitting station to a geostationary satellite, a satellite placed at an altitude of 35 786 km in a circular geosynchronous equatorial orbit such that its period of revolution around the Earth equals the period of rotation of Earth about its axis (24h), and its direction of revolution is the same as Earth's rotational direction, i.e., from west to east. The satellite thus positioned appears stationary or permanently fixed at one spot to an observer on Earth; hence known as a geostationary satellite. Three such satellites separated by 120° of longitude in space are sufficient to establish a communication network capable of providing communication across the world by receiving space waves from ground stations and relaying them back to receivers on Earth. Only exclusions are small circular regions at the poles.

5.2.5.4 Range of terrestrial communication with space waves

On the Earth, the space waves are not diffracted around hills or tall buildings on the route. So they are blocked by such terrains or structures. Terrestrial space wave range is restricted by curvature of Earth and the heights of the transmitting and receiving antennas. The greater the antenna heights, the larger is the distance covered before the antennas become invisible to each other. LOS terrestrial communication links through space waves have ranges equalling visual horizon distances of 48–64 km.

5.2.5.5 Path losses

The waves experience larger losses than ground or sky waves during movement through the atmosphere.

5.2.5.6 Antenna size for space waves

Since the antenna size is inversely proportional to the operational frequency, small-size antennas suffice at space wave frequencies. 1–20 cm long antennas are adequate at microwave frequencies.

5.2.5.7 Bandwidth of signal transmitted by space waves

The high frequencies of space waves, particularly microwaves facilitate the conveyance of vast amounts of data by virtue of the enormous bandwidths available at these high frequencies.

Table 5.1 presents the salient features of ground, sky and space waves for communication.

5.3 Atmospheric attenuation of space waves beyond 30 GHz microwaves towards terahertz waves

5.3.1 Absorption of microwaves by moisture and oxygen

Microwaves are absorbed by moisture, oxygen and other gases in the atmosphere. Absorption occurs due to the existence of absorption lines or frequency bands in the elements constituting the atmospheric gases. It takes place at the resonance frequencies of the gas molecules. The absorption peak of water vapor occurs at 22.254 GHz so that humidity adversely affects the microwaves. The absorption peak of oxygen occurs at 60 GHz so that oxygen molecules begin absorbing microwaves at this frequency.

Table 5.1. Ground, sky and space waves.

Parameter	Type of wave: Ground	Sky	Space
Frequency	500 kHz to 2 MHz, maximum up to 3 MHz	3 MHz to 30 MHz	30 MHz to 300 MHz and microwaves (300 MHz to 300 GHz)
Propagation	Move along the surface of the Earth	Repeated reflections between Earth's surface and ionosphere provides intercontinental communication	Move from one antenna to next through the troposphere, providing terrestrial communication, LOS links 48–64 km. Use three geostationary satellites for around-the-world communication for satellite communication
Losses	Low atmospheric losses	More atmospheric losses than ground waves	Higher losses than sky waves
Antenna size	Big antennas	Smaller antennas than ground waves	Smaller antennas than sky waves
Bandwidth	Small bandwidth	Larger bandwidth than ground waves	Largest bandwidth of the three

The molecule of water vapor is a permanent electric dipole. It undergoes rotational transitions by absorption of microwave energy. The oxygen molecule has a permanent magnetic dipole moment owing to which transitions occur between molecular fine structure levels of permitted rotational states. As the magnetic transitions of oxygen molecule are 10^{-4} times weaker than rotational transitions of water molecule, the impact of water vapor is often more pronounced than oxygen close to resonant frequencies, despite the lower water vapor content of the atmosphere relative to oxygen. The attenuation of microwaves with respect to distance of propagation is expressed in terms of an attenuation or absorption constant, also called specific attenuation, in units of dB/km. In practice, around 40–60 GHz, the microwave signals undergo significant attenuation so that communication range in the atmosphere is restricted to a few kms only.

5.3.2 Transmission windows for communication of sub-THz and THz waves

In the terahertz spectrum, there are openings through which the electromagnetic waves can propagate by undergoing fewer losses than adjoining portions of the spectrum. These slots are known as low-loss transmission windows. These windows are shown in table 5.2, which is compiled with data from ITU-R (2016) (pressure 1013.25 hPa, temperature 15 °C, water-vapor density 7.5 g m^{-3}) and Rahaman *et al* (2019).

5.3.3 Signal transmission in outer space

Note that this kind of attenuation does not take place for communication between two points in outer space because vacuum conditions exist. So signal deterioration due to atmospheric effects is not encountered. In these situations, the signal spreads uniformly in all directions and weakens as inverse square of distance in accordance with Coulomb's law.

5.4 Free space path loss

5.4.1 Definition and physical meaning

A parameter of relevance to the discussions here is the FSPL of a radio signal during its journey on a LOS path under ideal conditions through an obstacle-free space without suffering any reflection, diffraction, scattering, absorption in moisture or gases, buildings, walls or vegetation on the way, or by multipath interference in which the waves travel by two or more paths, e.g., a direct wave and a reflected wave, and engage in constructive or destructive interference. Expressed as a power ratio between two isotropic radiators,

$$\mathrm{FSPL} = \frac{\text{Transmitted power}}{\text{Received power}} = \frac{P_T}{P_R} \tag{5.27}$$

it provides an approximate theoretical estimate about the strength of the signal at a particular distance from the transmitter. It is thus helpful in predicting the performance of an RF link.

5.4.2 Derivation of expression for FSPL from the Friis formula

The Friis transmission formula is a fundamental equation in antenna theory used to calculate the power obtained at a receiver antenna from a transmitter antenna radiating electromagnetic waves at a frequency f and separated by a distance d from the receiver antenna. The waves spreading out in a spherical wavefront from an isotropic transmitter antenna are uniformly distributed over the surface of the sphere. If the radius of spherical surface is R, the transmitted power density S_{T} is related to transmitted power P_{T} as

$$S_T = \frac{P_T}{4\pi R^2} \tag{5.28}$$

Table 5.2. Low-loss transmission windows for propagation of electromagnetic waves in the sub-THz and THz frequency ranges.

Sl. No.	Frequency range (THz)	Width of the window (THz)	Atmospheric attenuation @dB/km	Atmospheric attenuation @ 10 dB/distance	Envisaged application distance	Reference
1.	<0.05		0.2 dB km^{-1}	10 dB/50 km	Very long distance	ITU-R 2016
2.	<0.10		15 dB km^{-1}	10 dB/0.667 km	Medium distance	
3.	0.10–0.15	0.05	2 dB km^{-1}	10 dB/5 km	Long distance	
4.	0.22–0.32	0.10	7dB km^{-1}	10 dB/1.43 km	Long distance	
5.	<0.300		30 dB km^{-1}	10 dB/0.333 km	Medium distance	
6.	0.330–0.370	0.04	8 dB km^{-1}	10 dB/1.25 km	Very short distance	
7.	<0.35		40 dB km^{-1}	10 dB/0.250 km	Short distance	
8.	0.390–0.440	0.05	18 dB km^{-1}	10 dB/0.555 km	Medium distance	
9.	<0.50		380 dB km^{-1}	10 dB/26.32 m	Device-to-device	
10.	0.625–0.725	0.10	60 dB km^{-1}	10 dB/166.7 m	Indoor	
11.	0.780–0.910		80 dB/km	10 dB/125 m	Indoor	
12.	0.940–0.960	0.02	130 dB km^{-1}	10 dB/76.92 m	Device-to-device	
13.	1.210–1.410	0.20	150dB km^{-1}	10 dB/66.7 m	Device-to-device	Rahaman *et al* (2019)
14.	1.420–1.590	0.17	150dB km^{-1}	10 dB/66.7 m	Device-to-device	
15.	<1.0		1.8×10^{4} dB km^{-1}	10 dB/0.555 m	Device-to-device	ITU-R (2016)
16.	<3.0		4×10^{5} dB km^{-1}	10 dB/2.5 cm	Device-to-device	Rahaman *et al* (2019)

The power received by the antenna of effective aperture area or aperture A_{E} is given by

$$P_R = S_T A_{\mathrm{E}} \tag{5.29}$$

Combining equations (5.28) and (5.29) we get

$$P_R = \frac{P_T}{4\pi R^2} A_{\mathrm{E}} \tag{5.30}$$

From the theory of antenna,

$$A_{\mathrm{E}} = \frac{\lambda^2 G}{4\pi} \tag{5.31}$$

where G symbolizes the gain of the antenna and λ is the wavelength of the signal.

Let us consider an isotropic antenna, then $G = 1$; hence

$$A_{\mathrm{E}} = \frac{\lambda^2}{4\pi} \tag{5.32}$$

If c stands for the velocity of electromagnetic waves, their frequency ν is given by

$$\nu = \frac{c}{\lambda} \tag{5.33}$$

Applying equation (5.33), equation (5.32) reduces to

$$A_{\mathrm{E}} = \left(\frac{c}{\nu}\right)^2 \frac{1}{4\pi} = \frac{c^2}{4\pi\nu^2} \tag{5.34}$$

Substituting for A_{E} from equation (5.34) in equation (5.30),

$$P_R = \frac{P_T}{4\pi R^2} \times \frac{c^2}{4\pi\nu^2} = P_T \left(\frac{c}{4\pi R\nu}\right)^2 \tag{5.35}$$

from which

$$\frac{P_T}{P_R} = \left(\frac{4\pi R\nu}{c}\right)^2 = \left(\frac{4\pi D\nu}{c}\right)^2 \tag{5.36}$$

where we have replaced R by the distance D between the transmitter and receiver antennas for notational convenience.

Since from equation (5.27), the left-hand side of equation (5.36) is FSPL, we have

$$\mathrm{FSPL} = \left(\frac{4\pi D\nu}{c}\right)^2 \tag{5.37}$$

To express FSPL in decibels, we write

$$\begin{aligned}
\text{FSPL} &= 10\log_{10}\left\{\left(\frac{4\pi D\nu}{c}\right)^2\right\} = 20\log_{10}\left(\frac{4\pi D\nu}{c}\right) \\
&= 20\log_{10}\left(\frac{4\pi}{c}\right) + 20\log_{10}D + 20\log_{10}\nu \\
&= 20\log_{10}D + 20\log_{10}\nu + 20\log_{10}\left(\frac{4 \times 3.141\ 59}{2.9979 \times 10^8}\right) \\
&= 20\log_{10}D + 20\log_{10}\nu + 20\log_{10}\left(\frac{4 \times 3.141\ 59 \times 10^{-8}}{2.9979}\right) \\
&= 20\log_{10}D + 20\log_{10}\nu + 20\log_{10}(4.1917 \times 10^{-8}) \\
&= 20\log_{10}D + 20\log_{10}\nu + 20\log_{10}(4.1917) - 8 \times 20 \times \log_{10}10 \\
&= 20\log_{10}D + 20\log_{10}\nu + 20 \times 0.62239 - 8 \times 20 \times 1 \\
&= 20\log_{10}D + 20\log_{10}\nu + 12.4478 - 160 \\
&= 20\log_{10}D + 20\log_{10}\nu - 147.55
\end{aligned} \tag{5.38}$$

Example 5.4

(i) Calculate the FSPL for communication between two stations at a distance of 1 km apart using: (a) radio waves at frequency 100 kHz, (b) microwaves at frequency 1 GHz (c) THz link at 1.2 THz, (d) IR link at 3×10^{13} Hz, (e) visible light link at 5×10^{14} Hz, (f) x-ray link at 2×10^{17} Hz and (g) gamma ray link at 3×10^{23} Hz.

(ii) Determine the energies of these radiations.

Solution: From equation (5.38) for FSPL, we have

$$\begin{aligned}
\text{FSPL} &= 20\log_{10}(1000) + 20\log_{10}\nu - 147.55 \\
&= 20\log_{10}(10^3) + 20\log_{10}\nu - 147.55 \\
&= 3 \times 20\log_{10}(10) + 20\log_{10}\nu - 147.55 \\
&= 3 \times 20 \times 1 + 20\log_{10}\nu - 147.55 \\
&= 20\log_{10}\nu + 60 - 147.55 = 20\log_{10}\nu - 87.55
\end{aligned} \tag{5.39}$$

This equation will be applied to calculate the FSPL at different frequencies.

(a)

$$\begin{aligned}
|\text{FSPL}|_{100\ \text{kHz}} &= 20\log_{10}(1 \times 10^5) - 87.55 \\
&= 20\log_{10}(1) + 20\log_{10}(10^5) - 87.55 \\
&= 20 \times 0 + 20 \times 5\log_{10}10 - 87.55 \\
&= 0 + 100 \times 1 - 87.55 = 12.45\ \text{dB}
\end{aligned} \tag{5.40}$$

(b)

$$\begin{aligned}|\mathrm{FSPL}|_{1\ \mathrm{GHz}} &= 20\log_{10}(1\times 10^{9}) - 87.55\\ &= 20\log_{10}(1) + 20\log_{10}(10^{9}) - 87.55\\ &= 20\times 0 + 20\times 9\log_{10}10 - 87.55\\ &= 20\times 0 + 20\times 9\times 1 - 87.55 = 0 + 180 - 87.55 = 92.45\ \mathrm{dB}\end{aligned} \tag{5.41}$$

(c)

$$\begin{aligned}|\mathrm{FSPL}|_{1.2\ \mathrm{THz}} &= 20\log_{10}(1.2\times 10^{12}) - 87.55\\ &= 20\log_{10}(1.2) + 20\log_{10}(10^{12}) - 87.55\\ &= 20\times 0.079\,18 + 20\times 12\log_{10}10 - 87.55\\ &= 1.5836 + 240\times 1 - 87.55 = 154.0336\ \mathrm{dB}\end{aligned} \tag{5.42}$$

(d)

$$\begin{aligned}|\mathrm{FSPL}|_{3\times 10^{13}\ \mathrm{Hz}} &= 20\log_{10}(3\times 10^{13}) - 87.55\\ &= 20\log_{10}(3) + 20\log_{10}(10^{13}) - 87.55\\ &= 20\times 0.477\,12 + 20\times 13\log_{10}10 - 87.55\\ &= 9.5424 + 260\times 1 - 87.55 = 181.9924\ \mathrm{dB}\end{aligned} \tag{5.43}$$

(e)

$$\begin{aligned}|\mathrm{FSPL}|_{5\times 10^{14}\ \mathrm{Hz}} &= 20\log_{10}(5\times 10^{14}) - 87.55\\ &= 20\log_{10}(5) + 20\log_{10}(10^{14}) - 87.55\\ &= 20\times 0.698\,97 + 20\times 14\log_{10}10 - 87.55\\ &= 13.9794 + 280\times 1 - 87.55 = 206.4294\ \mathrm{dB}\end{aligned} \tag{5.44}$$

(f)

$$\begin{aligned}|\mathrm{FSPL}|_{2\times 10^{17}\ \mathrm{Hz}} &= 20\log_{10}(2\times 10^{17}) - 87.55\\ &= 20\log_{10}(2) + 20\log_{10}(10^{17}) - 87.55\\ &= 20\times 0.301\,03 + 20\times 17\log_{10}10 - 87.55\\ &= 6.0206 + 340\times 1 - 87.55 = 258.4706\ \mathrm{dB}\end{aligned} \tag{5.45}$$

(g)

$$\begin{aligned}|\mathrm{FSPL}|_{3\times 10^{23}\ \mathrm{Hz}} &= 20\log_{10}(3\times 10^{23}) - 87.55\\ &= 20\log_{10}(3) + 20\log_{10}(10^{23}) - 87.55\\ &= 20\times 0.477\,121 + 20\times 23\log_{10}10 - 87.55\\ &= 9.5424 + 460\times 1 - 87.55 = 381.99\mathrm{dB} \approx 382\ \mathrm{dB}\end{aligned} \tag{5.46}$$

(i) The energy E is related to frequency ν as

$$E = h\nu = 4.135\,67 \times 10^{-15}\text{eV s} \times \nu \tag{5.47}$$

(a)

$$\begin{aligned} |E|_{100\text{ kHz}} &= 4.13567 \times 10^{-15}\text{ eV s} \times 100 \times 10^{3}\text{ s}^{-1} \\ &= 4.13567 \times 10^{-15+2+3} \\ &= 4.13567 \times 10^{-10}\text{ eV} \end{aligned} \tag{5.48}$$

(b)

$$\begin{aligned} |E|_{1\text{ GHz}} &= 4.13567 \times 10^{-15}\text{ eV s} \times 1 \times 10^{9}\text{s}^{-1} \\ &= 4.13567 \times 10^{-15+9} \\ &= 4.13567 \times 10^{-6}\text{ eV} \end{aligned} \tag{5.49}$$

(c)

$$\begin{aligned} |E|_{1.2\text{THz}} &= 4.13567 \times 10^{-15}\text{eV s} \times 1.2 \times 10^{12}\text{s}^{-1} \\ &= 4.9628 \times 10^{-15+12} = 4.9628 \times 10^{-3}\text{eV} \end{aligned} \tag{5.49a}$$

(d)

$$\begin{aligned} |E|_{3\times10^{13}\text{Hz}} &= 4.13567 \times 10^{-15}\text{eV s} \times 3 \times 10^{13}\text{Hzs}^{-1} \\ &= 4.13567 \times 10^{-15+13} = 12.407 \times 10^{-2}\text{eV} \\ &\qquad\qquad\qquad\qquad = 1.2407 \times 10^{-1}\text{eV} \end{aligned} \tag{5.49b}$$

(e)

$$\begin{aligned} |E|_{5\times10^{14}\text{ Hz}} &= 4.13567 \times 10^{-15}\text{ eV s} \times 5 \times 10^{14}\text{ s}^{-1} \\ &= 4.13567 \times 5 \times 10^{-15+14} \\ &= 20.67835 \times 10^{-1}\text{ eV} = 2.068\text{ eV} \end{aligned} \tag{5.50}$$

(f)

$$\begin{aligned} |E|_{2\times10^{17}\text{ Hz}} &= 4.13567 \times 10^{-15}\text{eV s} \times 2 \times 10^{17}\text{ s}^{-1} \\ &= 4.13567 \times 2 \times 10^{-15+17} \\ &= 8.27134 \times 10^{2}\text{ eV} = 827.134\text{ eV} \end{aligned} \tag{5.51}$$

(g)

$$\begin{aligned} |E|_{3\times10^{23}\text{ Hz}} &= 4.13567 \times 10^{-15}\text{ eV s} \times 3 \times 10^{23}\text{ s}^{-1} \\ &= 4.13567 \times 3 \times 10^{-15+23} \\ &= 12.40701 \times 10^{8}\text{ eV} = 1.241 \times 10^{9}\text{ eV} \\ &= \frac{1.241 \times 10^{9}}{10^{6}}\text{ MeV} = 1241\text{ MeV} \end{aligned} \tag{5.52}$$

Comment: With increase in frequency, the FSPL increases and so also the energy of the signal.

Example 5.5 The FSPL suffered in a communication link working at 3 THz is 110 dB. The source-to-observer distance is x. Find the value of x. With reference to table 5.2, compare FSPL with atmospheric loss at x distance.

Solution: From equation (5.38),

$$\begin{aligned} 110\text{ dB} &= 20\log_{10}(x) + 20\log_{10}(3 \times 10^{12}) - 147.55 \\ &= 20\log_{10}(x) + 20\log_{10}(3) + 20\log_{10}(10^{12}) - 147.55 \\ &= 20\log_{10}(x) + 20\log_{10}(3) + 20 \times 12 \times \log_{10}(10) - 147.55 \\ &= 20\log_{10}(x) + 20 \times 0.47712 + 20 \times 12 \times 1 - 147.55 \\ &= 20\log_{10}(x) + 9.5424 + 240 - 147.55 \\ &= 20\log_{10}(x) + 101.9924 \end{aligned} \tag{5.53}$$

or

$$20\log_{10}(x) = 110 - 101.99 = 8.01 \tag{5.54}$$

or

$$\log_{10}(x) = \frac{8.01}{20} = 0.4005 \tag{5.55}$$

$$\therefore x = 10^{0.4005} = 2.515\text{ m} \tag{5.56}$$

Verification: Right-hand side of equation (5.53) is

$$\begin{aligned} \text{RHS} &= 20\log_{10}(x) + 101.99 \\ &= 20\log_{10}(2.515) + 101.99 \\ &= 20 \times 0.4005 + 101.99 = 8.01 + 101.99 = 110 \\ &= \text{Left} - \text{hand side} \end{aligned} \tag{5.57}$$

Hence true.

From table 5.2, the atmospheric loss at ~ 3 THz is 10 dB/2.5 cm = 10 dB/0.025 m, so that the atmospheric loss at 2.515 m distance is = 10 dB × 2.515 m/0.025 m = 1006 dB. Hence atmospheric loss is 1006/110 = 9.145 times the FSPL.

Example 5.6 A communication link is designed to operate at 0.7 THz in the THz window 0.625–0.725 THz between two points 200 m apart. Find the FSPL and atmospheric loss for the link. Which of the two losses is an issue of concern?

Solution: By equation (5.38), the FSPL at 0.7 THz and 200 m distance is

$$\begin{aligned} \text{FSPL} &= 20\log_{10}D + 20\log_{10}\nu - 147.55 \\ &= 20\log_{10}200 + 20\log_{10}(0.7 \times 10^{12}) - 147.55 \\ &= 20 \times 2.301 + 20\log_{10}(0.7) + 20\log_{10}(10^{12}) - 147.55 \\ &= 46.02 + 20 \times -0.1549 + 12 \times 20\log_{10}(10) - 147.55 \\ &= 46.02 - 3.098 + 240 \times 1 - 147.55 = 286.02 - 150.648 \\ &= 135.372\text{ dB} \end{aligned} \tag{5.58}$$

From table 5.2, the atmospheric loss is 10 dB/166.7 m = 0.059 988 dB m^{-1}. At 200 m, it is 200 × 0.059 988 dB = 11.9976 ~ 12 dB. The FSPL is evaluated in equation (5.58).

$$\therefore \frac{\text{FSPL}}{\text{Atmospheric loss}} = \frac{135.372}{12} = 11.281 \approx 11.3 \tag{5.59}$$

or

$$\text{FSPL} = 11.3 \times \text{Atmospheric loss} \tag{5.60}$$

Hence, the FSPL causes more concern than atmospheric loss.

5.5 Link budget analysis

5.5.1 Writing the link budget equation

In a communication system, it is often desired to calculate the strength of the signal received at the receiver. Then an equation is written down expressing the received power in terms of the transmitted power, all the gains provided, e.g., the transmitter and receiver antenna gains; and the losses incurred, such as the FSPL; the atmospheric loss; terrain-dependent loss, urban or rural, mountainous or plain; and losses from ground reflection, diffraction etc. This equation is called the link budget equation and the analysis is referred to as link budget analysis. The analysis tells whether a link can be established to meet the targeted specifications, or whether any apportioning of gains or losses is necessary to achieve the goals. It is thus a design aid to quantify whether a link is viable and to assess its performance. The link budget equation has the form

$$\begin{aligned}\text{Received power (dBm)} = \text{Transmitted power (dBm)} + \text{Gains (dB)} \\ - \text{Losses (dB)}\end{aligned} \tag{5.61}$$

where dBm stands for decibel-milliwatt, which is a logarithmic measure of power expressing the power in decibels compared to a reference power level of 1 mW.

5.5.2 dBm to Watts conversion

Power in Watts is obtained from power in dBm by the formula

$$\begin{aligned}P_{\text{Watts}} &= \frac{1\ \text{W}}{1000}.\ 10^{\frac{P_{\text{dBm}}}{10}} = \frac{1\ \text{W}}{10^3}.\ 10^{\frac{P_{\text{dBm}}}{10}} \\ &= 1\ \text{W}.\ 10^{\frac{P_{\text{dBm}}}{10}} \times 10^{-3} = 1\ \text{W}.\ 10^{\frac{P_{\text{dBm}}}{10}-3} \\ &= 1\ \text{W}.\ 10^{\frac{P_{\text{dBm}-30}}{10}}\end{aligned} \tag{5.62}$$

If $P_{\text{dBm}} = 0$ dBm, from equation (5.62)

$$\begin{aligned}P_{\text{Watts}} &= 1\ \text{W}.\ 10^{\frac{P_{\text{dBm}-30}}{10}} = 1\ \text{W}.\ 10^{\frac{0-30}{10}} \\ &= 1\ \text{W}.\ 10^{-3} = 1\ \text{mW}\end{aligned} \tag{5.63}$$

For $P_{\mathrm{dBm}} = 10$ dBm, from equation (5.62)

$$\begin{aligned} P_{\mathrm{Watts}} &= 1\ \mathrm{W}.\ 10^{\frac{10-30}{10}} = 1\ \mathrm{W}.\ 10^{\frac{-20}{10}} \\ &= 1\ \mathrm{W}.\ 10^{-2} = 10\ \mathrm{mW} \end{aligned} \tag{5.64}$$

From equation (5.62), a power of 20 dBm corresponds to

$$\begin{aligned} P_{\mathrm{Watts}} &= 1\ \mathrm{W}.\ 10^{\frac{20-30}{10}} = 1\ \mathrm{W}.\ 10^{\frac{-10}{10}} \\ &= 1\ \mathrm{W}.\ 10^{-1} = 100\ \mathrm{mW} \end{aligned} \tag{5.65}$$

From equation (5.62), a power of 30 dBm becomes

$$\begin{aligned} P_{\mathrm{Watts}} &= 1\ \mathrm{W}.\ 10^{\frac{30-30}{10}} = 1\ \mathrm{W}.\ 10^{0} \\ &= 1\ \mathrm{W}.1 = 1\ \mathrm{W} \end{aligned} \tag{5.66}$$

and so on.

5.5.3 Logarithmic form of the link budget equation

Considering the contributions from transmitter and receiver losses together with antenna gains as well as the free-space and atmospheric losses, equation (5.38) is changed to a logarithmic equation

$$\begin{aligned} \text{Power received (dBm)} = {} & \text{Power transmitted (dBm)} + \text{Gain of transmitter antenna (dBi)} \\ & - \text{Transmitter losses (dB)} - \text{FSPL (dB)} - \text{Atmospheric loss (dB)} \\ & + \text{Gain of receiver antenna (dBi)} - \text{Receiver losses (dB)} \end{aligned} \tag{5.67}$$

where dBi is an abbreviation for decibel isotropic meaning decibels with respect to a theoretical isotropic radiator, defined as a point source which radiates and receives power spherically equally in all spatial directions with unit gain. The transmitter and receiver losses account for any losses in cables or connectors.

5.5.4 Link margin

A link margin M is defined as the difference between received power and sensitivity of reception:

$$M = \text{Power received (dBm)} - \text{Receive sensitivity (dBm)} \tag{5.68}$$

The receive sensitivity is the weakest received signal strength that the receiver is capable of identifying and processing. To ensure a reliable link, the link margin must be positive and $\geqslant$10 dB otherwise the link will be impractical.

5.5.5 Symbolic form of the logarithmic link budget equation

Replacing each term in equation (5.67) by its symbol, the equation becomes

$$\begin{aligned} P_{RX}(\mathrm{dBm}) = {} & P_{TX}(\mathrm{dBm}) + G_{TX}(\mathrm{dBi}) - L_{TX}(\mathrm{dB}) \\ & - \mathrm{FSPL}(\mathrm{dB}) - L_{\mathrm{Atmospheric}}(\mathrm{dB}) + G_{RX}(\mathrm{dBi}) - L_{RX}(\mathrm{dB}) \end{aligned} \tag{5.69}$$

Also, equation (5.68) is written as

$$M(\text{dBm}) = P_{RX}(\text{dBm}) - \text{Sensitivity}_{RX}(\text{dBm}) \tag{5.70}$$

Example 5.7 Ascertain the feasibility of a 1 km range terahertz link working at 0.4 THz along a LOS path through the atmosphere between a transmitting station radiating at 10 dBm with antenna gain 5 dBi and a receiver with identical antenna gain and receive sensitivity −80 dBm. Transmitter and receiver losses are 2 dB each. If the link is not possible, suggest a possible solution.

Solution: Applying equation (5.38)

$$\begin{aligned}\text{FSPL} &= 20\log_{10}(1000) + 20\log_{10}(0.4 \times 10^{12}) - 147.55\\ &= 20\log_{10}(10^{3}) + 20\log_{10}(0.4) + 20\log_{10}(10^{12}) - 147.55\\ &= 20 \times 3 \times \log_{10}(10) - 20 \times 0.3979 + 20 \times 12\log_{10}(10) - 147.55\\ &= 60 \times 1 - 7.958 + 20 \times 12 \times 1 - 147.55\\ &= 60 - 7.958 + 240 - 147.55 = 300 - 155.508 = 144.492\\ &\approx 144.5\text{dB}\end{aligned} \tag{5.71}$$

From table 5.2,

$$L_{\text{Atmospheric}} = 10 \text{ dB for } 0.555 \text{ km} = 18.018 \text{ dB for } 1 \text{ km} \tag{5.72}$$

Given for transmitter side:

$$P_{TX} = 10 \text{ dBm} \tag{5.73}$$

$$G_{TX} = 5 \text{ dBi} \tag{5.74}$$

$$L_{TX} = 2 \text{ dB} \tag{5.75}$$

From equations (5.71) and (5.72), let us rewrite

$$\text{FSPL} = 144.5 \text{ dB} \tag{5.76}$$

$$L_{\text{Atmospheric}} = 18.018 \text{ dB} \tag{5.77}$$

Given for receiver side:

$$G_{RX} = 5 \text{ dBi} \tag{5.78}$$

$$L_{RX} = 2 \text{ dB} \tag{5.79}$$

Putting the values in equations (5.69) and (5.70),

$$\begin{aligned}P_{RX} &= 10 \text{ dBm} + 5 \text{ dBi} - 2 \text{ dB} - 144.5 \text{ dB} - 18.018 \text{ dB} + 5 \text{ dBi} - 2 \text{ dB}\\ &= 20 \text{ dB} - 166.518 = -146.518 \text{ dBm}\end{aligned} \tag{5.80}$$

$$\begin{aligned} M &= P_{RX}\,(\text{dBm}) - \text{Sensitivity}_{RX}\,(\text{dBm}) \\ &= -146.518\text{ dBm} - (-80\text{ dBm}) = -146.518\text{ dBm} + 80\text{ dBm} = -66.518\text{ dBm} \end{aligned} \tag{5.81}$$

Negative value of link margin indicates that the link is untenable. One way to create the link is to increase the antenna gains. We work backwards to calculate the antenna gains required to achieve $M = 10$ dBm. Let the required antenna gain be x for transmitter or receiver. From equations (5.70) and (5.69) we get

$$\begin{aligned} M\,(\text{dBm}) &= P_{RX}\,(\text{dBm}) - \text{Sensitivity}_{RX}\,(\text{dBm}) \\ &= P_{TX}\,(\text{dBm}) + G_{TX}\,(\text{dBi}) \\ &\quad - L_{TX}\,(\text{dB}) - FSPL\,(\text{dB}) - L_{\text{Atmospheric}}\,(\text{dB}) \\ &\quad + G_{RX}\,(\text{dBi}) - L_{RX}\,(\text{dB}) - \text{Sensitivity}_{RX}\,(\text{dBm}) \end{aligned} \tag{5.82}$$

Putting the values in equation (5.82),

$$10 = 10 + x - 2 - 144.5 - 18.018 + x - 2 - (-80) \tag{5.83}$$

or

$$10 = 2x + 90 - 166.518 \tag{5.84}$$

or

$$2x = 10 + 166.518 - 90 = 86.518 \tag{5.85}$$

or

$$x = 86.518/2 = 43.259\text{ dB} \tag{5.86}$$

Let us calculate the diameter of the required antenna. Recall that we had put $G = 1$ in equation (5.31). Because the gain has a finite value here, re-inserting the gain G in equation (5.34) we have,

$$A_E = \frac{c^2 G}{4\pi\nu^2} \tag{5.87}$$

Considering a parabolic reflector antenna

$$G = 10\log_{10}\left\{\eta\left(\frac{\pi D}{\lambda}\right)^2\right\} \tag{5.88}$$

where η is the efficiency factor and D is the diameter of the parabolic reflector.

Equation (5.88) is expressed in terms of frequency as

$$G = 10\log_{10}\left\{\eta\left(\frac{\pi D}{\frac{c}{\nu}}\right)^2\right\} = 10\log_{10}\left\{\eta\left(\frac{\pi^2 D^2 \nu^2}{c^2}\right)\right\} \tag{5.89}$$

or

$$\frac{G}{10} = \log_{10}\left\{\eta\left(\frac{\pi^2 D^2 \nu^2}{c^2}\right)\right\} \tag{5.90}$$

or

$$\eta\left(\frac{\pi^2 D^2 \nu^2}{c^2}\right) = 10^{\frac{G}{10}} \tag{5.91}$$

or

$$D^2 = \frac{10^{\frac{G}{10}}}{\eta} \times \frac{c^2}{\pi^2 \nu^2} \tag{5.92}$$

$$\begin{aligned} \therefore D = \frac{c}{\pi\nu}\sqrt{\frac{10^{\frac{G}{10}}}{\eta}} &= \frac{2.9979 \times 10^8}{3.14159 \times 0.4 \times 10^{12}}\sqrt{\frac{10^{\frac{43.259}{10}}}{0.5}} \\ &= 2.386 \times 10^{8-12}\sqrt{\frac{10^{4.3259}}{0.5}} = 2.386 \times 10^{-4}\sqrt{\frac{21178.7342}{0.5}} \\ &= 2.386 \times 10^{-4}\sqrt{42357.4684} \\ &= 2.386 \times 10^{-4} \times 205.809 = 491.06 \times 10^{-4}\ \text{m} = 0.049106\ \text{m} \end{aligned} \tag{5.93}$$

where we have taken $\eta = 0.5$.

Note 1: dBm can be added to dB or dBi but dBm cannot be added to dBm. This becomes evident when we look at dBs in proper perspective. dB and dBi are unitless numbers but dBm is in milliwatts. But all dBs have a logarithmic nature. When we are adding logarithms of two numbers we are actually multiplying the original numbers because

$$\log A + \log B = \log AB \tag{5.94}$$

i.e., addition of logarithms becomes multiplication of linear numbers. dB represents a number and dBm is a power value in mW. So the number can be multiplied with power value to get the new power in mW. But the same statement looks confusing when we add a dB to a dBm value. Let us clarify.

Suppose we add 1 dB to 1 dBm. If the ratio of two power levels is x,

$$1\ \text{dB} = 10\log_{10}(\text{x}) \tag{5.95}$$

from which

$$\log_{10}(\text{x}) = \frac{1}{10} = 0.\ 1 \tag{5.96}$$

or,

$$\text{x} = 10^{0.1} = 1.\ 259 \tag{5.97}$$

So 1 dB corresponds to simply a number $x = 1.259$ with no unit.

Let us take a number y such that

$$1\ \text{dBm} = 10\log_{10}\{y\ (\text{mW})\} \tag{5.98}$$

or,

$$\log_{10}\{y\ (\text{mW})\} = \frac{1}{10} = 0.1 \tag{5.99}$$

or,

$$\{y\ (\text{mW})\} = 10^{0.1} = 1.259 \tag{5.100}$$

y is in milliwatts; it is not unitless.

From equations (5.95), (5.98), (5.97) and (5.100)

$$\begin{aligned}1\ \text{dB} + 1\ \text{dBm} &= 10\log_{10}(x) + 10\log_{10}\{y\ (\text{mW})\}\\ &= 10\{\log_{10}(x) + \log_{10}\{y\ (\text{mW})\}\}\\ &= 10\log_{10}[(x)\{y\ (\text{mW})\}] = 10\log_{10}\{xy(\text{mW})\}\\ &= 10\log_{10}(xy)\ \text{dBm} = 10\log_{10}(1.259 \times 1.259)\text{dBm}\\ &= 10\log_{10}(1.585)\ \text{dBm} = 10 \times 0.2\ \text{dBm} = 2\text{dBm}\end{aligned} \tag{5.101}$$

Note 2: dBm cannot be added to dBm because both are powers and addition of two powers in logarithmic scale means their multiplication in linear scale leading to a product in $(\text{mW})^2$, which is meaningless.

Note 3: Receive sensitivity is a negative quantity because it represents a small value of power. The receiver should be capable of detecting low levels of power. Since the logarithm of a fraction is negative, the negative sensitivity indicates a power level less than 1 mW, e.g., suppose the power value for −80 dBm is P mW. Then

$$-80\ \text{dBm} = 10\log_{10}\{\text{P}\ (\text{mW})\} \tag{5.102}$$

or,

$$\log_{10}\{\text{P}\ (\text{mW})\} = -\frac{80}{10} = -8 \tag{5.103}$$

or

$$\text{P}\ (\text{mW}) = 10^{-8} \tag{5.104}$$

or

$$\text{P} = 10^{-8}\ \text{mW} \tag{5.105}$$

Example 5.8 The FSPL increases with frequency and becomes a significant component of the total loss at high frequencies. An LOS link is operating through the atmosphere at 1.5 THz in the 1.420–1.590 THz frequency band. The specific attenuation due to atmosphere is 10 dB at 66.7 m. What is FSPL at 66.7 m? By what factor is it higher than atmospheric loss?

Solution: Applying equation (5.38)

$$\begin{aligned}\text{FSPL} &= 20\log_{10}(66.7) + 20\log_{10}(1.5 \times 10^{12}) - 147.55 \\ &= 20 \times 1.8241 + 20\log_{10}(1.5) + 20\log_{10}(10^{12}) - 147.55 \\ &= 36.482 + 20 \times 0.1761 + 20 \times 12 \times \log_{10}(10) - 147.55 \\ &= 36.482 + 3.522 + 20 \times 12 \times 1 - 147.55 \\ &= 280.004 - 147.55 = 132.454 \text{ dB}\end{aligned} \tag{5.106}$$

Hence at 66.7 m,

$$\text{FSPL}/L_{\text{Atmospheric}} = 132.45/10 = 13.245 \tag{5.107}$$

Example 5.9 A 0.1 THz bandwidth LOS communication link at 0.85 THz is made between two stations A and B 100 m apart. Transmitting station A has a transmitter with power 25 dBm, and antenna gain 57 dBi. Receiving station B has an antenna with gain 55 dBi. Transmission loss is 3 dB and the receiver noise figure is 4 dB. Taking into account Johnson noise, calculate the Shannon limit on the channel capacity.

Solution: To calculate the Shannon limit, the SNR is required for which the signal and the noise at the receiver must be evaluated. First, we find the signal power at the receiver by taking into account all the gains and losses. Then we determine the noise power at the receiver. Finally, we find the SNR at the receiver, and from it the Shannon limit.

By equation (5.38), we have

$$\begin{aligned}\text{FSPL} &= 20\log_{10}(100) + 20\log_{10}(0.85 \times 10^{12}) - 147.55 \\ &= 20\log_{10}(10^2) + 20\log_{10}(0.85) + 20\log_{10}(10^{12}) - 147.55 \\ &= 20 \times 2 \times \log_{10}(10) - 20 \times 0.070\,58 + 20 \times 12\log_{10}(10) - 147.55 \\ &= 40 \times 1 - 1.4116 + 20 \times 12 \times 1 - 147.55 \\ &= 40 - 1.4116 + 240 - 147.55 = 280 - 148.9616 = 131.0384 \\ &\approx 131 \text{ dB}\end{aligned} \tag{5.108}$$

From table 5.2, we write

$$L_{\text{Atmospheric}} = 10 \text{ dB for } 125 \text{ m} = 8 \text{ dB for } 100 \text{ m} \tag{5.109}$$

So, we take

$$L_{\text{Atmospheric}} = 8 \text{ dB} \tag{5.110}$$

Station A:

$$P_{TX} = 25 \text{ dBm} \tag{5.111}$$

$$G_{TX} = 57 \text{ dBi} \tag{5.112}$$

$$L_{TX} = 3 \text{ dB} \tag{5.113}$$

Station B:

$$G_{RX} = 55 \text{ dBi} \tag{5.114}$$

Noise figure (F) is a number used to express the performance of a receiver. It measures the degradation in SNR as the signal passes through the receiver. It is the ratio of SNR at the input to that at the output of a receiver, i.e., $F = \text{SNR}_{\text{Input}}/\text{SNR}_{\text{Output}}$. Here, the receiver noise figure is

$$\text{RX Noise Figure } (F) = 4 \text{ dB} \tag{5.115}$$

Thermal agitation of electrons leads to Johnson noise. The Johnson noise is proportional to the bandwidth B of the communication link and the temperature T of the surroundings. Thermal noise power is given by

$$\begin{aligned} N(\text{Watts}) &= k_B T B \\ &= 1.381 \times 10^{-23} \times 300 \times 0.1 \times 10^{12} \\ &= 41.43 \times 10^{-23+12} \\ &= 41.43 \times 10^{-11} \text{ W} \\ &= 41.43 \times 10^{-11} \times 10^{3} \text{ mW} \\ &= 41.43 \times 10^{-8} \text{ mW} \end{aligned} \tag{5.116}$$

where k_B is the Boltzmann constant. Thermal noise power is expressed in dBm as

$$\begin{aligned} N \text{ (dBm)} &= 10\log_{10}(41.43 \times 10^{-8}) = 10\{\log_{10}(41.43) + \log_{10}(10^{-8})\} \\ &= 10\{1.6173 - 8\log_{10}(10)\} = 10\{1.6173 - 8 \times 1\} \\ &= -10 \times 6.3827 = -63.83 \text{ dBm} \end{aligned} \tag{5.117}$$

From equations (5.115) and (5.117), the noise at the receiver is

$$\begin{aligned} \text{Noise}_{RX} &= F\ (dB) + N \text{ (dBm)} = 4 + (-63.83) = 4 - 63.83 \\ &= -59.83 \text{ dBm} \end{aligned} \tag{5.118}$$

Signal at receiver is

$$\begin{aligned} \text{Signal}_{RX} &= P_{TX} \text{ (dBm)} + G_{TX} \text{ (dBi)} - L_{TX}\text{(dB)} - FSPL \text{ (dB)} \\ &\quad - L_{\text{Atmospheric}} \text{ (dB)} + G_{RX} \text{ (dBi)} \\ &= 25 \text{ dBm} + 57 \text{ dBi} - 3 \text{ dB} - 131 \text{ dB} - 8 \text{ dB} + 55 \text{ dBi} \\ &= 137 \text{ dBm} - 142 \text{ dB} = -5 \text{ dBm} \end{aligned} \tag{5.119}$$

where values from equations (5.111)–(5.114), (5.108) and (5.110) are put.

From equation (5.118), the noise at the receiver is rewritten

$$\text{Noise}_{RX} = -59.83 \text{ dBm} \tag{5.120}$$

Applying equations (5.119) and (5.120), the SNR is obtained as

$$\begin{aligned}\text{SNR (dB)} &= -5\text{ dBm} - (-59.83\text{ dBm}) = -5\text{ dBm} + 59.83\text{ dBm} \\ &= 54.83\text{ dB}\end{aligned} \tag{5.121}$$

From equations (5.4) and (5.121),

$$\begin{aligned}C &= 3.32B\log_{10}(1 + RMS\text{ SNR}) = 3.32B\log_{10}\left(1 + 10^{\frac{\text{SNR(dB)}}{10}}\right) \\ &= 3.32 \times 0.1 \times 10^{12} \times \log_{10}\left(1 + 10^{\frac{54.83}{10}}\right) \\ &= 0.332 \times 10^{12} \times \log_{10}(1 + 10^{5.483}) \approx 0.332 \times 10^{12} \times \log_{10}(10^{5.483}) \\ &= 0.332 \times 10^{12} \times 5.483 \times \log_{10}(10) \\ &= 1.82 \times 10^{12} \times 1 = 1.8 \times 10^{12} = 1.8\text{ Tbps}\end{aligned} \tag{5.122}$$

Example 5.10 A bidirectional link $P \leftrightarrow Q$ is made between two points P and Q. Both P and Q have transmission and reception capabilities. The distance between P and Q is 300 m. The link works at a frequency of 1.5 THz. The transmission power of point P is 21 dBm. The antenna gain is 47 dBi. The receive sensitivity of P is −83 dBm. The losses at P are 3 dB. The transmitting power of Q is 27 dBm and the antenna gain is 53 dBi. The receive sensitivity at Q is −73.7 dBm. The losses at Q are 2.3 dB. Check for feasibility of links $P \rightarrow Q$ and $Q \leftarrow P$ if the atmospheric loss is 10 dB/66.7 m.

Solution: FSPL and atmospheric loss are the same in both directions $P \rightarrow Q$ and $Q \leftarrow P$:

By equation (5.38),

$$\begin{aligned}\text{FSPL} &= 20\log_{10}(300) + 20\log_{10}(1.5 \times 10^{12}) - 147.55 \\ &= 20\log_{10}(3 \times 10^{2}) + 20\log_{10}(1.5) + 20\log_{10}(10^{12}) - 147.55 \\ &= 20\log_{10}(3) + 20\log_{10}(10^{2}) + 20 \times 0.176\,09 + 20 \times 12\log_{10}(10) - 147.55 \\ &= 20 \times 0.477\,12 + 20 \times 2\log_{10}(10) + 3.5218 + 20 \times 12 \times 1 - 147.55 \\ &= 9.5424 + 20 \times 2 \times 1 + 3.5218 + 240 - 147.55 \\ &= 9.5424 + 40 + 3.5218 + 240 - 147.55 = 293.0642 - 147.55 \\ &= 145.5142 \approx 145.5\text{ dB}\end{aligned} \tag{5.123}$$

From table 5.2, at 1.5 THz operational frequency

$$L_{\text{Atmospheric}} = 10\text{ dB for }66.7\text{ m} = 44.9775\text{ dB for }300\text{ m} \tag{5.124}$$

So, we take

$$L_{\text{Atmospheric}} = 44.98\text{ dB} \tag{5.125}$$

We tabulate the gains and losses for communication links in directions *P* to *Q* and *Q* to *P* in the form of link budget tables. This tabulation helps to get a clear perception of the possibility of practicality of the link or otherwise.

Link budget: *P*→*Q*

TX Power of *P*	21 dBm
Antenna gain of P	47 dBi
Losses of P	−3 dB
Antenna gain of *Q*	53 dBi
Losses of *Q*	−2.3 dB
Total gain	115.7 dB

FSPL at 300 m	−145.5 dB
$L_{\text{Atmospheric}}$	−44.98 dB
Total loss	−190.48

Total gain	115.7 dB
Total loss	−190.48
Received signal level	−74.78 dB

Received signal level	−74.78 dB
Receive sensitivity of *Q*	−73.7 dBm
Link margin	−1.08 dB

Value of link margin is negative. Hence link *P*→Q is not achieved.

Link budget: *Q*→*P*

TX Power of *Q*	27 dBm
Antenna gain of *Q*	53 dBi
Losses of *Q*	−2.3 dB
Antenna gain of *P*	47 dBi
Losses of *P*	−3 dB
Total gain	121.7 dB

FSPL at 300 m	−145.5 dB
$L_{Atmospheric}$	−44.98 dB
Total loss	−190.48

Total gain	121.7 dB
Total loss	−190.48
Received signal level	−68.78 dB

Received signal level	−68.78 dB
Receive sensitivity of P	−83 dBm
Link margin	+14.22 dB

Link margin is positive and > 10 dB. Hence link Q→P is achieved.

5.6 Opportunities offered by terahertz waves for communications

In the preceding, we have perused the characteristics, merits and demerits of three widely practiced distinct communication modes of ground, sky, and space waves including satellite links. Our primary concerns were how far distant points on Earth can be accessed by each mode, the type of antennas for each mode and the bandwidths obtainable. We shall now examine where and how the terahertz range of frequencies fit in fulfilling our need of getting more spectrum against the higher frequencies of the electromagnetic spectrum. To this end, we must look at the full electromagnetic spectrum including gamma and x-rays, and ask what different frequency ranges have to offer.

5.6.1 Present allocation of the electromagnetic spectrum to communication signals

5.6.1.1 Radio-frequency and microwave spectra

We are aware that the electromagnetic spectrum we have addressed so far has already been distributed by international bodies. Its different portions are allocated to different applications, e.g., extremely low frequency (ELF: 3–300 Hz) to telegraph, and teletypewriter; voice frequency (VF: 300 Hz–3 kHz) to telephony circuits; very low frequency (VLF: 3–30 kHz) and low frequency (LF: 30–300 kHz) to maritime radio and navigation; medium frequency (MF: 300 KHz–3 MHz) to AM radio, aviation radio and navigation; high frequency (HF: 3–30 MHz) to shortwave radio; very high frequency (VHF: 30–300 MHz) to VHF TV and FM radio; ultrahigh frequency (UHF: 300 MHz to 3 GHz) to UHF TV, mobile phones, GPS, WI-FI, radar, and 4G; super high frequency (SHF: 3–30 GHz) to satellite communications and WI-FI; and extremely high frequency (EHF: 30–300 GHz) to radio astronomy, radar and satellite communications. The UHF+SHF+EHF comprise the microwave spectrum. WI-FI (not an acronym) represents a wireless networking technology. It provides internet connectivity to devices such as desktop and laptop computers through a wireless router. Printers, video cameras and smart phones are other devices covered.

Briefly, the radio frequency spectrum (3 Hz–300 MHz) is fully allocated. It aims at long-range communication. Any buildings and objects en route cannot block these waves. The bandwidth of signals is limited to a few MHz. The microwave spectrum (300 MHz–300 GHz) is devoted to medium-range communication such as radar and WI-FI.

5.6.1.2 Quest for new unused spectrum

Both the radio-frequency and microwave spectra are already reserved for defined applications. The spectra are overused, jam-packed and congested, and there is little scope for assigning further applications to them. But there is a continuous demand from customers for better quality of service (QoS) by increasing the bandwidth. The capacity can be expanded to some extent by improvement of SNR by employing suitable modulation and coding schemes, and other advanced techniques. But SNR minimization will reach physical limits. Then the only answer is to look for more empty, unutilized spectrum.

Since the higher the frequency, the more the available bandwidth, we begin to explore for spectrum from the highest frequency of electromagnetic spectrum and see the possibilities of spectrum utilization moving down the frequency ladder towards lower frequencies. Since the spectrum is for human use, it is necessary to first satisfy ourselves with safety of users.

5.7 Biosafety and human health concerns on electromagnetic radiation exposure

Let us focus our attention on the effects of radiation from different parts of the electromagnetic spectrum on the human body and living matter. The biological effects of electromagnetic radiation are tabulated in table 5.3 for each segment of the electromagnetic spectrum.

5.7.1 Gamma and x-ray communication spectra

We can now comment on the pros and cons and feasibility of gamma and x-ray communication systems as high data carrying capacity vehicles. Gamma rays carry the highest energy of all types of electromagnetic waves and occupy the first position. They are followed by x-rays at the second position. Then come the UV, visible, IR, THz, microwaves and radio waves. Gamma rays energies lie in the MeV range, optical radiation a few eV and radio waves very low around 10^{-8} eV (Example 5.4). Because gamma rays have the highest frequencies amongst all electromagnetic waves, and the frequencies decrease in the same order as the energies, the data carrying capability decreases from the highest values for gamma rays to the lowest

Table 5.3. Interaction of electromagnetic waves with the human body and biological matter.

Sl. No.	Portion of electromagnetic spectrum	Effects on human body/living beings
1.	Radio waves	Human body is transparent to radio waves. The effect is mainly thermal. Prolonged exposure at high intensities heats the tissues. Exposure to RF fields at low frequencies does not pose a health risk.
2.	Microwaves	Human body is almost transparent. A small amount of radiation absorbed rotates molecules and has a heating effect. Microwaves cause tissue heating. Can destroy living tissue when the temperature of the tissue exceeds 43°C. Exposure to >20 mW cm^{-2} of body surface is harmful. High levels produce painful burns. Eyes are particularly vulnerable to their harmful effects.
3.	Infrared rays	Vibrates molecules and causes heating. Prolonged exposure can damage the lens, cornea and retina of the eye. Can cause cataracts. Corneal ulcers and retinal burns may also occur.
4.	Visible light	Strongly absorbed by human body promoting electrons to higher energy levels but does not cause ionization.
5.	Ultraviolet rays	Very strongly absorbed by human body moving electrons to higher energy levels. Causes ionization at high-frequency portion. Has a damaging effect on soft tissues of the skin and eyes. Premature aging of skin may result. Effects on eyes include retinal damage, cataracts, macular degeneration and photokeratitis.
6.	THz waves	Nonionizing radiation. Low-intensity pulses do not harm living organisms. Not so harmful to tissues and DNA as x-rays but short, powerful bursts can damage DNA.
7.	X-rays and gamma rays	Human body is transparent. Cause ionization. X-rays cause mutations in DNA leading to increased risk to cancer with cumulative exposure. Exposure to high x-ray levels can cause vomiting, fainting, hair loss, etc. Gamma rays can damage DNA, and cause mutations leading to cancer. They have high penetrating power requiring several cm thick lead sheets for stopping.

value for radio waves for the same transmission power. But practically instead of using the high energy, more data carrying gamma rays, the lower energy, less data carrying waves are used for communication. Why?

The reasons are not far to seek. Gama rays and x-rays and UV radiation also are ionizing radiations. Particularly, gamma rays and x-rays are lethal to human beings and other living organisms. They damage the tissues and the DNA. They can cause cancers of various types. So their use for terrestrial communication is not encouraging. Further they interact with air molecules and any objects in their path. They suffer absorption so that their energies fall rapidly with distance. Path losses due to attenuation are enormous. So they are not ideal candidates as communication waves on Earth as they appear prima facie.

But they can be used for communication in deep space where their hazardous effects can be safely handled. Here also they have the drawbacks that they are difficult to produce as well as put to applications because they can only be modulated mechanically using mirrors and shutters. Such sophistication and difficulties have prevented their utilization in communication despite the gigantic data capacities available. Microwaves are better suited for communication in foggy, cloudy, and snowy environments without any biological harming effects. Radio waves can be used at kilowatts and megawatts of power without any adverse effects. Even these high power levels are harmless. Generation of equivalent powers of gamma rays or x-rays is riddled with complexities of the nuclear processes involved.

If better production and manipulation technologies are developed then exploitation of gamma rays and x-rays for communication in space will receive a boost. It is envisaged that the x-ray communication (XCOM) is useful in penetrating the Martian dust storms and also the hot plasma sheath that causes a communication blackout when the astronaut-carrying capsule of an incoming space vehicle re-enters the Earth's atmosphere and the enormous frictional heat generated suddenly raises the temperature of the capsule jacket. XCOM can also convey data across a physically shielded enclosure. A bit error rate $<10^{-4}$ is realizable using x-rays of energy > 150 keV after transit through 20 mm iron. X-rays of energy > 250 keV enable good communication across 30 mm iron (Mu *et al* 2020). Radio waves are unable to do so. Additionally, x-rays can be transmitted in less divergent, tightly focused, collimated beams over large distances offering wide bandwidths but requiring less power. The adoption of any portion of the electromagnetic spectrum for communication therefore calls for carefully balancing their up- and downsides for the specific situation at hand.

5.7.2 Visible light, infrared and terahertz spectra

The visible light spectrum (430–790 THz, width = 360 THz) is 1200 times larger than the radio frequency spectrum (3 kHz–300 GHz, width = 299.999 997 GHz). The IR spectrum (300 GHz–430 THz, width=429.7 THz) is 1432.33 times larger than RF spectrum. Both these spectra can provide enormously large free bandwidths. The terahertz spectrum (0.1–10 THz, width=9.9 THz) is only 33 times larger than the RF spectrum. None of these three types of radiation can pass through walls or doors; hence they provide secure communication with assured privacy. All are LOS links. Path loss increases with frequency in the order terahertz to IR/visible. The IR/visible communication is suitable only for short-range, static communication such as indoor use. The upper end of the terahertz band is used for short-range, static communication similar to the IR/visible spectrum. However, the lower end of the terahertz band can be used for mobile communication with constrained movement and for long-range communication resembling the microwave spectrum.

Limitations of visible/IR communication must be indicated. The visible light/IR communication is highly sensitive to interference from noise due to ambient sources such as incandescent bulbs, fluorescent lamps and light emitting diodes, and the natural light source in the form of sunlight (Kahn and Barry 1997, Matheus *et al* 2019). The intensity of IR radiation/visible light cannot be indiscriminately increased due to its adverse effect on the human eye. Therefore, it seems logical to look at a frequency band intermediate between microwave and IR/visible light bands. This band will share both the advantages and disadvantages of the microwave and IR/visible light bands, and hence provide an optimal solution. This is the terahertz band inheriting the advantages of both IR/visible light and microwave bands to balance their respective disadvantages.

A thorough comparison of contenders for wireless communication is made in table 5.4 (Elayan *et al* 2020).

5.8 Salient features of terahertz communication

5.8.1 Motivation for utilization of terahertz band for communication

The main reason for accessing the terahertz band is that the radio-frequency and microwave bands were fully assigned to specific tasks, and there was a pressing demand for new bands of wider bandwidth for providing faster data transfer and large throughput (Singh and Sicker 2019). The gamma ray and x-ray frequencies were found to be full of health-related risks for communication in a human environment. The visible light/IR bands were considered, and are being used in a restricted way. The terahertz band has so far remained unutilized/under-utilized and so there is an upsurge of interest in this band. Availability of large chunks of free spectrum has encouraged all to seek its support for expansion of telecommunication networks.

5.8.2 Viable bandwidths

It is immediately apparent that the full width of the terahertz band = 9.9 THz cannot be used as a single large piece of spectrum because the terahertz waves are subject to atmospheric attenuation. So, the communication will be achieved through transmission windows. If we look at these windows in table 5.2, we find that the widths vary from 0.02 THz to 0.2 THz. In addition, fog, humidity and other climatic conditions severely influence the performance of the link and therefore cannot be ignored. All these factors must be taken into account in choosing the appropriate spectrum range for communication.

5.8.3 Provision of interference-free communication

Unlike radio waves, terahertz waves will not cross from one room or cubicle to an adjoining room or cubicle because they are blocked by the walls. So the low penetration capability allows undisturbed communication in neighboring rooms

Table 5.4. Using Millimeter, terahertz, infrared and visible light waves for communication.

	Type of wave			
Parameter	Millimeter	Terahertz	Infrared	Visible
Frequency range	30 GHz–300 GHz	0.1–10 THz	300 GHz–430 THz	430–790 THz
Maximum data rate	10 Gbps; higher than microwave (1 Gbps)	100 Gbps	10 Gbps	10 Gbps
Distance covered	Short and medium; selecting frequencies between absorption curve peaks, i.e., in the valleys minimizes atmospheric loss	Short and medium; selecting frequencies between absorption curve peaks, i.e., in the valleys minimizes atmospheric loss	Short	Short
Effect of weather conditions	Sensitive to humidity, rain, fog but less than terahertz	Sensitive to humidity, rain, fog but less than IR/ visible light	Sensitive to humidity, rain, fog	Sensitive to humidity, rain, fog
Topology of network	Point to multi-point	Point to multi-point	Point to point	Point to point
Sources of noise	Thermal	Thermal	Artificial ambient light and sunlight; scintillations and variation of signal intensity	Artificial ambient light and sunlight; scintillations and variation of signal intensity
Security	Moderate	High	High	High
Effect on human eye	No	No	Yes; imposes safety limits on eye exposure	Yes; imposes safety limit on eye exposure

even if the same spectrum portion is used. The inter-room disturbance can be further prevented by suitable antenna design because terahertz wave communication is highly directional with pencil beams propagating along LOS paths. Thus the same spectrum portion can be repeatedly used in neighboring localities without any interference. This arrangement permitting spectral reuse benefits from the short-range communication capability in an enclosed space. It is not possible with radio waves.

5.8.4 Security of communication link

Communications across limited distances are advantageous for establishing secure linkages. Over and above this, making them resistant to jamming can further prevent from mischievous attacks. Chances of eavesdropping are minimized because THz links use highly directional antennas and any person trying to get unauthorized access must be on the same narrow beamwidth as used for THz link, for which probability is low. So the THz link is averse to such attempts. Confinement of link to small regions such as rooms also makes it robust against eavesdropping.

5.8.5 Creation of ultra-dense networks

Noting that the antenna size diminishes with wavelength and the high-frequency terahertz waves have small wavelengths, nanosize antennas may be fabricated. Using such antennas, the network density can be increased. Such networks will not entangle with the adjacent networks, bearing in mind the short distances covered by terahertz waves in the atmosphere, a characteristic, which endows the ability to provide interference-friendly communication.

5.8.6 Mobile fronthaul improvement and backhaul sustenance

Mobile fronthaul is the front-end interfacing part of a cellular network connecting macrocells to small cells, i.e., cloud or centralized radio access network, an architecture of baseband units, to remote standalone radio heads called the small cells.

Mobile backhaul is the back end interfacing part of a network connecting the core network with radio access network. It uses wireless communication to fetch data from an end user to a major node in the network such as the internet.

Present fronthaul operation can be improved by offloading the highly demanding users to the THz spectrum and thereby making the spectrum more free from crowding. Thus the throughput can be increased by proper resource allocation.

Existing backhaul is based on optical fibers. Terahertz waves can provide a wireless backhaul. The influence of weather conditions cannot be overlooked. Taking note of this loss mechanism of signal, a closely-knit mesh of antennas interconnecting short-distance points can be woven to build a terahertz wireless backhaul. This will remain unaffected by atmospheric losses. The size of the backhaul will increase with distance between the server and the end user. As a consequence, wireless backhaul will be unable to support communication across a large distance, and must be supplemented by alternative options. To meet the challenge of communicating over larger distances, mixed or hybridized propagation media such as optical fiber-THz wireless-optical fiber as well as THz wireless-optical fiber-THz wireless links can act as useful options.

5.9 Challenges posed/implementational difficulties of terahertz communication, and possible solutions

5.9.1 High path attenuation

As we have indicated several times, terahertz waves suffer enfeeblement of power density as they propagate through the atmosphere or free space. The path loss depends on atmospheric conditions such as dry or moist air and the distance between the transmitter and receiver. The complexity of path loss is aggravated by the fact that in certain transmission windows of frequency ranges, the absorption is small while in the remaining frequency ranges, there are peaks of absorption where the waves are intensely absorbed and so the signal reaching the receiver is very weak. Avoidance of absorption peaks is possible at the expense of losing continuity of band so that the width of the allowed band is reduced.

5.9.2 Low penetration ability

The benefit of THz waves in being blocked by objects such as walls and surfaces is gainfully applied in THz imaging and spectroscopy to detect safety threats posed by people carrying arms and ammunition aboard public transport and using them for unlawful activities. But the same boon becomes a curse for communication because it shortens the communication distance. A LOS link can provide cost-effective communication. Non-LOS communication using reflection mirrors and focusing lenses or a wired backbone providing interconnections between wireless terahertz access points will require expensive infrastructural augmentation.

5.9.3 Necessity of high-gain antennas

To counter the path losses, the obvious strategy seems to be raising the transmitter power and the antenna gain because the received power is inversely proportional to the path loss and directly proportional to the transmitter power and antenna gain. But as frequency is increased, the transmitter power is also raised. It is difficult to build high power sources of terahertz radiation. This makes us more inclined towards increasing antenna gain as a viable solution than trying other options.

Directivity and beamwidth are two vital parameters related to the gain of an antenna. Focusing the radiation on a smaller area enables an antenna to achieve higher gain. The area is measured in degrees of beamwidth. So the smaller the area and hence the smaller the beamwidth, the higher is the gain of the antenna. The antenna gain varies inversely as beamwidth. A directional antenna with a narrow beamwidth gives more gain than an omnidirectional antenna. Therefore, for working with THz waves, the design of antenna has moved from omnidirectional approach to directional and pencil beams.

The shift in perspective from omnidirectional to unidirectional antennas brings with it alignment problems. The receiver antenna must be accurately aligned with the transmitter antenna for efficient signal reception. This alignment is easily done for static networks but becomes difficult in a mobile scenario. The procedure of alignment of the antenna with the beam is time consuming, and the time needed will increase rapidly in a dense network, thereby lowering the throughput and offsetting the advantages derived with terahertz waves.

5.9.4 Frequency of outages

Owing to the low penetration power, the need of high-gain directional antennas, and the precision necessary in beam alignment, terahertz networks are liable to more chances of failure. These outages are brought about by blockages, and the blockages occur when an opaque object intercepts the path of the terahertz wave obstructing the line-of-path transmission or there is a change in direction of the antenna. Particularly, the mobile applications are more susceptible to outages. Management of mobility is therefore essential to reduce the number of outages.

5.10 Optically- and electrically-driven terahertz wave modulators

The discussion on terahertz communication will remain incomplete without talking about modulation techniques. Demodulation or detection of terahertz waves has been discussed at several places earlier in the book.

5.10.1 Dependence of interaction of terahertz waves with a material on its electrical conductivity

Terahertz waves are reflected by metals but transmitted through dielectrics. In other words, the interaction of terahertz waves with a material depends on the density of carriers in it and hence the conductivity of the material. This difference of interaction of terahertz waves with materials is applied for modulating them. As we know, semiconductors or regions in a semiconductor device can be made conducting or insulating by varying the free carrier concentration. This variation can be done not only by doping the semiconductor but also by optical or electrical driving/pumping/tuning (figure 5.2).

5.10.2 Optical pumping

During optical pumping the semiconductor is irradiated with a laser beam. If the energy of the beam exceeds the bandgap of the semiconductor, the electrons are dislodged from their shells creating electron–hole pairs. This process called photo-ionization increases the density of free carriers in the semiconductor thereby increasing its conductivity. The laser beam used for optical pumping can be either in the form of pulses of energy bursts or as a continuous wave. Both types of beams are effective in causing photo-ionization but the pulse mode provides a larger density of free carriers and therefore provides more conductivity enhancement. This happens because the transient power of a continuous wave laser is much smaller than that of a femtosecond laser.

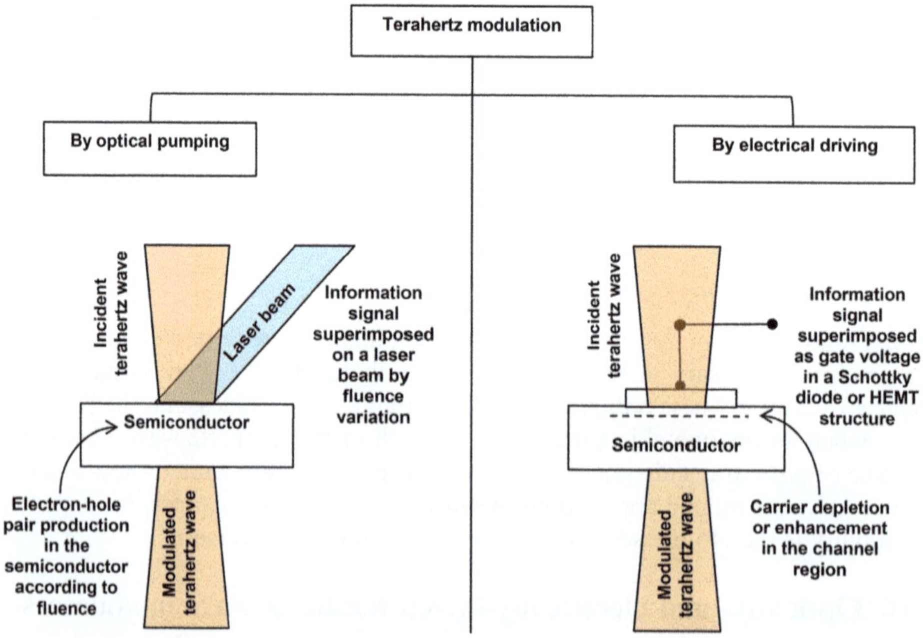

Figure 5.2. Terahertz wave modulation by optical pumping and electrical driving.

5.10.3 Electrical driving

In electrical driving, the free carrier concentration in the semiconductor is varied by applying a gate voltage to convert a semiconductor region with high carrier concentration to one of low carrier concentration, e.g., by applying a negative gate bias to an N-channel field-effect transistor, we can repel the electrons away from the region underneath the gate dielectric to produce a depletion region. Thus we create a channel for providing a high conductivity region and deplete the channel to get a low conductivity region.

5.10.4 Terahertz modulation by optical pumping

Now let us see how to apply optical pumping for modulation of terahertz waves. The information signal is used to vary the fluence of a laser beam. The laser beam falls on a semiconductor. Simultaneously, the terahertz wave to be modulated is incident on the semiconductor. The laser beam creates electron–hole pairs in the semiconductor in accordance with the fluence changes. The generated carriers alter the conductivity of the semiconductor. The conductivity changes affect the reflectance and transmittance of the terahertz beam striking the semiconductor. Hence the variations in the laser beam fluence are impressed on the terahertz wave transmitted through the semiconductor. If the conductivity of semiconductor is raised, the intensity of transmitted terahertz wave decreases. Conversely, if the conductivity of semiconductor is reduced, the transmitted wave intensity rises. Thus the transmitted terahertz wave serves as a carrier signal of the message in the information signal supplied to the laser for optical pumping.

5.10.5 Terahertz modulation by electrical driving

The electrical driving is applied in the same way for terahertz modulation. The information signal is used to bias the gate. Depending on the message, the gate either produces a depletion zone or a conducting channel in the semiconductor. The terahertz wave to be modulated is also falling on the semiconductor at the same time. When the terahertz wave passes through a low-conductivity depletion zone, it undergoes less reflection and the intensity of transmitted terahertz wave is high. The reverse happens when the terahertz wave passes through a high-conductivity, highly reflecting zone. The intensity of transmitted terahertz wave is low. So the information in the gate bias is imprinted on the terahertz wave enabling it to serve as a carrier of the message.

5.11 Optically-pumped spatial modulator

The spatial light modulator (SLM) used is of the liquid crystal type (Okada and Tanaka 2011). The SLM electrically controls the optical properties of a nematic liquid crystal to produce a variable amplitude/phase mask to shape spatially resolved ultrashort laser pulses. The phase-varied femtosecond laser pulses are focused on a silicon prism of high resistivity material ~100 ohm-cm (figure 5.3). These are the optical pumping pulses constituting the information signal. They have a wavelength of 800 nm. This wavelength is sufficient to excite the electrons to the conduction band. The holes are left behind in the valence band. Thus electron–hole pairs are generated for electrical conduction.

Upon illumination of the silicon wafer with the laser pulses from the SLM, a pattern of free carriers is produced in the silicon wafer depending on the relative intensities of the light beam in the information signal. Following this pattern of photo-generated carriers, the conductivity of the silicon prism and hence its transparency to terahertz waves varies at different positions of the silicon prism. The terahertz wave to be modulated is also made to fall on the silicon prism using a non-coaxial arrangement. As the transmissibility of the silicon prism follows the message in the information signal, the information is engraved on the terahertz wave transmitted by the silicon prism. With the information encoded in the terahertz beam, the function of a spatial terahertz modulator is realized.

5.12 Electrically-driven modulator using two-dimensional electron gas (2DEG)

This device of figure 5.4 utilizes a high-electron-mobility transistor (HEMT)-type structure in which a 2DEG is formed at the interface between $Al_{0.3}Ga_{0.7}As$ and pristine (undoped) GaAs layers (Kleine-Ostmann *et al* 2004).

In a HEMT, band offset between a wide bandgap semiconductor and a narrow bandgap semiconductor causes transference of electrons from the wider bandgap

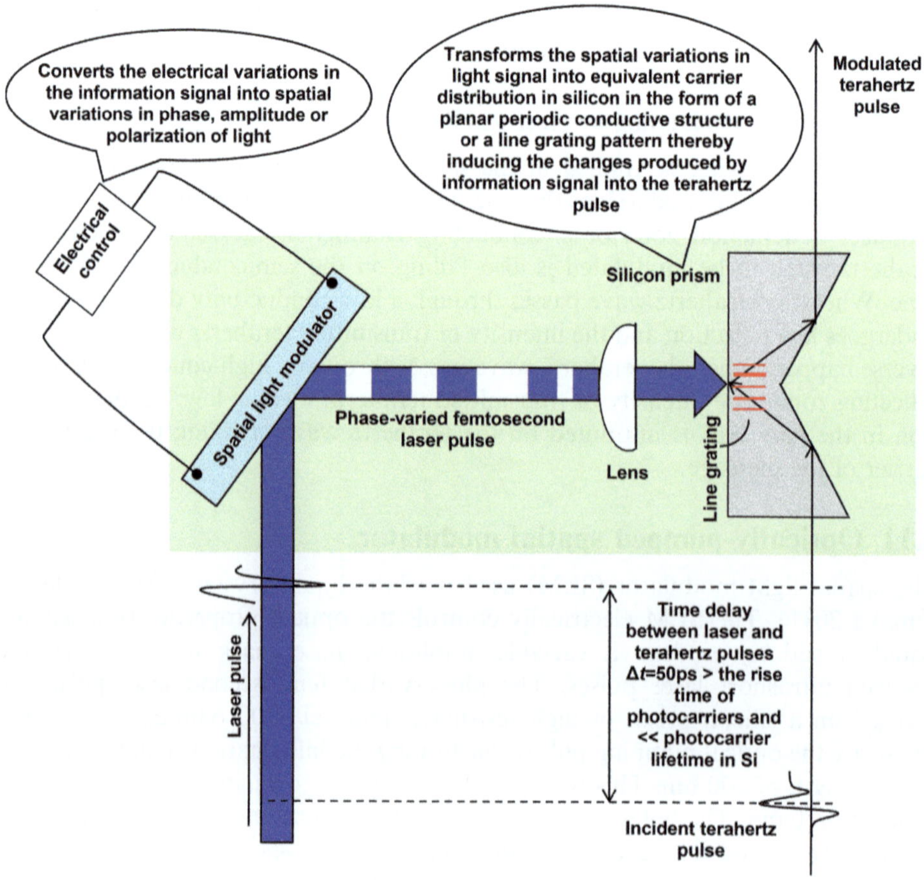

Figure 5.3. Modulation of a terahertz pulse by induction of a planar line grating pattern mimicking the information signal changes at the total internal reflection surface of a silicon prism.

semiconductor to the narrow bandgap semiconductor. The free electrons are concentrated in the form of a 2DEG of a few nanometers thickness in the vicinity of the AlGaAs/GaAs heterojunction. This type of doping is called modulation doping. Modulation doping differs from the conventional doping in the respect that the electrons are separated from the donor impurity atoms. Consequently, scattering of electrons from donor impurity ions known as ionized impurity scattering is eliminated. In heavily-doped semiconductors, the ionized impurity scattering reduces the mobility of electrons. But its absence here increases the mobility of the electrons. So a high conductivity region exists at the AlGaAs/GaAs interface. This region presents high reflectivity to terahertz waves and allows very little transmission.

For modulation, the terahertz waves are made to fall on the device at the AlGaAs/GaAs region. The information signal is applied to the gate of the device as the gate voltage. Under zero bias, the AlGaAs/GaAs region is conducting. So, the intensity of transmitted terahertz wave (the modulated wave) is low. But when the

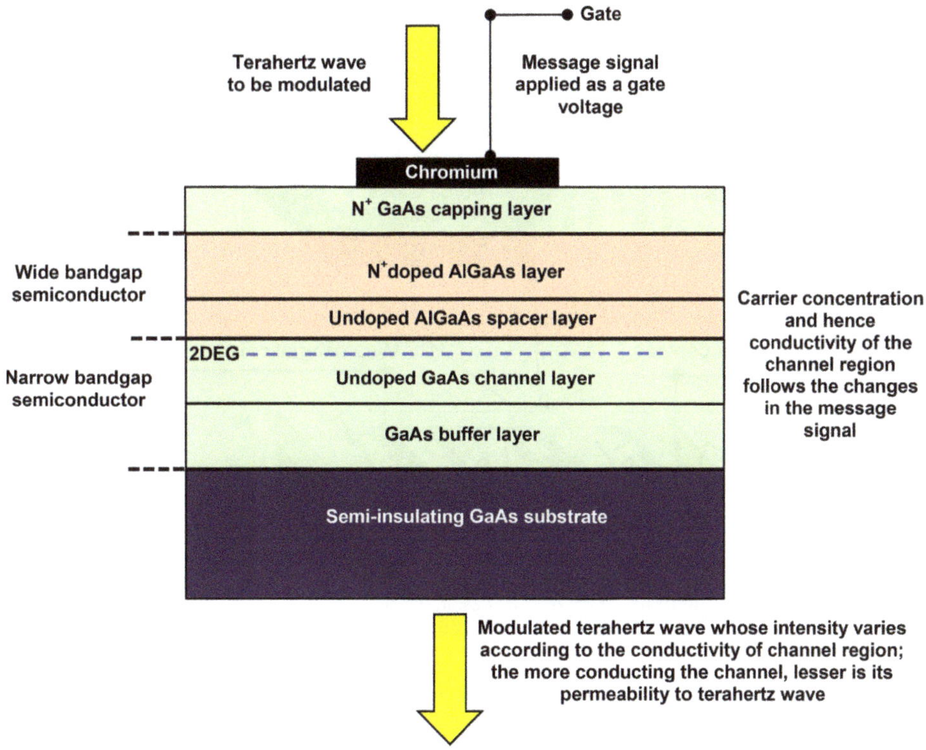

Figure 5.4. Terahertz wave modulation by varying the conductivity of the 2DEG in a HEMT through gate-fed information signal and thereby influencing the penetration through and absorption of terahertz radiation in the channel region.

gate bias becomes negative, the electrons are repelled away from the AlGaAs/GaAs interface. The depletion of charges renders this region insulating. Consequently, the intensity of transmitted terahertz wave increases. Thus the transmitted terahertz wave follows the variations of the gate signal, and the message on the gate signal is written on the terahertz wave, resulting in its modulation. The maximum modulation depth achieved is 3% over the frequency range 0.1–2.0 THz.

5.13 Electrically-driven metamaterial-based modulator

Metamaterials are artificial materials designed to exhibit desired electromagnetic responses. Using these materials, resonant structures can be fabricated to work over a large portion of electromagnetic spectrum spanning radio-frequency through terahertz to infrared regions from commonly used materials in microelectronic technology.

In figure 5.5, a metamaterial working on an electrical analogue of split-ring resonator is used (Chen *et al* 2006). It acts as a parallel inductance-capacitance (L-C) circuit. The two rings form inductive loops. The split gap provides the capacitance. An array of gold resonator elements constituting the metamaterial is fabricated on an N-type GaAs layer of thickness 1 μm and doping concentration 1.9×10^{16} cm^{-3}

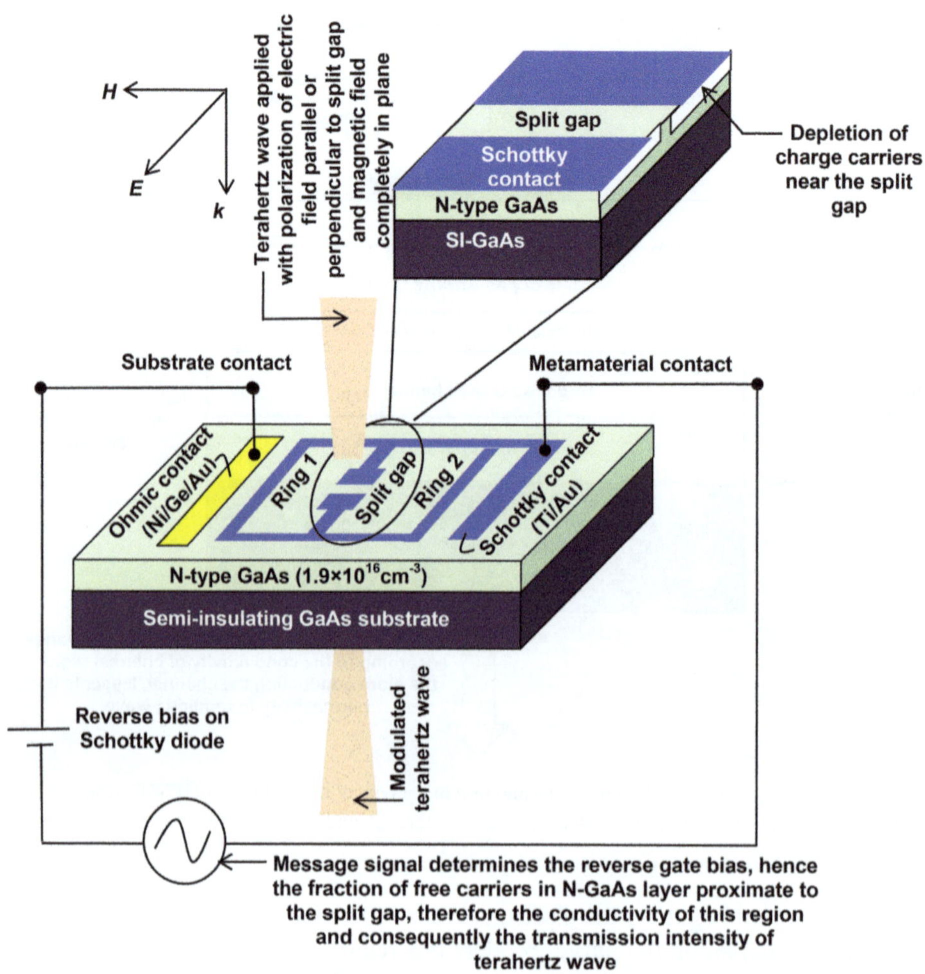

Figure 5.5. Modulation of terahertz wave by metamaterial device.

grown on a semi-insulating GaAs substrate. The metal and the N-type GaAs layer form a Schottky junction, and the interconnected resonators act as a metal gate.

The information signal is applied as a bias to this metal gate to control the conductivity the N-type GaAs layer adjoining the split gaps through displacement of free charge carriers in it. When the applied bias is zero, the N-type GaAs layer at the split gap shows high conductivity. The free carriers in the substrate short out the capacitive effects at the split gaps. So, terahertz waves experience difficulty in transmission through the N-type GaAs layer. This is the state of dampened resonance response with the resonance subdued due to the lossy N-type GaAs layer. Resonance condition is established when a reverse bias is applied to the metamaterial gate. Depletion of charge carriers occurs in the N-type GaAs layer near the split gaps. The resultant decrease of conductivity increases the intensity of transmitted terahertz wave. As the transmission of the terahertz wave falling on the split gaps is controlled by the voltage applied between the Schottky and ohmic contacts, the terahertz wave is modulated in intensity by the information content of the Schottky contact/ohmic contact voltage. The modulation of transmission of terahertz waves is 50% at 0.72 THz, meaning that the change in relative intensity of transmission is 50% at this frequency.

5.14 Graphene modulators

The low modulation depths achieved in semiconductor-based terahertz modulators forced researchers to search for new materials. Metamaterial-semiconductor modulators provided high-speed modulation but these devices had narrow bandwidths associated with their fixed structures. The complexity of fabrication was also disadvantageous.

Unique electrical, optical and mechanical properties of graphene, a zero bandgap two-dimensional semiconductor have made it a popular material amongst device engineers. Graphene supports a high carrier concentration ~1×10^{14} cm^{-2}. It displays an extraordinarily high carrier mobility ~20 000 cm^2 V^{-1} s^{-1} for both electrons and holes. Unlike semiconductors, holes are as agile in graphene as electrons. The similar mobility values for electrons and holes owe their origin to the symmetrical conical band structure of graphene. Graphene therefore makes available a much broader range of tunable conductivity for modulation than possible in semiconductors dependent on electrons only. Moreover, graphene can be easily integrated with semiconductors to reap benefits by combining them in hybrid structures.

5.14.1 Optically-pumped graphene modulator

Optical absorption in graphene causes transitions of carriers in the same energy band (intraband) and from one energy band to another band (interband) (Sensale-Rodriguez *et al* 2012). Hence optical conductivity $\sigma(\omega)$ due to absorption of waves of frequency ω by graphene is the sum of two terms $\sigma_{\text{Intraband}}(\omega)$ and $\sigma_{\text{Interband}}(\omega)$ expressed as

$$\sigma(\omega) = \sigma_{\text{Intraband}}(\omega) + \sigma_{\text{Interband}}(\omega) \tag{5.126}$$

In the infrared/visible light frequency range, interband transitions dominate while at terahertz frequencies, intraband transitions are dominant. Then graphene acts like a conductive film. The optical conductivity of graphene follows its electrical conductivity. It is described by the Drude model as

$$\sigma(\omega) = \frac{\sigma_{\text{DC}}(E_{\text{F}})}{1 + \omega^2\tau^2} \tag{5.127}$$

where σ_{DC} is the DC electrical conductivity, a function of the Fermi level E_{F}; and τ is the momentum relaxation time for scattering of carriers. The equation implies that absorption of terahertz waves by graphene can be varied by changing its DC electrical conductivity or Fermi level. For $\omega\tau \ll 1$, equation (5.127) reduces to

$$\sigma(\omega) \approx \sigma_{\text{DC}}(E_{\text{F}}) \tag{5.128}$$

For $\omega >> 1$, equation (5.127) tells us

$$\sigma(\omega) < \sigma_{DC}(E_F) \tag{5.129}$$

A trilayer hybrid terahertz modulator (figure 5.6) consists of: (i) a P-type silicon substrate, (ii) a titanium dioxide interlayer and (iii) a single-layer graphene top layer: graphene/TiO_2/P-type Si (Wei *et al* 2019). The resistivity of the silicon substrate is 1–10 Ω-cm. Its thickness is 500 μm. The 10-nm thick TiO_2 layer is deposited over the Si substrate by immersion in 0.1 M $TiCl_4$ solution. Graphene is grown on copper by chemical vapor deposition. It is transferred over TiO_2 by wet etching to form the trilayer modulator of area 1 cm^2.

The information signal is a light beam from a semiconductor laser at a wavelength of 808 nm. The information signal and the terahertz wave to be modulated are simultaneously incident on the graphene surface. The terahertz beam of diameter 3 mm overlaps the laser beam of spot diameter 5 mm.

A P-N heterojunction is formed between P-type silicon and TiO_2 films. This P-N junction has a built-in electric field due to which the free carriers generated in silicon by laser excitation are injected into graphene. As a result, the Fermi level of graphene is shifted to a higher conduction band. The consequent increase in electrical conductivity of graphene is accompanied by a reduction of transmittance of terahertz waves.

The changes in electrical conductivity of graphene along with those in its terahertz wave transmissibility are inscribed in the terahertz wave. So the intensity

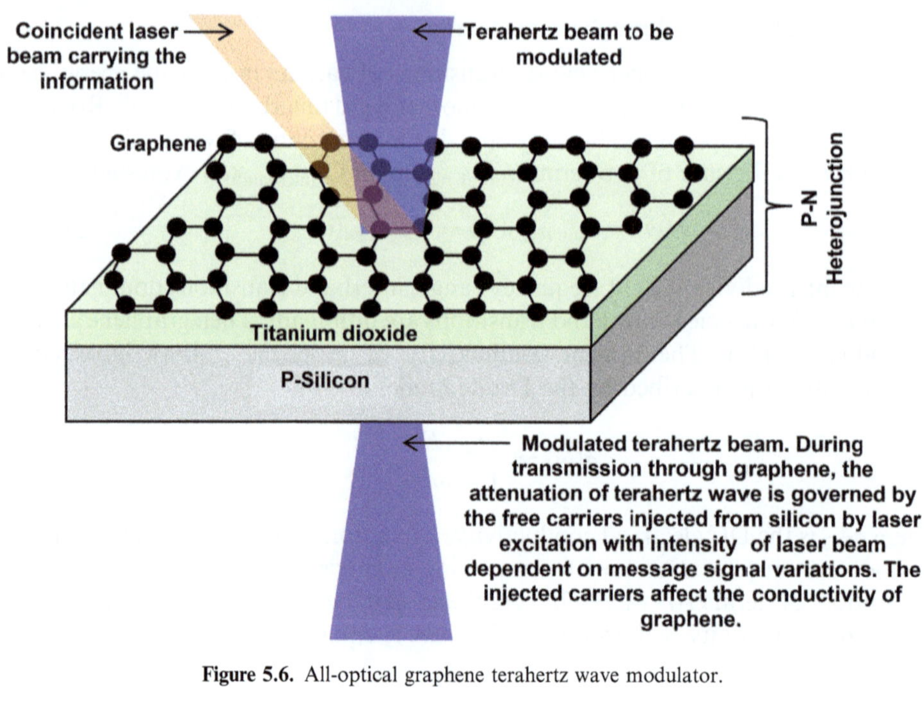

Figure 5.6. All-optical graphene terahertz wave modulator.

of terahertz wave shows variations in response to the laser signal, and its modulation is implemented. Broadband modulation is obtained in the frequency range 0.3–1.7 THz. The depth of modulation is 88% for pumping with 808 nm laser at 1.4 W, and decreases to 33% at 0.2 W. At 0.2 W, the modulation depth for the graphene/TiO_2/Si is 2.5 times that of graphene/Si, which is the factor giving the extent of improvement by using TiO_2 over graphene/Si. At 0.2 W, the modulation depth for the graphene/TiO_2/Si is six times that of bare Si, which measures the improvement by including graphene/TiO_2 over Si.

5.14.2 Electrically-driven graphene modulator

Copper-catalyzed chemical vapor deposition is carried out in a quartz tube using methane and hydrogen to grow monolayer graphene (Sensale-Rodriguez *et al* 2012). The grown graphene sheets are transferred to SiO_2/P-type silicon substrates using polymethyl methacrylate and wet etching process for removal of underlying metal (figure 5.7). These sheets have 1 cm^2 area. The device has two top contacts to graphene. The bottom gate contact is ring-shaped. The doping concentration of

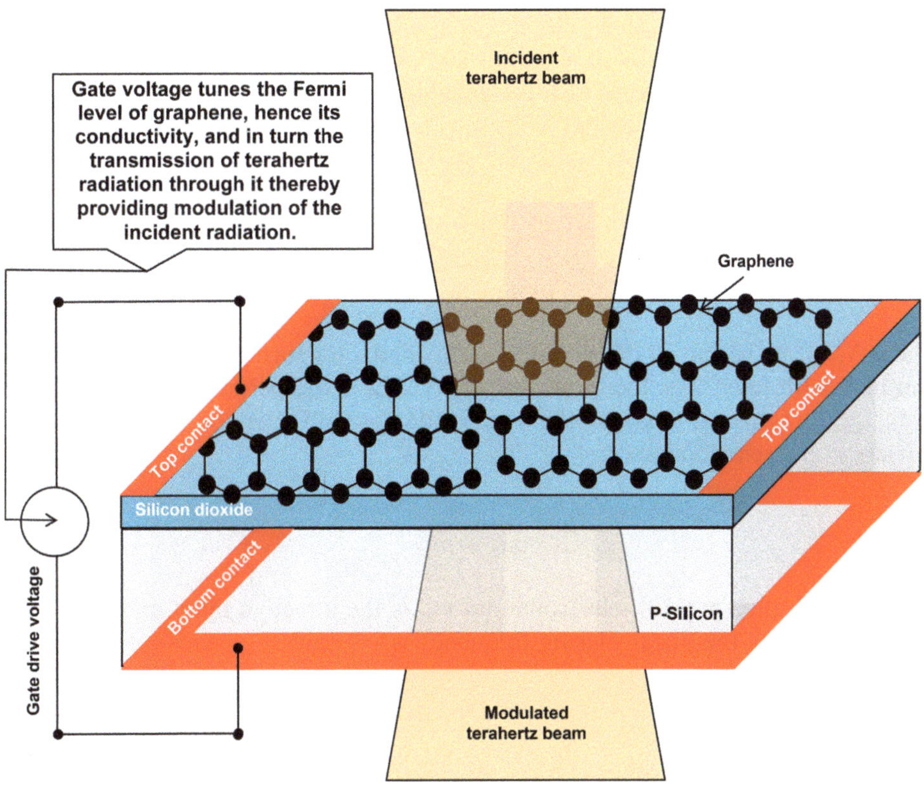

Figure 5.7. Electrically operated graphene terahertz wave modulator.

P-type silicon is kept low ~ 1×10^{15} cm^{-3}. A high doping density will render it conducting resulting in severe absorption of terahertz waves.

The information signal is applied as a gate voltage between the top graphene contact and the bottom ring contact. At the same time, the terahertz wave to be modulated is incident on the graphene film. The gate voltage tunes the Fermi level of graphene. By tuning the Fermi level, we tune the density of states for intraband transitions in graphene, hence the conductivity of graphene and thereby the transmittance of terahertz radiation through graphene. At a gate bias at which the conductivity of graphene is maximum, the transmittance is minimum and conversely. The method yields a modulation depth of 16% in the frequency range 0.57–0.63 THz.

Example 5.11 In a terahertz wave modulator using silicon as the semiconductor, a free carrier density of 1×10^{17} cm^{-3} is produced by optical pumping with laser. Calculate the plasma frequency for silicon given that dielectric constant of silicon is 11.6, effective mass of electrons in silicon is 0.27 times the electron rest mass and the effective mass of holes in silicon is 0.37 times the electron rest mass.

Solution: Plasma frequency is the frequency above which the metal starts behaving like a dielectric with the real part of its permittivity becoming positive; below this frequency it shows metallic behavior with negative value of real part of permittivity. The plasma angular frequency ω_p of a material is given by

$$\omega_p = \sqrt{\frac{N_{Pump}q^2}{\varepsilon_0\varepsilon_{Semiconductor}\mu}} \tag{5.130}$$

where N_{Pump} is the carrier density at the semiconductor surface generated by optical pumping with laser, q is the electronic charge, ε_0 is the permittivity of free space, $\varepsilon_{Semiconductor}$ is permittivity of the semiconductor and μ is the effective mass of carriers in the semiconductor:

$$\mu = \frac{m_e m_h}{m_e + m_h} \tag{5.131}$$

m_e is the effective mass of electrons and m_h is the effective mass of holes. From equation (5.130), the linear plasma frequency is

$$f_p = \frac{1}{2\pi}\sqrt{\frac{N_{Pump}q^2}{\varepsilon_0\varepsilon_{Semiconductor}\mu}} \tag{5.132}$$

Here $N_{\text{Pump}} = 1 \times 10^{17}\ \text{cm}^{-3} = 1 \times 10^{17} \times 10^{6}\ \text{m}^{-3}$, $\varepsilon_{\text{Semiconductor}} = 11.6$, $m_e = 0.27\ m_0$, $m_h = 0.37\ m_0$. Putting the given values in equation (5.132),

$$
\begin{aligned}
f_p &= \frac{1}{2\pi}\sqrt{\frac{N_{\text{Pump}} q^2}{\varepsilon_0 \varepsilon_{\text{Semiconductor}} \mu}} \\
&= \frac{1}{2 \times 3.141\,59} \times \sqrt{\frac{1 \times 10^{17} \times 1 \times 10^{6} \times (1.602 \times 10^{-19})^2}{8.854 \times 10^{-12} \times 11.6 \times \left(\frac{0.27m_0 \times 0.37m_0}{0.27m_0 + 0.37m_0}\right)}} \\
&= \frac{1}{6.2832} \times \sqrt{\frac{1 \times 10^{17} \times 1 \times 10^{6} \times 10^{-38} \times 2.5664}{8.854 \times 10^{-12} \times 11.6 \times \left(\frac{0.27m_0 \times 0.37m_0^2}{0.64m_0}\right)}} \\
&= \frac{1}{6.2832} \times \sqrt{\frac{2.5664 \times 10^{17+6-38}}{8.854 \times 10^{-12} \times 11.6 \times \left(\frac{0.27 \times 0.37m_0}{0.64}\right)}} \\
&= \frac{1}{6.2832} \times \sqrt{\frac{2.5664 \times 10^{-15}}{8.854 \times 10^{-12} \times 11.6 \times \left(\frac{0.27 \times 0.37 \times 9.109 \times 10^{-31}}{0.64}\right)}} \\
&= \frac{1}{6.2832} \times \sqrt{\frac{2.5664 \times 10^{-15}}{\frac{8.854 \times 10^{-12-31} \times 11.6 \times 0.27 \times 0.37 \times 9.109}{0.64}}} \\
&= \frac{1}{6.2832} \times \sqrt{\frac{2.5664 \times 10^{-15}}{146.0339 \times 10^{-43}}} = \frac{1}{6.2832} \times \sqrt{\frac{2.5664}{146.0339}} \times 10^{-15+43} \\
&= \frac{1}{6.2832} \times \sqrt{0.017\,574 \times 10^{28}} = \frac{1}{6.2832} \times 0.132\,567 \times 10^{14} \\
&= 0.021\,099 \times 10^{14}\text{Hz} = 2.1099 \times 10^{12}\ \text{Hz} \approx 2.1\ \text{THz}
\end{aligned}
\tag{5.133}
$$

Example 5.12 Find the modulation depth of a terahertz wave modulator in which the maximum transmittance is 0.9 and the minimum transmittance is 0.45.

Solution: The modulation depth or modulation index m measures the extent of excursion or deviation of a chosen parameter of modulated terahertz wave with respect to the value of this parameter for unmodulated terahertz wave. If we take the intensity of the terahertz wave as the parameter then

$$
\begin{aligned}
m &= \frac{\text{Intensity of unmodulated terahertz wave} - \text{Intensity of modulated terahertz wave}}{\text{Intensity of unmodulated terahertz wave}} \\
&= \frac{\text{Intensity}_{\text{Unmodulated}} - \text{Intensity}_{\text{Modulated}}}{\text{Intensity}_{\text{Unmodulated}}}
\end{aligned}
\tag{5.134}
$$

Considering that the unmodulated wave corresponds to zero optical pumping, hence minimum conductivity of semiconductor giving maximum transmission and vice versa, equation (5.134) may be rewritten in terms of transmittances of the semiconductor as

$$m = \frac{\text{Maximum transmittance of the semiconductor} - \text{Minimum transmittance of the semiconductor}}{\text{Maximum transmittance of the semiconductor}}$$
$$= \frac{0.9 - 0.45}{0.9} = \frac{0.45}{0.9} = 0.5 \tag{5.135}$$

5.15 Discussion and conclusions

Terahertz communication is full of opportunities as an enabling technology for next-generation wireless communication.

One area where use of terahertz band is envisaged for static applications aims at removing the already overloaded spectrum and making it available for mobile networks. Another area where terahertz communication can make a dent is providing interference-free short-range communication in small rooms or cubicles. The low penetration power of terahertz radiation prevents communication in one room from disturbing that in the adjoining room. Outdoor use of terahertz communication is affected by weather conditions such as rain, snow or fog, and is therefore challenging. Careful prior assessment of path losses is essential. No less challenging is use of terahertz waves for wireless mobile network. As opposed to UV, x-rays or gamma rays, terahertz waves have no injurious effect on human vision.

References

Chen H-T, Padilla W J, Zide J M O, Gossard A C, Taylor A J and Averitt R D 2006 Active terahertz metamaterial devices *Lett. Nat.* **444** 597–600

Elayan H, Amin O, Shihada B, Shubair R M and Alouini M-S 2020 Terahertz band: the last piece of RF spectrum puzzle for communication systems *IEEE Open J. Commun. Soc.* **1** 1–32

Forouzan B A and Fegan S C 2007 *Data Communications and Networking* (New York: McGraw Hill Higher Education) pp 86–7

Hartley R V L 1928 July Transmission of information (Presented at the Int. Congress of Telegraphy and Telephony, Lake Como, Italy, September 1927) *Bell Syst. Tech. J.* **7** 535–63

ITU-R 2016 Rec. ITU-R P.676–11: Attenuation by atmospheric gases, P Series Radiowave propagation p 5 https://www.itu.int/dms_pubrec/itu-r/rec/p/R-REC-P.676-11-201609-S!!PDF-E.pdf (Accessed on 2nd December 2021)

Kahn J M and Barry J R 1997 Wireless infrared communications *Proc. IEEE* **85** 265–98

Kleine-Ostmann T, Dawson P, Pierz K, Hein G and Koch M 2004 Room-temperature operation of an electrically driven terahertz modulator *Appl. Phys. Lett.* **84** 3555–7

Matheus L E M, Vieira A B, Vieira L F M, Vieira M A M and Gnawali O 2019 Visible light communication: concepts, applications and challenges *IEEE Commun. Surv. Tutor.* **21** 3204–37

Mu J, Tang X, Liu Y, Hang S, Li H, Zhou W, Dang P and Lai S 2020 High penetration X-ray communication under physical shielding *J. X-ray Sci. Technol.* **28** 187–96

Okada T and Tanaka K 2011 Photo-designed terahertz devices *Sci. Rep.* **1** 1–5

Nyquist H 1928 Certain topics in telegraph transmission theory *Trans. Am. Inst. Electr. Eng.* **47** 617–44. Reprint as classic paper *Proc. IEEE* Vol. 90 Feb 2002

Rioul O and Magossi J C 2014 On Shannon's formula and Hartley's rule: beyond the Mathematical Coincidence *Entropy* **16** 4892–910

Rahaman M H, Bandyopadhyay A, Christopher S, Ray K P and Pal S 2019 THz communication technology in India: present and future *Def. Sci. J.* **69** 510–6

Sensale-Rodriguez B, Yan R, Kelly M M, Fang T, Tahy K, Hwang W S, Jena D, Liu L and Xing H G 2012 Broadband graphene terahertz modulators enabled by intraband transitions *Nat. Commun.* **3** 1–7

Singh R and Sicker D 2019 Beyond 5G: THz Spectrum Futures and Implications for Wireless Communication *30th European Conf. of the Int. Telecommunications Society (ITS): Towards a Connected and Automated Society (Helsinki, 16–19 June, 2019)* p 30

Shannon C E 1949 January Communication in the presence of noise *Proc. Inst. Radio Eng.* **37** 10–21

Shannon C E 1948 A mathematical theory of communication *Bell Syst. Tech. J.* **27** 623–56

Shannon C E and Weaver W 1949 *The Mathematical Theory of Communication* (Champaign, IL: University of Illinois Press) Paper bound edition 1963, 1964, 125pp

Stallings W 2007 *Data and Computer Communications* (Upper Saddle River, NJ: Pearson Education) pp 92–3

Wei M, Zhang D, Li Y, Zhang L, Jin L, Wen T, Bai F and Zhang H 2019 High-performance all-optical terahertz modulator based on graphene/TiO_2/Si trilayer heterojunctions *Nanoscale Res. Lett.* **14** 1–6

IOP Publishing

Chapter 6

Spectroscopy

Beginning with the preliminaries of spectroscopy, the time and frequency domain representations of waves are explained and their interrelation is examined. The basic concepts and components of a terahertz time domain spectroscopy (TDS) system are described. The paths traversed by the pump and the probe beams as well as the electrical signals are traced out. Terahertz spectroscopy shares several features with its two bordering spectroscopies, microwave and infrared spectroscopies, and therefore with the rotational aspects of the former and vibrational aspects of the latter, which are brought out by a mathematical analysis. A detailed comparison of terahertz TDS with infrared spectroscopies is made. Far infrared spectroscopy is inferior to terahertz TDS below 3 THz. Fourier transform infrared spectroscopy responds to intramolecular vibrations while terahertz spectroscopy is sensitive to intermolecular vibrations.

In this chapter, we delve into the intricacies of terahertz spectroscopy in relation to well-known spectroscopic techniques and look into the requirement of a new spectroscopy when several similar techniques are already existent. Terahertz spectroscopy is often combined with imaging to investigate the external features and chemical constitution of materials and objects, which will be pursued in the next chapter on imaging.

For easy comprehension, we shall start by defining some commonly used terms (Pavia *et al* 2015).

6.1 Spectra and spectroscopies

6.1.1 Spectrum

This is a recording of variation of intensity or energy of electromagnetic radiation with its frequency.

6.1.2 Spectroscopy

This is a popular exploratory tool widely used for chemical analysis in material science for compositional identification and quantification. It works on the interaction between electromagnetic radiation and materials as a function of frequency or wavelength of the waves used (table 6.1). The instrument used to break the electromagnetic radiation into components of different frequencies and to measure or record the spectra is called a spectrometer.

Looking at the type of radiant energy used for excitation, spectroscopy is named as microwave, terahertz, infrared, UV–visible, x-rays or gamma-ray spectroscopy. Because each type of radiation from microwaves to gamma rays has a different frequency and hence energy, it interacts with materials differently so that vital information about materials is revealed from the spectra obtained at those energies (table 6.2).

Depending on the nature of interaction, spectroscopy is classified into absorption, reflection or scattering, and emission modes accordingly as the energy is absorbed by the material, reflected or scattered from it or emitted by the material itself (table 6.3).

According to the method of obtaining the spectra used in the spectrometer, spectroscopy is divided into two main subheads: dispersive and Fourier transform categories (table 6.4).

In dispersive absorption spectroscopy, the incoming radiation is dispersed into its frequency components by passage through dispersive elements such as prisms or gratings. The material is showered with radiations of different frequencies using a monochromator. The intensity of radiation emerging after absorption by the

Table 6.1. Interaction of electromagnetic waves with matter.

Sl. No.	Name of waves	Consequences of interaction
1.	Radio waves	Penetrate insulating materials such as wood, bricks, and concrete but are stopped by conducting materials like metals and water.
2.	Microwaves	Cause rotation and torsion of molecules.
3.	Terahertz waves	Strongly absorbed by polar molecules but can easily penetrate non-polar ones. Cause rotation, torsion, vibration, translation of molecules and their conformational states. Show properties of both microwaves and infrared radiations.
4.	Infrared rays	Produce vibrations of molecules.
5.	UV–visible light waves	Bring about electronic transitions between energy levels.
6.	X-rays	Cause ionization (photoionization) and x-ray production.
7.	Gamma rays	Responsible for photoeffects (photoelectric and photonuclear), Crompton scattering, and electron-positron pair production.

Table 6.2. Features of some classes of spectroscopies.

Sl. No.	Name of class	Main features
1.	Microwave spectroscopy	Also called rotational spectroscopy, it provides information regarding polar molecules, i.e., molecules having a permanent dipole moment, e.g., water, ammonia, hydrogen chloride, hydrogen bromide, nitric oxide, etc in the gas phase because in liquid or solid phases, molecular rotation is hindered by intermolecular interactions.
2.	THz spectroscopy	A wide range of materials such as polar liquids, gases, biological samples, semiconductors, superconductors, etc can be analyzed. Strong temporal and spatial coherences of terahertz waves enable accurate determination of refractive indices and absorption coefficients of materials.
3.	IR spectroscopy	Also called vibrational spectroscopy, it is used to determine functional groups in molecules in solid, liquid or gaseous states. IR active molecules include benzene, carbonyl group, HCl, CO, NO, etc while O_2, H_2, Cl_2 are IR inactive.
4.	UV-vis spectroscopy	Also called electronic spectroscopy, it is mostly used for analyzing samples in liquid forms or as solutions. Solid samples or suspensions can also be analyzed. Concentrations of analytes in the gas, liquid, or solid phase can be measured. Optical and electronic properties of materials can be characterized.

Table 6.3. Principles of spectroscopy.

Sl. No.	Name of spectroscopy	Basic underlying principle
1.	Absorption or reflection/scattering spectroscopies	The material is struck with a beam of electromagnetic energy of the concerned type and the effect of this stimulating radiation is recorded with respect to the frequency components of the incident beam.
2.	Emission spectroscopy	The material is energized by heating with a flame, spark, arc or plasma and the radiation produced is studied in relation to frequency.

Table 6.4. Spectroscopies according to spectra recording methods.

Sl. No.	Name of spectroscopy	Method of spectra recording
1.	Dispersive absorption spectroscopy	Uses dispersive elements (prisms or gratings).
2.	Fourier transform spectroscopy	Interferometry and application of Fourier transform to interferogram data.

material is measured at each frequency with respect to that of radiation passing through a reference or blank sample. When the frequency of incident radiation matches with the vibrational frequency of a chemical bond, absorption occurs, thereby revealing its identity.

In Fourier transform spectroscopy, an interferogram traces intensity of light with respect to optical path difference (Smith 1996). A Michelson–Morley interferometer is used to generate the interferogram. The interferogram is a pattern of alternately bright and dark fringes. Application of Fourier transform to the interferogram data yields the spectrum. Performing calculations with a digital computer, the experimental results for a wide range of frequencies are simultaneously deduced, whereas the monochromator gives the result for one frequency or a narrow frequency band only at a time.

6.1.3 Time and frequency domains

6.1.3.1 Time domain curve

This is a plot of amplitude of the wave against time. Curves are very complex if there are multiple signals.

6.1.3.2 Frequency domain curve

This is a plot of amplitude of the wave against frequency. Complexity is much less if there are multiple signals.

6.1.3.3 Conversion from time domain to frequency domain representation

In the time domain, amplitude of the wave is plotted against time. The resulting graph is called the waveform. Figure 6.1(a) shows a typical time-domain plot. Every waveform can be decomposed into a series of simple sinusoidal waves. These sinusoidal waves are the components or constituents of the waveform. They have different amplitudes, frequencies and phases. The process of decomposition of the waveform is implemented using a mathematical tool called the Fourier transform. The process is called Fourier analysis or spectrum analysis. The name 'spectrum analysis' arises from the fact that the objective of the process is to generate the equivalent plot to the waveform in frequency domain showing the amplitude and phase of the component waves with respect to frequency. The plot of amplitude versus frequency is known as a spectrum.

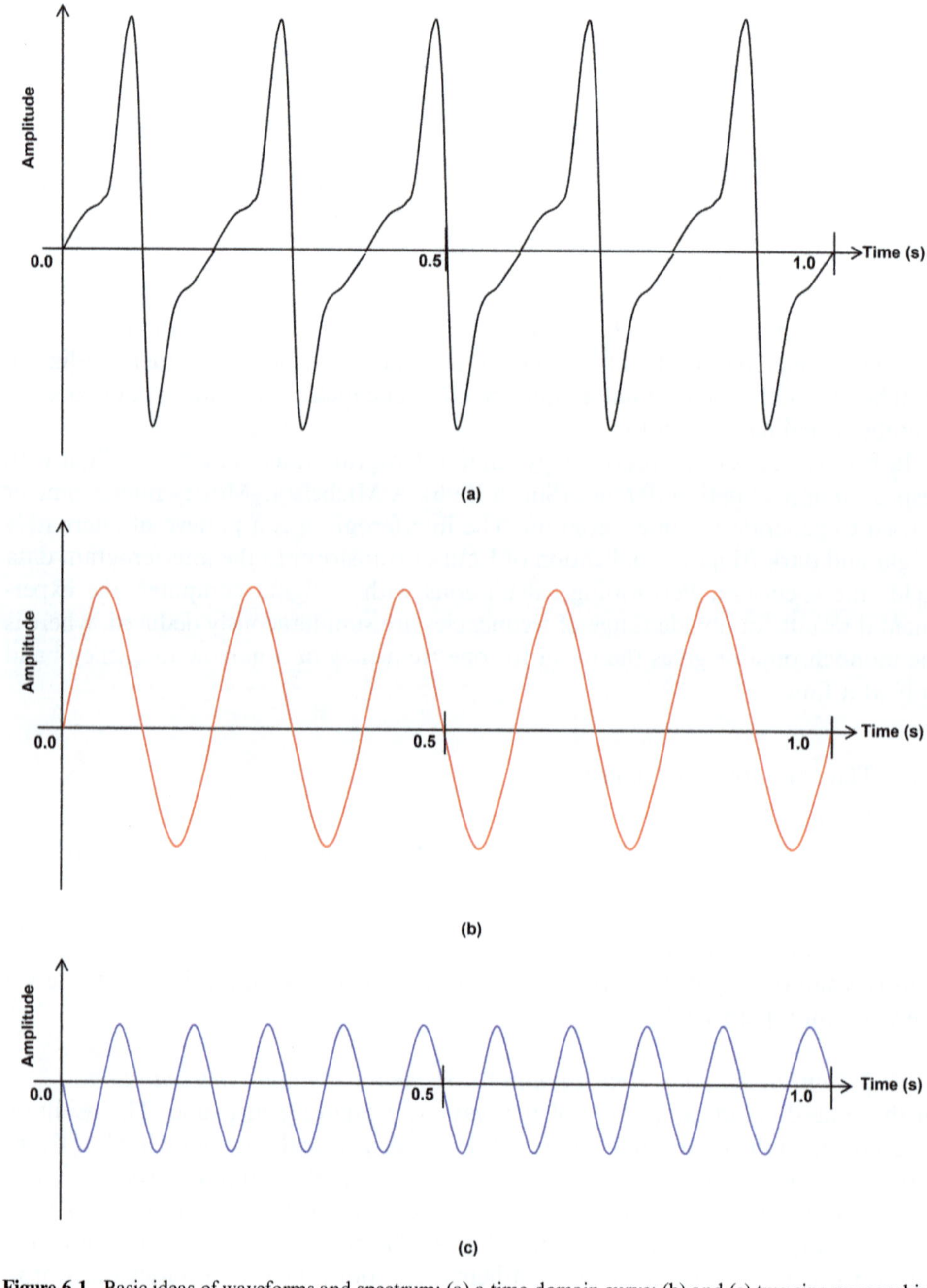

Figure 6.1. Basic ideas of waveforms and spectrum: (a) a time-domain curve; (b) and (c) two sine waves which have been combined together to produce the time-domain curve of (a); and (d) frequency spectrum equivalent to the curve of (a).

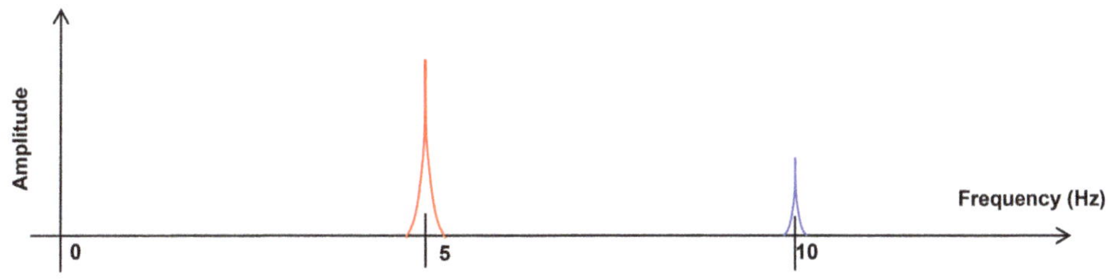

(d)

Figure 6.1. (Continued.)

Figures 6.1(b) and (c) illustrate the sine waves present in the waveform of figure 6.1(a). As can be seen, these sine waves have frequencies of 5 Hz and 10 Hz. Look at figures 6.1(a)–(c) carefully. It is clear that the waves of figures 6.1(b) and (c) can be superimposed to represent the wave of figure 6.1(a). At time instants where the crest of the wave in figure 6.1(b) coincides with the crest of the wave in figure. 6.1(c), constructive interference takes place between the two waves and the amplitude of the composite wave at these instants becomes larger. Similarly, at those instants of time at which the crest of one wave falls on the trough of another, the two waves interfere destructively so that the amplitude of the composite wave becomes smaller than the amplitudes of the interfering waves.

The defining equation for the Fourier transform of a function of time $E(t)$ is (Cho 2018)

$$E(\omega) = \int_{-\infty}^{+\infty} E(t)\exp(-j\omega t)dt \tag{6.1}$$

where ω is the angular frequency. This is the forward Fourier transform or analysis equation.

The Fourier transform consists of a real part x and an imaginary part y, and is represented by the complex number $E(\omega)$:

$$E(\omega) = x(\omega) + jy(\omega) \tag{6.2}$$

This complex number is written in polar form with a magnitude and phase as

$$\begin{aligned} E(\omega) &= |E(\omega)| \exp\{j\phi(\omega)\} = |E(\omega)| \{\cos\phi(\omega) + j\sin\phi(\omega)\} \\ &= |E(\omega)| \cos\phi(\omega) + j\ |E(\omega)| \sin\phi(\omega) \end{aligned} \tag{6.3}$$

where $|E(\omega)|$ is the continuous amplitude spectrum of the wave and angle $\phi(\omega)$ is its continuous phase spectrum. $|E(\omega)|$ tells us about the relative presence of sine waves in $E(t)$ while $\phi(\omega)$ indicates the lining up of different sine waves to form $E(t)$ to see how they are offset with respect to each other.

$$x = \text{Real part of } E(\omega) = |E(\omega)| \cos\phi(\omega) \tag{6.4}$$

$$y = \text{Imaginary part of } E(\omega) = |E(\omega)| \sin\phi(\omega) \tag{6.5}$$

from which

$$\frac{|E(\omega)| \sin \phi(\omega)}{|E(\omega)| \cos \phi(\omega)} = \frac{y(\omega)}{x(\omega)} \tag{6.6}$$

or

$$\tan \phi(\omega) = \frac{y(\omega)}{x(\omega)} \tag{6.7}$$

$$\therefore \phi(\omega) = \tan^{-1}\left\{\frac{y(\omega)}{x(\omega)}\right\} \tag{6.8}$$

The amplitudes of the two sine waves present in the waveform of figure 6.1(a) are given by

$$|A_1(\omega)| = \sqrt{\{x_1(\omega)\}^2 + \{y_1(\omega)\}^2} \tag{6.9}$$

$$|A_2(\omega)| = \sqrt{\{x_2(\omega)\}^2 + \{y_2(\omega)\}^2} \tag{6.10}$$

The phases of the waves are

$$\phi_1(\omega) = \tan^{-1}\left\{\frac{y_1(\omega)}{x_1(\omega)}\right\} \tag{6.11}$$

$$\phi_2(\omega) = \tan^{-1}\left\{\frac{y_2(\omega)}{x_2(\omega)}\right\} \tag{6.12}$$

Figure 6.1(d) shows the spectrum obtained by applying the Fourier transform on the waveform of figure 6.1(a). It is immediately obvious that the complicated waveform of figure 6.1(a) contained hidden information about frequency components present inside it. This information is revealed by the spectrum.

The contents of figures 6.1(a)–(d) are pictorially displayed in a single diagram shown in figure 6.2. The time and frequency domains are in fact two perspectives of the same scenario. They are views obtained when placing the eyes in the directions indicated in the diagram. The original information is preserved when interconverting between the two domains. But certain aspects of the circumstance which were not easily seen in the time domain become strikingly noticeable in the frequency domain.

6.2 Scheme of terahertz TDS

Owing to the non-availability of straightforward absorption measurement methods, terahertz spectroscopy is not performed by the traditional dispersive or Fourier transform approaches. Instead, measurements are done in the time domain.

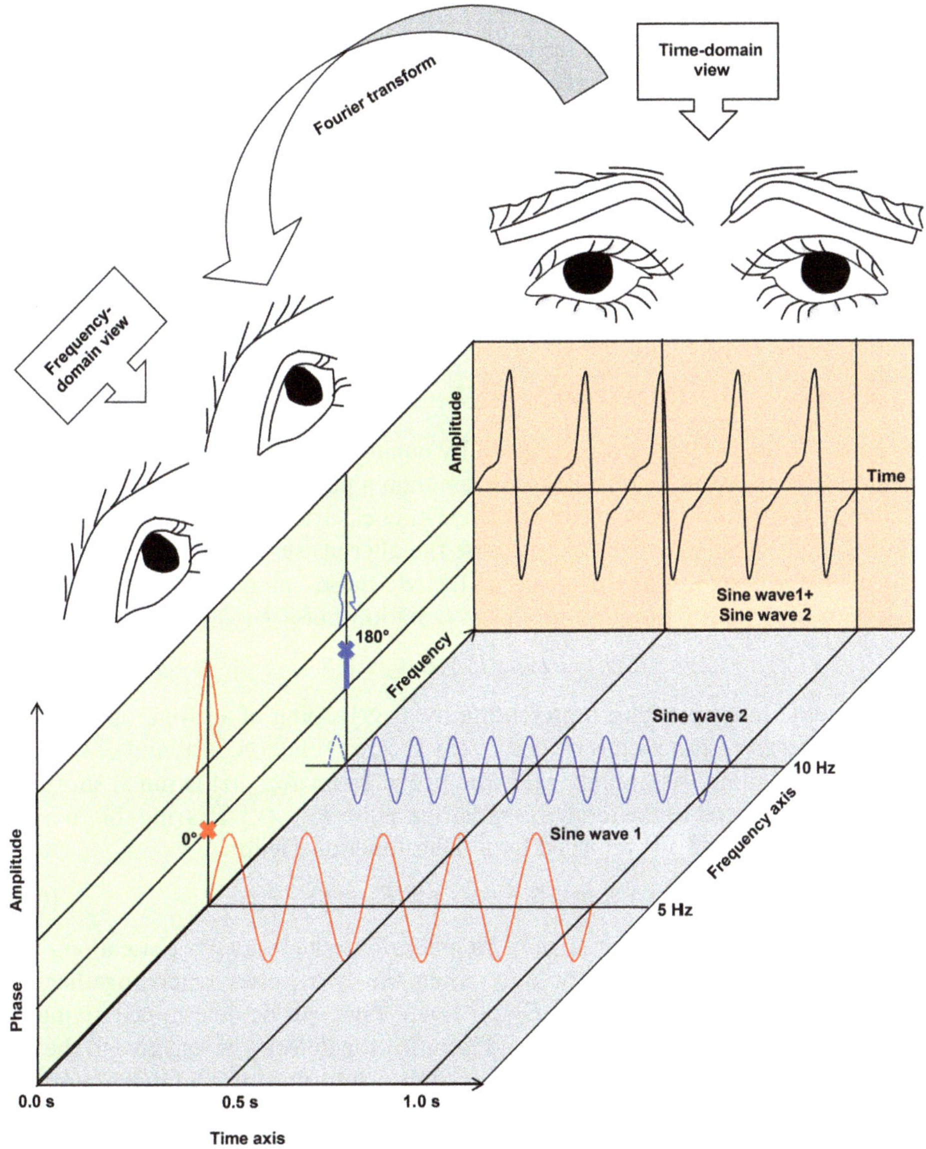

Figure 6.2. Visualization of the frequency domain version (spectrum) of a time-domain pattern (waveform). Direction of viewing is indicated by the eye positions. Note that the positive slope of sine wave 2 passes through the horizontal axis before time $t = 0$. This means that sine wave 2 has shifted to the left with respect to sine wave 1 giving a leading phase angle for sine wave 2, which is taken as positive angle and the amount of shift towards the left determines the phase angle ϕ, which is here π radians or 180°.

6.2.1 THz-TDS concept

6.2.1.1 The convolution operation

In THz-TDS, the magnitude of the transient terahertz electric field is in the range of $10–10^2$ V cm^{-1}. The time duration of this electric field is extremely short, around a few picoseconds (Neu and Schmuttenmaer 2018). The rise and fall times of available electrical detectors are mostly of the order of nanoseconds and can be pushed up to at best several hundreds or tens of picoseconds. Hence, they cannot be used for accurate measurements of electric field with sub-picosecond resolution. The path adopted to solve this problem is to resort to optical techniques. A near infrared optical pulse of 100 fs duration is beamsplit. The beams thus obtained propagate along two different paths. They are used both for production and detection of the terahertz electric field.

A femtosecond (fs) IR laser pulse is used to sample the unknown terahertz electric field. This IR laser pulse is known as the readout pulse or probe beam. An instantaneous signal $S(t)$ is obtained only on simultaneous arrival of the terahertz radiation pulse $E_{\mathrm{THz}}(t)$ and the IR probe beam of intensity profile $I_{\mathrm{Probe}}(t)$ so that

$$S(t) \propto I_{\mathrm{Probe}} E_{\mathrm{THz}}(t) \tag{6.13}$$

where t is the time. The signal $S(t)$ given by equation (6.13) is the signal of interest which must be measured with resolution less than a picosecond. As already said and to reiterate further, the slow speed of the existing electrical detectors failing to cope with these fast signals compels us to devise the alternative scheme which entails the performance of convolution operation on the two pulses, namely the IR probe beam $I_{\mathrm{Probe}}(t)$ and the longer duration terahertz radiation pulse $E_{\mathrm{THz}}(t')$, written as

$$S(t') \propto I_{\mathrm{Probe}}(t) \circledast E_{\mathrm{THz}}(t') \tag{6.14}$$

The convolution operation representing the overlapping of a given function with another as the given function is shifted over the other function, is a kind of blending of the two functions. As already said, the probe beam $I_{\mathrm{Probe}}(t)$ is much shorter in duration as compared to the terahertz radiation pulse $E_{\mathrm{THz}}(t')$. Bearing this in mind, the probe beam $I_{\mathrm{Probe}}(t)$ is replaced by a delta function yielding

$$S(t') \propto \delta(t) \circledast E_{\mathrm{THz}}(t') = E_{\mathrm{THz}}(t') \tag{6.15}$$

Due to the relativity shorter span of IR pulse than the terahertz pulse along with exclusive responsivity of detector only when the two pulses reach together, the variation of terahertz electric field $E_{\mathrm{THz}}(t')$ with time can be determined giving the terahertz field as a function of time. Additionally, the detector is sensitive to the sign of the electric field $E_{\mathrm{THz}}(t')$ also. Thus complete information about the temporal dependence of the amplitude of terahertz electric field is obtained. The capability of provision of both amplitude and phase of the terahertz electric field is a significant advantage favoring THz-TDS with respect to Fourier transform infrared red spectroscopy in which knowledge of the intensity of the terahertz electric field represented by the square of its amplitude $E^2(t)$ is obtained.

6.2.1.2 Signal measurements at different time points

Equation (6.15) gives the signal $S(t')$ at a single time point t'. For measurements of signal at all the points t_1', t_2', t_3', ..., the probe beam $\delta(t)$ is delayed with respect to the terahertz electric field $E_{\mathrm{THz}}(t')$. For this purpose, the IR laser beam is resolved into two components. One of these components is the pump beam employed for generation of the terahertz radiation pulse. The other component is the readout or probe beam. The path length traversed by one of the beams is taken to be longer than the path of the other beam. If the length of the path is s and the time taken to cover this distance is t, the time t and the distance s are related through the velocity of laser beam, which is velocity c of light:

$$t = \frac{s}{c} \tag{6.16}$$

Since

$$\begin{aligned} c &= 2.9979 \times 10^8\,\mathrm{ms}^{-1} = \frac{2.9979 \times 10^8}{10^{15}}\frac{\mathrm{m}}{\mathrm{fs}} = 2.9979 \times 10^{-7}\frac{\mathrm{m}}{\mathrm{fs}} \\ &= 2.9979 \times 10^{-7} \times 10^6 \frac{\mu\mathrm{m}}{\mathrm{fs}} \\ &= 2.9979 \times 10^{-1}\,\mu\mathrm{mfs}^{-1} = 0.29979\,\mu\mathrm{mfs}^{-1} \end{aligned} \tag{6.17}$$

a distance of 0.29979 μm corresponds to a delay time of 1 fs; hence 1 μm distance means 1/0.29979=3.33 fs. For round trip propagation, the total delay time is double this amount (6.66 fs). Computer-controlled positioning stages called delay lines exploit this translation of fs time duration to μm scale spatial movement to provide the required time delays for measurements at various instants of time.

6.2.1.3 Avoiding moisture-induced vitiation of measurements

Noting the severity of absorption and weakening of terahertz radiation by the water vapor present in the atmosphere, careful precaution is necessary to mitigate moisture influence. The procedure followed involves confinement of the path of the terahertz beam within a closed vessel containing dry air. The vessel is either purged with dry air or nitrogen, or completely evacuated prior to measurements.

6.2.1.4 Fourier transformation

The transient electric field obtained from the measurements is transformed from the time domain to the frequency domain by Fourier transformation using equation (6.1). This transformation gives the spectral information about the terahertz pulse:

$$\text{Electric field (Time)} \rightarrow \text{Fourier Transform} \rightarrow \text{Electric field (Frequency)} \tag{6.18}$$

6.2.1.5 Discrete Fourier transform (DFT)

In order to work with a digital computer, the continuous-time signal is approximated with a finite-duration sequence containing a finite number of discrete samples from the original continuous-time signal. So, instead of the continuous-time Fourier transform, its discrete version, the DFT is used. The DFT $E(\omega_\mathrm{k})$ of a signal $e(t_\mathrm{n})$ replaces the infinite integral in equation (6.1) with a finite sum as (Smith 2007)

$$E(\omega_\mathrm{k}) = \sum_{n=0}^{N-1} e(t_\mathrm{n}) \exp(-j\omega_\mathrm{k} t_\mathrm{n}) \text{ where } k = 0,\ 1,\ 2,\ 3,\ \ldots\ldots(N-1) \tag{6.19}$$

where $e(t_n)$ is the amplitude of the input electric field at an instant of time t_n; the symbol $t_n = nT$ denotes the nth sampling instant, n is an integer $\geqslant 0$; T is the sampling interval or sampling period; $E(\omega_k)$ is the frequency spectrum of the electric field $e(t_n)$ at frequency ω_k; the symbol ω_k stands for $k\Omega$, the kth frequency sample (radians per second); $\Omega = 2\pi/NT$ is the radian-frequency sampling interval; $1/T$ is the sampling rate; N is the number of time samples = number of frequency samples, i.e., the number of samples in the data frame of DFT.

6.2.1.6 *Fast Fourier transform (FFT)*

Determination of DFT is a computationally intensive procedure. The FFT is not any new transform but an efficient algorithm that increases the computational speed of DFT by reducing the number of multiplications and addition operations necessary taking advantage of its symmetry and periodicity properties (Rao *et al* 2010). In opposition to the N^2 complex multiplications required for DFT evaluation, $(N/2)\log_2 N$ complex multiplications give the FFT, revealing its computational efficiency.

Example 6.1 By what factor is the number of mathematical operations reduced if computation of DFT is carried out with FFT algorithm for the case in which the number of samples in the data frame $N = 5000$.

Solution: Number of mathematical operations in DFT computation is

$$\text{Number(DFT)} = N^2 = (5000)^2 = 2.5 \times 10^7 \tag{6.20}$$

while the number of operations in FFT computation is

$$\begin{aligned}\text{Number(FFT)} &= \frac{N}{2}\log_2 N = \frac{N\log_{10}N}{2\ \log_{10}2} = \frac{5000}{2}\frac{\log_{10}5000}{\log_{10}2} \\ &= 2500\frac{3.69897}{0.30103} = 30719.28 = 3.072 \times 10^4\end{aligned} \tag{6.21}$$

where formula for the change of base of logarithm: $\log_b a = \log_d a/\log_d b$ has been applied.

The reduction factor of mathematical operations is obtained as

$$\begin{aligned}\text{Reduction factor} &= \frac{\text{Number(DFT)}}{\text{Number(FFT)}} \\ &= \frac{2.5 \times 10^7}{3.072 \times 10^4} = 0.8138 \times 10^3 = 813.8\end{aligned} \tag{6.22}$$

6.2.2 THz-TDS instrumentation system

The main parts of a terahertz TDS system are (figure 6.3): a femtosecond laser, a terahertz emitter, a terahertz detector; beam splitter, mirrors, lenses and other

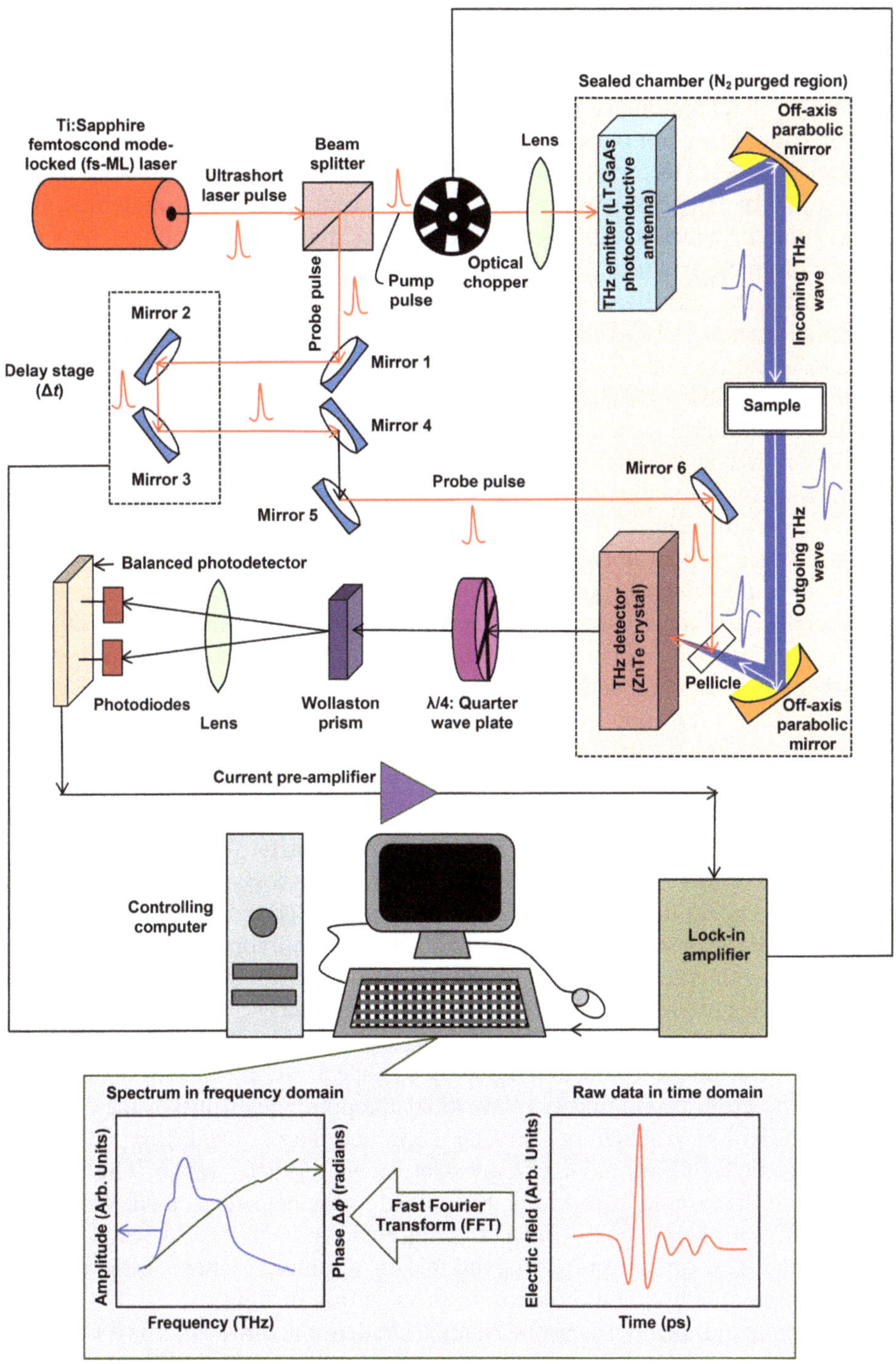

Figure 6.3. Setup of coherent terahertz time-domain spectroscopy (THz-TDS) system in transmission mode. Optical beam paths are represented by red lines and arrows while terahertz beam paths are indicated by white arrows against a blue background. The system uses LT-GaAs photoconductive antenna as terahertz emitter and ZnTe electro-optic crystal as terahertz detector.

components for laser beam and terahertz wave optics; a chopper, a delay stage and a computer (Fischer *et al* 2005). Regarding the terahertz emitter and detector used in spectroscopy, a commonly used terahertz emitter is the photoconductive antenna. Another example is a nonlinear crystal which produces terahertz radiation by optical rectification. The detector is a photoconductive antenna or electro-optic crystal. Electro-optic or Pockel's effect, observed in non-centrosymmetric materials, is the phenomenon in which the refractive index of a material is modified under the influence of an electric field.

To explain the operation, we suppose that a photoconductive antenna is used for terahertz generation and ZnTe crystal for THz detection, as shown in figure 6.3. We call the photoconductive antenna as the THz generator or the THz emitter and the ZnTe detector crystal as the THz detector. In practice, photoconductive antennas can be used as both THz emitter and detector. Similarly, ZnTe crystals can be employed for THz generation as well as detection. So different emitter/detector combinations can be used.

As already mentioned, terahertz spectroscopy works through probing the properties of a material by pulses of terahertz waves having ultra-short duration. A pump beam-probe beam instrumentation is used employing a pump laser beam and a probe laser beam, which is a split-off component of the pump beam itself (figure 6.3). The procedure for generation of the pump and probe beams involves directing the beam from a mode-locked laser on a beam splitter. Mode-locking is a technique of production of a train of pulses of extremely short duration using a laser source. These pulses have duration in the range from subpico- to femtoseconds. The paths followed by the two beams and the resulting electrical signals are as follows:

(i) Pump beam and electrical signal path: femtosecond laser–beam splitter–chopper-lens–THz emitter–parabolic mirror-sample–parabolic mirror–pellicle–THz detector–quarter wave plate–Wollaston prism–lens–photodiodes–photodetector–current pre-amplifier–lock-in amplifier–computer;
(ii) Probe beam and electrical signal path: femtosecond laser–beam splitter–mirror 1–delay stage (mirrors 2 and 3)–mirrors 4,5,6–pellicle–THz detector–quarter wave plate–Wollaston prism–lens-photodiodes–photodetector–current pre-amplifier–lock-in amplifier–computer.

Tracing the beam paths through the setup of a spectroscopy system, we find that a laser pulse of femtosecond duration known as the pump beam strikes the emitter of terahertz radiation (the emitter crystal supposed here), producing a pulse of electromagnetic radiation having frequencies in the terahertz range. The terahertz pulse falls on the sample from where it is guided to the detector crystal. The electric field of this terahertz pulse induces a birefringence in the detector crystal. Birefringence is an optical property exhibited by a material whose refractive index depends on polarization and direction of propagation of light.

Unlike the pump beam, the probe beam is incident on a moving delay stage from which it impinges on the detector. The birefringence induced in the detector crystal by the terahertz pulse from the sample causes a phase retardation in the probe pulse which has a pre-aligned polarization. This phase retardation varies linearly with the electric field of the terahertz pump pulse.

Using the delay stage, the probe pulse is time delayed by varying amounts with respect to the pump pulse. In this way, the phase retardation of the probe pulse is measured with respect to time, enabling the extraction of electric field of the terahertz pump pulse on the time scale. In essence, a sampling of the electric field of the terahertz pump pulse is accomplished with varying time. From this sampling of the impulsive or transient response in time, the frequency spectrum is derived by fast Fourier transformation.

The above detection scheme resolves the temporal profile of the terahertz pulse giving both amplitude and phase of the electric field with high signal-to-noise ratio.

6.3 Rotational and vibrational aspects of terahertz spectroscopy

After learning about the setup of THz-TDS system, let us explore the rotational and vibrational aspects in this spectrosopy. In section 6.1.2, table 6.2, it was mentioned that three forms of spectroscopies, namely rotational or microwave spectroscopy (Brown and Carrington 2003), vibrational or infrared spectroscopy (Chalmers and Griffiths 2002) and electronic or UV-vis spectroscopy (Perkampus 1992) are immediately evident judging from the different nature of interactions of electromagnetic waves with matter in accordance with their energies. Further, THz spectroscopy filling the void between microwave and infrared spectroscopies shares features of its two bordering spectroscopies, and is concerned with molecular transitions between quantized rotational states as well as those between quantified vibrational states. In terahertz spectroscopy, we shall come across both types of transitions. Therefore, taking examples of diatomic molecules, prominent aspects of rotational and vibrational spectroscopies will be presented in the exercises below. As a diatomic molecule contains two atoms, we begin with the concept of reduced mass of two particles because it is essential for understanding the subsequent mathematical treatments.

Example 6.2 Reduced mass notion: Show that the equations of motion describing two mutually interacting particles can be transformed into an equation for the motion of a single particle whose mass equals the product of masses of the two particles divided by the sum of masses.

Solution: Let the position vectors of the particles of masses m_1 and m_2 at a given instant of time t be $\mathbf{r}_1$ and $\mathbf{r}_2$ with respect to an inertial frame of reference and origin O (figure 6.4). Then

$$\begin{aligned} F_{21} &= \text{Force acting on particle of mass } m_1 \text{ due to particle of mass } m_2 \\ &= m_1 \frac{d^2\mathbf{r}_1}{dt^2} \end{aligned} \tag{6.23}$$

and

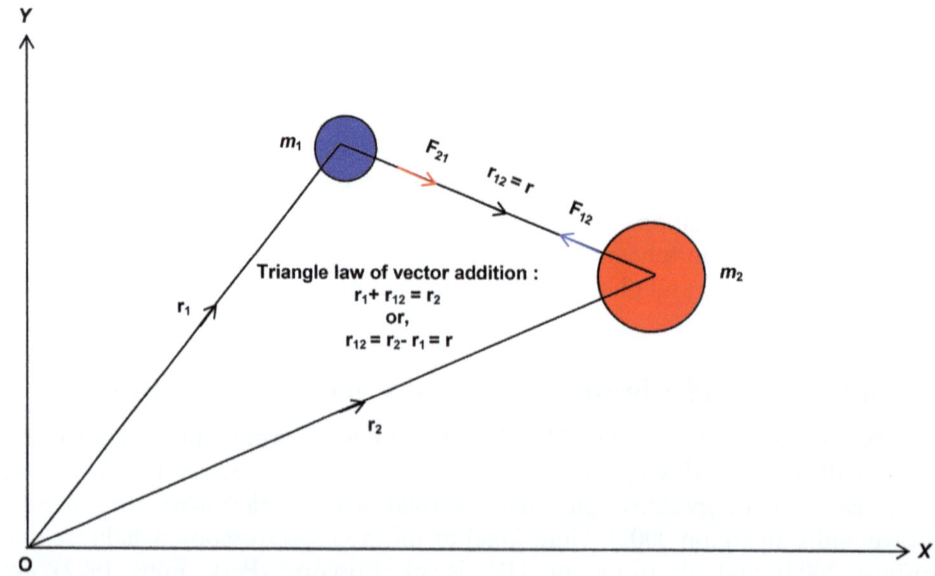

Figure 6.4. Reduced mass concept illustrating that two particles of masses m_1 and m_2 with position vectors $\mathbf{r_1}$ and $\mathbf{r_2}$ and exerting forces F_{12} and F_{21} on each other can be analyzed by replacing the two-particle system with a single particle of reduced mass μ.

$$\begin{aligned} F_{12} &= \text{Force acting on particle of mass } m_2 \text{ due to particle of mass } m_1 \\ &= m_2 \frac{d^2\mathbf{r}_2}{dt^2} \end{aligned} \tag{6.24}$$

Equation (6.23) gives

$$\frac{d^2\mathbf{r}_1}{dt^2} = \frac{F_{21}}{m_1} \tag{6.25}$$

while equation (6.24) yields

$$\frac{d^2\mathbf{r}_2}{dt^2} = \frac{F_{12}}{m_2} \tag{6.26}$$

Subtracting equation (6.25) from equation (6.26),

$$\frac{d^2\mathbf{r}_2}{dt^2} - \frac{d^2\mathbf{r}_1}{dt^2} = \frac{F_{12}}{m_2} - \frac{F_{21}}{m_1} \tag{6.27}$$

Also, from figure 6.4

$$\mathbf{r}_{12} = \mathbf{r}_2 - \mathbf{r}_1 \tag{6.28}$$

$$\therefore \frac{d^2\mathbf{r}_{12}}{dt^2} = \frac{d^2\mathbf{r}_2}{dt^2} - \frac{d^2\mathbf{r}_1}{dt^2} \tag{6.29}$$

or

$$\frac{d^2\mathbf{r}_2}{dt^2} - \frac{d^2\mathbf{r}_1}{dt^2} = \frac{d^2\mathbf{r}_{12}}{dt^2} = \frac{d^2\mathbf{r}}{dt^2} \tag{6.30}$$

replacing $\mathbf{r}_{12}$ by $\mathbf{r}$.

Looking at equations (6.27) and (6.30), we can write

$$\frac{d^2\mathbf{r}}{dt^2} = \frac{F_{12}}{m_2} - \frac{F_{21}}{m_1} \tag{6.31}$$

According to Newton's third law of motion,

$$F_{12} = -F_{21} = F \tag{6.32}$$

so that equation (6.31) reduces to

$$\frac{d^2\mathbf{r}}{dt^2} = \frac{F_{12}}{m_2} + \frac{F_{12}}{m_1} = \frac{F}{m_2} + \frac{F}{m_1} = F\left(\frac{1}{m_2} + \frac{1}{m_1}\right) = F\left(\frac{m_1 + m_2}{m_1 m_2}\right) \tag{6.33}$$

or

$$F = \left(\frac{m_1 m_2}{m_1 + m_2}\right)\frac{d^2\mathbf{r}}{dt^2} = \mu\left(\frac{d^2\mathbf{r}}{dt^2}\right) \tag{6.34}$$

where

$$\frac{m_1 m_2}{m_1 + m_2} = \mu = \text{Reduced mass of particles of masses } m_1 \text{ and } m_2 \tag{6.35}$$

Equation (6.34) resembles the equation of motion of a single particle of mass μ at a distance $\mathbf{r}$ from one of the particles. This particle is called the center of mass (COM) of the two particles. Thus the two-particle problem is reduced to an easier one-particle problem in which a particle of mass μ is moving about the COM of the two particles. COM is the unique imaginary point at which the entire mass of the system can be assumed to be located for describing its behavior.

Ex.6.3 Rotational spectroscopy: (a) Apply the rigid rotor model to derive an expression for the frequency separation between two rotational states of a diatomic molecule. (b) The absorption frequencies of two successive rotational energy levels of HCl molecule are 2.498 THz (83.32 cm^{-1}) and 3.122 THz (104.13 cm^{-1}). Find the rotational constant and moment of inertia of the molecule and also the bond length in Å.

Solution: (a) A diatomic molecule is composed of two atoms of masses m_1 and m_2. The molecule rotates about the COM of the two atoms with a frequency ν with the first atom rotating in a circle of radius r_1 about the COM and the second atom rotating in a circle of radius r_2 about the COM (figure 6.5). The COM lies at the point of intersection of r_1 and r_2.

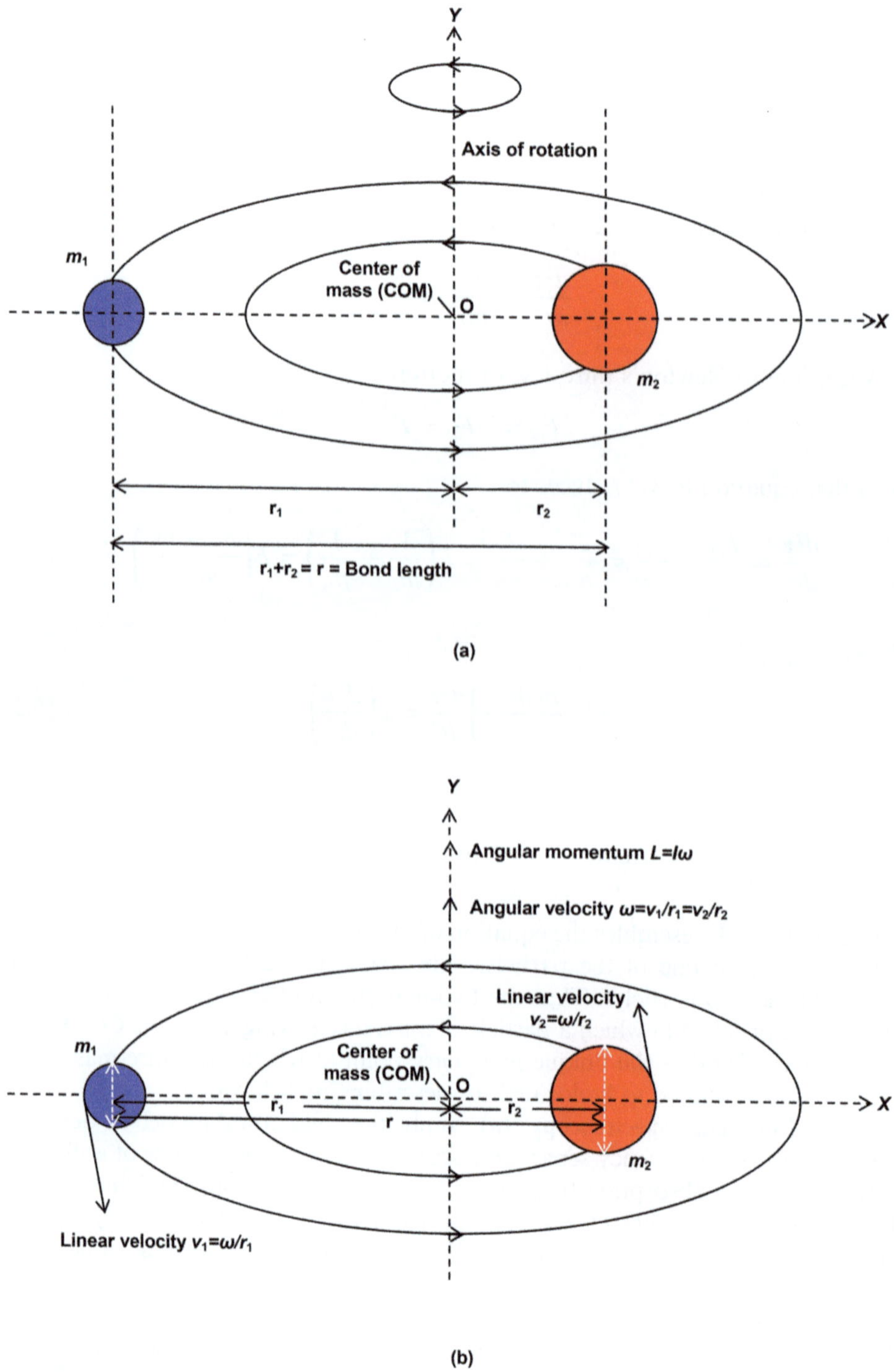

Figure 6.5. Molecular rotation: (a) The rigid rotor model representing the rotation of a diatomic molecule about a vertical axis. (b) Simplified version of (a) showing the linear and angular velocities and angular momentum.

Applying the reduced mass concept elaborated in example 6.2, the system can be described by the fixed distance between the two atoms given by

$$r = r_1 + r_2 \tag{6.36}$$

instead of the individual radii of rotation r_1, r_2, which are interrelated according to the COM as

$$m_1 r_1 = m_2 r_2 \tag{6.37}$$

From equation (6.36)

$$r_2 = r - r_1 \tag{6.38}$$

while from equation (6.37)

$$r_2 = \frac{m_1 r_1}{m_2} \tag{6.39}$$

Since the left-hand sides of equations (6.38) and (6.39) are the same,

$$r - r_1 = \frac{m_1 r_1}{m_2} \tag{6.40}$$

or

$$r_1 + \frac{m_1 r_1}{m_2} = r \tag{6.41}$$

or

$$r_1\left(1 + \frac{m_1}{m_2}\right) = r \tag{6.42}$$

or

$$r_1\left(\frac{m_2 + m_1}{m_2}\right) = r \tag{6.43}$$

$$\therefore r_1 = \frac{m_2}{m_1 + m_2} r \tag{6.44}$$

Similarly,

$$r_2 = \frac{m_1}{m_1 + m_2} r \tag{6.45}$$

Kinetic energy of atom of mass m_1 moving with linear velocity v_1 is

$$T_1 = \frac{1}{2} m_1 v_1^2 = \frac{1}{2} m_1 (\omega r_1)^2 = \frac{1}{2} m_1 \omega^2 r_1^2 = \frac{1}{2} m_1 r_1^2 \omega^2 \tag{6.46}$$

and the kinetic energy of atom of mass m_2 moving with linear velocity v_2 is

$$T_2 = \frac{1}{2}m_2 v_2^2 = \frac{1}{2}m_2(\omega r_2)^2 = \frac{1}{2}m_2\omega^2 r_2^2 = \frac{1}{2}m_2 r_2^2\omega^2 \tag{6.47}$$

where the relation

$$v = \omega r \tag{6.48}$$

between linear velocity v, angular velocity ω and radius r of circle has been used.

Adding together equations (6.46) and (6.47), the total kinetic energy of the system is

$$\begin{aligned} T &= T_1 + T_2 = \frac{1}{2}m_1 r_1^2\omega^2 + \frac{1}{2}m_2 r_2^2\omega^2 = \frac{1}{2}(m_1 r_1^2 + m_2 r_2^2)\omega^2 \\ &= \frac{1}{2}I\omega^2 = \frac{I\omega^2}{2} \end{aligned} \tag{6.49}$$

since

$$m_1 r_1^2 + m_2 r_2^2 = \text{Moment of inertia of the system of particles} = I \tag{6.50}$$

Putting the values of r_1, r_2 from equations (6.44), (6.45) into equation (6.50)

$$\begin{aligned} I &= m_1\left(\frac{m_2}{m_1 + m_2}r\right)^2 + m_2\left(\frac{m_1}{m_1 + m_2}r\right)^2 \\ &= \frac{m_1^2}{m_1}\left(\frac{m_2}{m_1 + m_2}r\right)^2 + \frac{m_2^2}{m_2}\left(\frac{m_1}{m_1 + m_2}r\right)^2 \\ &= \frac{1}{m_1}\left\{m_1^2\left(\frac{m_2}{m_1 + m_2}r\right)^2\right\} + \frac{1}{m_2}\left\{m_2^2\left(\frac{m_1}{m_1 + m_2}r\right)^2\right\} \\ &= \frac{1}{m_1}\left(\frac{m_1 m_2}{m_1 + m_2}r\right)^2 + \frac{1}{m_2}\left(\frac{m_1 m_2}{m_1 + m_2}r\right)^2 \\ &= \left(\frac{1}{m_1} + \frac{1}{m_2}\right)\left(\frac{m_1 m_2}{m_1 + m_2}r\right)^2 = \left(\frac{m_1 + m_2}{m_1 m_2}\right)\frac{(m_1 m_2 r)^2}{(m_1 + m_2)^2} \\ &= \left(\frac{m_1 m_2}{m_1 + m_2}\right)r^2 = \mu r^2 \end{aligned} \tag{6.51}$$

where

$$\frac{m_1 m_2}{m_1 + m_2} = \mu = \text{Reduced mass of the two-atom system} \tag{6.52}$$

Hence,

$$I = \mu r^2 \tag{6.53}$$

The moment of inertia I of the system is related to its angular momentum L. The equation connecting I with L is:

$$L = I\omega \tag{6.54}$$

Equations (6.49) and (6.54) enable us to write

$$T = \frac{I\omega^2}{2} = \frac{I^2\omega^2}{2I} = \frac{(I\omega)^2}{2I} = \frac{L^2}{2I} \tag{6.55}$$

The Hamiltonian of the system is

$$H = T + V = T + 0 = T = \frac{L^2}{2I} \tag{6.56}$$

because in the absence of electric or magnetic fields, the energy is independent of orientation of the system, making the potential energy component to be zero. The angular momentum or moment of momentum of a particle defined as the vector product of its position vector **r** and linear momentum **p** as (figure 6.6)

$$\mathbf{L} = \mathbf{r} \times \mathbf{p} \tag{6.57}$$

is resolved into projections L_x, L_y, L_z on the three axes, and a new operator is defined in quantum mechanics in Cartesian co-ordinates as

$$\hat{L}^2 = \hat{L}_x^2 + \hat{L}_y^2 + \hat{L}_z^2. \tag{6.58}$$

Transformation from Cartesian to spherical polar co-ordinate system enables us to write

$$\begin{aligned}\hat{\boldsymbol{L}}^2 &= \text{Angular momentum operator} \\ &= -\hbar^2\left\{\frac{1}{\sin\theta}\frac{\partial}{\partial\theta}\left(\sin\theta\frac{\partial}{\partial\theta}\right) + \frac{1}{\sin^2\theta}\frac{\partial^2}{\partial\phi^2}\right\}\end{aligned} \tag{6.59}$$

Putting this value of $\hat{\boldsymbol{L}}^2$ from equation (6.59) in equation (6.56), the equation for Hamiltonian becomes

$$H = -\frac{\hbar^2}{2I}\left\{\frac{1}{\sin\theta}\frac{\partial}{\partial\theta}\left(\sin\theta\frac{\partial}{\partial\theta}\right) + \frac{1}{\sin^2\theta}\frac{\partial^2}{\partial\phi^2}\right\} \tag{6.60}$$

The wave function ψ is obtained by solving the Schrodinger's equation. This equation is written as

$$\hat{H}\psi = E\psi \tag{6.61}$$

where E is the total energy of the two-atom system.

Using equation (6.60), we can write equation (6.61) as

$$-\frac{\hbar^2}{2I}\left\{\frac{1}{\sin\theta}\frac{\partial}{\partial\theta}\left(\sin\theta\frac{\partial}{\partial\theta}\right) + \frac{1}{\sin^2\theta}\frac{\partial^2}{\partial\phi^2}\right\}\psi = E\psi \tag{6.62}$$

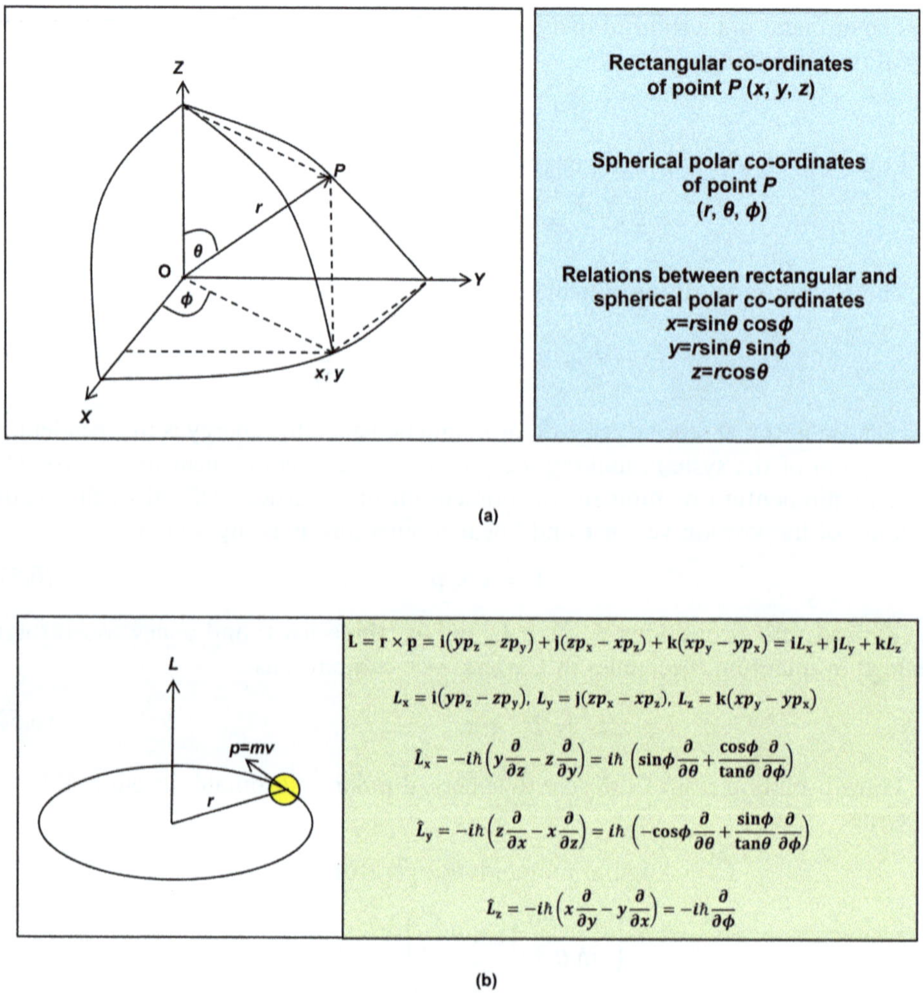

Figure 6.6. About co-ordinate systems relationships and angular momentum: (a) rectangular Cartesian co-ordinates (x, y, z) and spherical polar coordinates (r, θ, ϕ), and (b) formulae for angular momentum.

The wave function $\psi\ (\theta, \phi)$ is usually written as $Y\ (\theta, \phi)$. So equation (6.62) becomes

$$-\frac{\hbar^2}{2I}\left\{\frac{1}{\sin\theta}\frac{\partial}{\partial\theta}\left(\sin\theta\frac{\partial}{\partial\theta}\right) + \frac{1}{\sin^2\theta}\frac{\partial^2}{\partial\phi^2}\right\}Y(\theta, \phi) = EY(\theta, \phi) \tag{6.63}$$

The equation contains two variables θ and ϕ. To solve this equation, we write $Y(\theta, \phi)$ as a product

$$Y(\theta, \phi) = \Theta(\theta)\Phi(\phi) \tag{6.64}$$

Thus we get two equations by separating the variables, one in the variable Θ and the other in the variable Φ:

$$\frac{\sin\theta}{\Theta(\theta)}\frac{d}{d\theta}\left(\sin\theta\frac{d\Theta}{d\theta}\right)+\frac{2IE}{\hbar^2}\sin^2\theta=m^2 \tag{6.65}$$

and

$$\frac{1}{\Phi(\phi)}\frac{d^2\Phi}{d\phi^2}=-m^2 \tag{6.66}$$

where m is a constant. Solution of equation (6.66) gives

$$\Phi(\phi)=\frac{1}{\sqrt{2\pi}}\exp(im\phi) \tag{6.67}$$

where m is a quantum number $=0, \pm1, \pm2, \ldots$
Solution of equation (6.65) results in

$$\frac{2IE}{\hbar^2}=J(J+1) \tag{6.68}$$

where J is the rotational energy level with integral values $J = 0, 1, 2, 3, \ldots\ldots$. From equation (6.68), the energy is expressed as

$$E=\frac{\hbar^2}{2I}J(J+1) \tag{6.69}$$

The constant

$$\frac{\hbar^2}{2I}=B=\text{Rotational constant} \tag{6.70}$$

Hence, equation (6.69) becomes

$$E=BJ(J+1) \tag{6.71}$$

which is re-rewritten for rotational level J as

$$E_J=BJ(J+1) \tag{6.72}$$

Let us look at a transition from rotational energy level E_J to another rotational energy level $E_{J'}$. For the level E_J, we rewrite equation (6.72) and for $E_{J'}$. we have a new equation. Thus

$$E_J=BJ(J+1) \tag{6.73}$$

and

$$E_{J'}=BJ'(J'+1) \tag{6.74}$$

The difference of energies between the two levels is

$$E_J - E_{J'} = BJ(J+1) - BJ'(J'+1) \tag{6.75}$$

The selection rule allows

$$\Delta J = \pm 1 \tag{6.76}$$

For ΔJ=1, we have

$$\Delta J = J - J' = 1 \tag{6.77}$$

or

$$J' = J - 1 \tag{6.78}$$

Using equations (6.75) and (6.78), we write

$$\begin{aligned} E_J - E_{J'} &= BJ(J+1) - B(J-1)(J-1+1) = BJ(J+1) - B(J-1)(J) \\ &= BJ(J+1-J+1) = BJ(2) = 2BJ \end{aligned} \tag{6.79}$$

which is the energy of radiation absorbed to produce a transition between two rotational levels. By Planck's theory,

$$E_J - E_{J'} = h\nu \tag{6.80}$$

where h is Planck's constant and ν is the frequency. Hence, from equations (6.79) and (6.80)

$$h\nu = 2BJ \tag{6.81}$$

or

$$\nu = \frac{2BJ}{h} \tag{6.82}$$

When $J = 0$, $\nu = 0$, the molecule is not rotating. A non-rotating molecule is said to be in the ground state. When $J = 1$, the frequency of the first absorption line is obtained as

$$\nu_1 = \frac{2B}{h} \tag{6.83}$$

Similarly, for $J = 2$, the frequency of the second absorption line is

$$\nu_2 = \frac{2B(2)}{h} = \frac{4B}{h} \tag{6.84}$$

For the third absorption line, the frequency is

$$\nu_3 = \frac{2B(3)}{h} = \frac{6B}{h} \tag{6.85}$$

and so on. We find from equations (6.83)–(6.85) that

$$\nu_2 - \nu_1 = \frac{4B}{h} - \frac{2B}{h} = \frac{2B}{h} \tag{6.86}$$

$$\nu_3 - \nu_2 = \frac{6B}{h} - \frac{4B}{h} = \frac{2B}{h} \tag{6.87}$$

Thus the frequency separation between any two consecutive absorption lines of the rotational spectrum (figure 6.7) is always

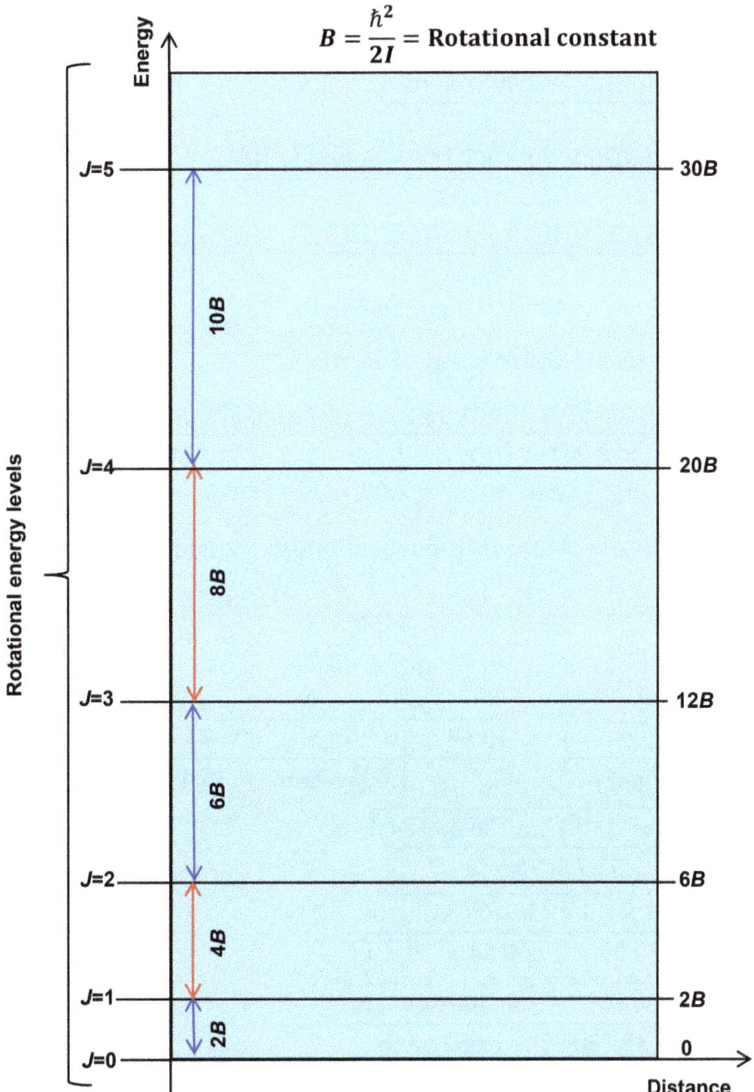

Figure 6.7. Rotational energy levels of a rigid rotor diatomic molecule.

$$\Delta\nu = \frac{2B}{h} \tag{6.88}$$

(b) For HCl molecules, the rotational constant B is calculated from equation (6.88) by putting the values of frequencies (3.122 and 2.498 THz), and Planck's constant $h = 6.626 \times 10^{-34}$ J s^{-1}:

$$3.122 \times 10^{12} - 2.498 \times 10^{12} = \frac{2B}{h} \tag{6.89}$$

or

$$\begin{aligned} B &= \frac{h(3.122 - 2.498) \times 10^{12}}{2} \\ &= \frac{(6.626 \times 10^{-34})(3.122 - 2.498) \times 10^{12}}{2} \mathrm{J} \times \mathrm{s} \times \mathrm{s}^{-1} \\ &= \frac{6.626 \times 10^{-34} \times 0.624 \times 10^{12}}{2} \mathrm{J} \\ &= 2.067 \times 10^{-34+12}\ \mathrm{J} = 2.067 \times 10^{-22}\ \mathrm{J} \end{aligned} \tag{6.90}$$

From equation (6.70), the moment of inertia is

$$\begin{aligned} I &= \frac{\hbar^2}{2B} = \frac{(1.05457 \times 10^{-34})^2}{2 \times 2.067 \times 10^{-22}} \frac{(\mathrm{J} \times \mathrm{s})^2}{\mathrm{J}} = \frac{(1.05457)^2 \times 10^{-68+22}}{2 \times 2.067} \mathrm{J\ s^2} \\ &= 0.269 \times 10^{-46}\ \mathrm{kg\ m^2\ s^{-2}\ s^2} = 2.69 \times 10^{-47}\ \mathrm{kg\ m^2} \end{aligned} \tag{6.91}$$

The bond length r is obtained from equations (6.52) and (6.53), which give

$$r^2 = \frac{I}{\mu} = \frac{I}{\frac{m_1 m_2}{m_1 + m_2}} = \frac{I}{\frac{m_\mathrm{H} m_\mathrm{Cl}}{m_\mathrm{H} + m_\mathrm{Cl}}} = \frac{I(m_\mathrm{H} + m_\mathrm{Cl})}{m_\mathrm{H} m_\mathrm{Cl}} \tag{6.92}$$

$$\begin{aligned} \therefore r &= \sqrt{\frac{I(m_\mathrm{H} + m_\mathrm{Cl})}{m_\mathrm{H} m_\mathrm{Cl}}} = \sqrt{\frac{2.69 \times 10^{-47}\mathrm{kgm^2}(1.007\,84 + 35.453)\mathrm{amu}}{1.007\,84\mathrm{amu} \times 35.453\mathrm{amu}}} \\ &= \sqrt{\frac{2.69 \times 10^{-47}\mathrm{kgm^2}(36.460\,84)}{35.730\,95\mathrm{amu}}} \\ &= \sqrt{\frac{2.69 \times 10^{-47}(36.460\,84)\mathrm{kgm^2}}{35.730\,95 \times 1.660\,54 \times 10^{-27}\mathrm{kg}}} \\ &= \sqrt{\frac{2.69 \times 10^{-47+27}(36.460\,84)\quad \mathrm{m^2}}{35.730\,95 \times 1.660\,54}} \\ &= 1.2857 \times 10^{-10}\,\mathrm{m} = \frac{1.2857 \times 10^{-10}}{1 \times 10^{-10}} \text{Å} = 1.286\text{Å} \end{aligned} \tag{6.93}$$

Example 6.4 Vibrational spectroscopy: (a) Apply the simple harmonic oscillator model and Hooke's law to derive the formula for frequency of vibration of a diatomic molecule in terms of force constant and reduced mass. (b) Absorption of terahertz radiation by molecular networks causes vibrations of weak intermolecular hydrogen bonds between molecular aggregates or macromolecules. If the absorption frequency is 2.5 THz, find the force constant for vibrations of hydrogen bond between oxygen atom of one molecule and hydrogen atom of another. The infrared stretching wave number of a free OH bond is 3600 cm^{-1}. What is the force constant of this bond?

Solution: (a) In mechanics, simple harmonic motion (SHM) is a periodic motion of an object that takes place under the influence of a restoring force, which is proportional to the displacement of the object from the equilibrium (or mean) position, and is always directed towards the equilibrium position. A representative example of SHM is the motion of a mass m suspended from a fixed support by a spring (figure 6.8). Suppose the mass is pulled downwards and released. The released mass repeatedly moves up and down executing a SHM. This motion obeys the Hooke's law of elasticity. This law states that: For small deformations of an elastic object, the deformation produced is directly proportional to the deforming force. In the case of the spring, the force F needed to stretch or compress the spring is proportional to the extension/shortening x of the spring:

$$F = -kx \tag{6.94}$$

where k is a constant. It is known as the spring constant or force constant of the spring. Its physical significance is that it gives an idea about the stiffness of the spring. The stiffer the spring, the larger is its force constant. The negative sign is introduced to imply that the restoring force exerted by the spring to return to equilibrium position acts in a direction opposite to the force causing extension of the spring.

Molecules are made up of atoms interconnected by bonds. Vibration of a molecule is the periodic motion of atoms constituting the molecule. This motion of molecules takes place with respect to each other. However, the COM of the molecules remains unperturbed. Molecular vibrations are of two types:

(i) stretching or compressing which involves the change of bond length, and
(ii) bending in which the bond angle varies.

Treating the atoms of a molecule as particles and the bonds as springs, a familiar analogy to the molecular vibrations is the simple harmonic oscillator. So, the simple harmonic approximation is applied to derive the formula for vibrational frequency of a molecule.

Let us examine thoroughly the SHM of a particle of mass m suspended from the free end of a spring whose top end is tied to a rigid support. Let the force constant of the spring be k. By Newton's second law of motion,

$$\text{Force} = \text{Mass} \times \text{Acceleration} \tag{6.95}$$

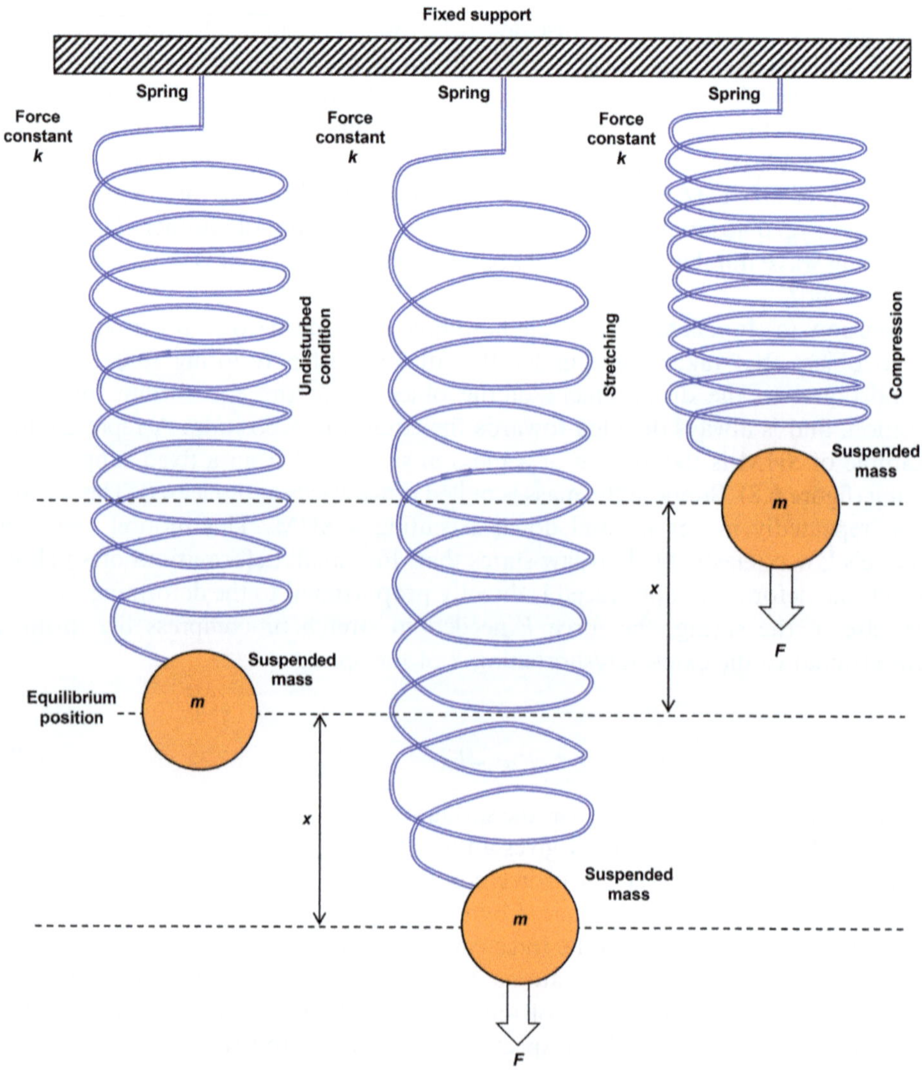

Figure 6.8. Harmonic oscillator model of molecular vibration.

or

$$F = m\frac{d^2x}{dt^2} \tag{6.96}$$

From equations (6.94) and (6.96) we get

$$m\frac{d^2x}{dt^2} = -kx \tag{6.97}$$

or

$$m\frac{d^2x}{dt^2} + kx = 0 \tag{6.98}$$

or

$$\frac{d^2x}{dt^2} + \left(\frac{k}{m}\right)x = 0 \tag{6.99}$$

It is a linear second-order differential equation. The solution of this equation is

$$x(t) = C\cos(\omega t) + D\sin(\omega t) = A\cos(\omega t - \phi) \tag{6.100}$$

where C and D are constants determined from the initial conditions

$$A = \sqrt{C^2 + D^2} = \text{Amplitude} \tag{6.101}$$

$$\omega = \sqrt{\frac{k}{m}} = \text{Angular frequency} = 2\pi \times \text{Linear frequency} = 2\pi\nu \tag{6.102}$$

and

$$\phi = \tan^{-1}\left(\frac{D}{C}\right) = \text{Phase angle} \tag{6.103}$$

Equation (6.102) is rewritten as

$$2\pi\nu = \sqrt{\frac{k}{m}} \tag{6.104}$$

$$\therefore \nu = \frac{1}{2\pi}\sqrt{\frac{k}{m}} \tag{6.105}$$

Here the motion of only one particle of mass m has been considered. The frequency is for a one-particle problem. But a diatomic molecule is a two-particle problem. To obtain the vibrational frequency of a diatomic molecule consisting of two atoms of masses m_1 and m_2 joined together by a chemical bond having a force constant k, we apply the reduced mass concept to convert the two-particle problem to a one-particle problem. Then equations (6.51) and (6.52) can be applied to the molecule. Arguing in this manner, equation (6.105) is modified to

$$\nu = \frac{1}{2\pi}\sqrt{\frac{k}{\mu}} \tag{6.106}$$

where μ is the reduced mass of the two atoms in the diatomic molecule. Therefore equation (6.106) gives the vibrational frequency of the molecule.

(b) Squaring both sides of equation (6.106) we get

$$\nu^2 = \frac{1}{4\pi^2}\left(\frac{k}{\mu}\right) \tag{6.107}$$

or

$$k = 4\pi^2\nu^2\mu \tag{6.108}$$

The reduced mass for hydrogen bonded oxygen and hydrogen atoms (O…H) is

$$\mu_{\mathrm{O...H}} = \frac{m_{\mathrm{O}} m_{\mathrm{H}}}{m_{\mathrm{O}} + m_{\mathrm{H}}} \tag{6.109}$$

Mass of oxygen atom is

$$m_{\mathrm{O}} = 15.999\mathrm{amu} = 15.999 \times 1.661 \times 10^{-27}\ \mathrm{kg} \tag{6.110}$$

Mass of hydrogen atom is

$$m_{\mathrm{H}} = 1.007\,84\mathrm{amu} = 1.007\,84 \times 1.661 \times 10^{-27}\ \mathrm{kg} \tag{6.111}$$

$$\begin{aligned} \mu_{\mathrm{O...H}} &= \frac{15.999 \times 1.661 \times 10^{-27}\ \mathrm{kg} \times 1.007\,84 \times 1.661 \times 10^{-27}\ \mathrm{kg}}{15.999 \times 1.661 \times 10^{-27}\ \mathrm{kg} + 1.007\,84 \times 1.661 \times 10^{-27}\ \mathrm{kg}} \\ &= \frac{15.999 \times 1.007\,84}{15.999 + 1.007\,84} \times \frac{(1.661 \times 10^{-27}\ \mathrm{kg})^2}{1.661 \times 10^{-27}\ \mathrm{kg}} \\ &= \frac{16.1244 \times 1.661 \times 10^{-27}\ \mathrm{kg}}{17.0068} = 1.5748 \times 10^{-27}\ \mathrm{kg} \end{aligned} \tag{6.112}$$

Since

$$\nu = 2.5\ \mathrm{THz} = 2.5 \times 10^{12}\ \mathrm{Hz} \tag{6.113}$$

Equation (6.108) gives

$$\begin{aligned} k_{\mathrm{O.....H}} &= 4 \times (3.14159)^2 \times (2.5 \times 10^{12})^2 \times 1.5748 \times 10^{-27} \\ &= 4 \times 9.8696 \times 6.25 \times 10^{24} \times 1.5748 \times 10^{-27} \\ &= 388.566 \times 10^{24-27} = 388.566 \times 10^{-3} = 0.3886\ \mathrm{Nm}^{-1} \end{aligned} \tag{6.114}$$

The given wave number of a free oxygen-to-hydrogen bond (OH) is converted to frequency as

$$\begin{aligned} 3600\ \mathrm{cm}^{-1} &= 3600 \times 2.9979 \times 10^{10}\ \mathrm{Hz} \\ &= 10\,792.44 \times 10^{10}\ \mathrm{Hz} = 1.0792 \times 10^{14}\ \mathrm{Hz} \end{aligned} \tag{6.115}$$

for which the force constant is

$$\begin{aligned} k_{\mathrm{OH}} &= 4 \times (3.14159)^2 \times (1.0792 \times 10^{14})^2 \times 1.5748 \times 10^{-27} \\ &= 4 \times 9.8696 \times 1.164\,67 \times 10^{28} \times 1.5748 \times 10^{-27} \\ &= 72.4082 \times 10^{28-27} = 724.082\ \mathrm{Nm}^{-1} \end{aligned} \tag{6.116}$$

When the hydrogen atom of a water molecule is not hydrogen bonded to the oxygen atom of a neighboring water molecule, the OH bond can be considered to be free. Here the oxygen and hydrogen atoms are linked together by the strong covalent bond, not by weak hydrogen bonding.

The ratio

$$\frac{k_{\mathrm{O.....H}}}{k_{\mathrm{OH}}} = \frac{0.3886}{724.082} = 5.367 \times 10^{-4} \tag{6.117}$$

from equations (6.114) and (6.116).

Note: Lutz *et al* (1982) observed that the unperturbed OH^- ions vibrated at a wave number of 3570 ± 10 cm^{-1} but the vibrational wave number decreased to 70 cm^{-1} for XH.... OH^- bonds. The ratio

$$\frac{\text{Force constant for wave number of XH....OH}^-\text{bonds}}{\text{Force constant for wave number of OH}^-\text{ions}} = \left(\frac{70}{3570}\right)^2 = (0.0196)^2 \tag{6.118}$$
$$= 3.84 \times 10^{-4}$$

since from equation (6.108), force constant is proportional to square of frequency, and frequency is proportonal to wave number, so force constant varies as square of wave number.

6.4 Comparison of THz time-domain with far infrared Fourier transform spectroscopic techniques

THz-TDS resembles far infrared Fourier transform spectroscopy in approach. A detailed comparison of terahertz spectroscopy with far infrared spectroscopy has been made by Han *et al* (2001). The signal-to-noise ratios of the two spectroscopic techniques differ with respect to frequency ranges. Between 10 GHz and 4 THz, the field amplitude SNR for terahertz spectroscopy is more than 10^4, whereas the SNR for far infrared Fourier transform spectroscopy is around 300. But the reverse happens at higher frequencies. At these frequencies, the SNR for far infrared spectroscopy has a higher value and also shows a broader spectrum. In addition, the time-gated coherent nature of terahertz spectroscopy makes it suitable for high-temperature processes. Thus at frequencies <3 THz, terahertz spectroscopy is superior, while at frequencies > 5 THz, far infrared spectroscopy gives better performance.

After comparing THz-TDS with far-infrared spectroscopy, let us also look at infrared spectroscopy in full IR frequency range. Differences between THz-TDS and FT-IR spectroscopies become evident when we examine the type of absorptions taking place in the two frequency ranges. FT-IR measurements are sensitive to intra-molecular vibrations in individual bonds, whereas THz-TDS recordings access the intermolecular vibrations, i.e., those occurring between different molecules. FT-IR absorptions are specific to functional groups while such absorptions are non-existent in the terahertz frequency range. THz-TDS is therefore more suitable for characterization of materials in solid state. Discrete transitions resulting from periodicity of crystalline lattices provide unique signatures in THz-TDS. This feature of THz-TDS imparts to it added special status for study of crystalline materials. It is extremely useful in distinguishing polymorphic materials. Polymorphs are solid materials having the same chemical composition, which are found in different crystalline forms. Additionally, the hydration of crystals and formation of salts can be tracked. However, THz-TDS is not limited in scope to crystalline materials. The stability of amorphous materials can also be predicted with THz-TDS.

6.5 Discussion and conclusions

The importance of THz-TDS arises from its sensitivity to resonances of rotational and vibrational modes of various molecules inaccessible to other techniques. Such availability of chemical signatures of several materials in the terahertz region results in exhibition of distinct absorption peaks. These attributes endow this spectroscopy with the ability to identify and quantify a vast body of materials broadening the scope of its applications enormously. By virtue of its burgeoning applications, it occupies a coveted position amidst the already established spectroscopic techniques. Particularly biomolecules like amino acids, mono- and disaccharides and large molecules like DNA, proteins and polysaccharides have evinced great interest.

A significant advantage of THz-TDS comes from the fact that the electric field of the pulse is measured whereby amplitude and phase information about the frequency components is obtained. From this information both the real and imaginary components of an optical constant can be determined without the need of measurements at frequencies outside usable bandwidth of the pulse avoiding the necessity for application of Kramers–Kronig relations which are bidirectional mathematical relations connecting the real and imaginary parts of an optical constant.

References

Brown J M and Carrington A 2003 *Rotational Spectroscopy of Diatomic Molecules* (Cambridge: Cambridge University Press) p 1046

Chalmers J M and Griffiths P R (ed) 2002 *Handbook of Vibrational Spectroscopy* (5 volume set) (Hoboken, NJ: Wiley) p 4000

Cho S 2018 *Fourier Transform and Its Applications using Microsoft EXCEL* (San Rafael, CA: Morgan & Claypool Publishers) pp 2-1–14

Fischer B, Hoffmann M, Helm H, Modjesch G and Uhd Jepsen P 2005 Chemical recognition in terahertz time-domain spectroscopy and imaging *Semicond. Sci. Technol.* **20** S246–53

Han P Y, Tani M, Usami M, Kono S, Kersting R and Zhang X-C 2001 A direct comparison between terahertz time-domain spectroscopy and far-infrared Fourier transform spectroscopy *J. Appl. Phys.* **89** 2357–9

Lutz H D, Eckers W and Haeuseler H 1982 OH stretching frequencies of solid hydroxides and of free OH^- ions *J. Mol. Struct.* **80** 221–4

Neu J and Schmuttenmaer C A 2018 An introduction to terahertz time domain spectroscopy (THz-TDS) *J. Appl. Phys.* **124** 231101-1–231101-14

Pavia D L, Lampman G M, Kriz G S and Vyvyan J R 2015 *Introduction to Spectroscopy* (Stamford, CA: Cengage Learning) p 784

Perkampus H H 1992 *UV–vis Spectroscopy and Its Applications* Springer Lab Manuals (Berlin: Springer) p 244

Rao K R, Kim D N and Hwang J J 2010 *Fast Fourier Transform: Algorithms and Applications* (Dordrecht: Springer Science+Business Media B.V) p 426

Smith J O 2007 *Mathematics of the Discrete Fourier Transform (DFT) with Audio Applications* (W3K Publishing) p 322

Smith B C 1996 *Fundamentals of Fourier Transform Infrared Spectroscopy* (Boca Raton, FL: CRC Press) pp 15–54

IOP Publishing

Practical Terahertz Electronics: Devices and Applications, Volume 2

Optical devices and applications

Vinod Kumar Khanna

Chapter 7

Imaging

Imaging modalities are classified. These include: transmission and reflective imaging, pulsed and continuous wave imaging, and close-range and stand-off imaging. Comparisons of THz imaging with millimeter wave imaging, infrared imaging and x-ray imaging are presented. Among closed-range imaging techniques, pulse imaging by raster scanning, and inverse synthetic aperture method, and continuous wave imaging by electro-optic detection are discussed. Two stand-off imaging techniques, viz., focal plane array and interferometric imaging are explained. Applications of imaging in security surveillance for explosives and illicit drugs, characterization of pharmaceutical products, clinical diagnosis, cancer detection, inspection of food and agri-products, and in non-destructive evaluation are described.

The capability of benign terahertz waves to acquire images of objects is a real 'feather in the cap' of terahertz electronics. In this chapter, we shall explore the capabilities of terahertz imaging leading to new opportunities.

7.1 Types of imaging

Imaging is the process of creation of visual representation of an object using electromagnetic waves in the terahertz frequency range. The imaging also includes the use of terahertz waves for detection and identification of chemical composition of the substances such as concealed explosives, chemical and biological warfare agents, illegitimate drugs and related items from their characteristic transmission/reflection spectra in this frequency range. So, the imaging has a dual role. It discloses the physical appearance or shape of the hidden object. Additionally, it aids in chemical recognition of the object. No sooner than any destructive or hazardous content is found, necessary action for obliterating the threat is initiated.

Imaging can be classified into transmission and reflection modes depending on whether terahertz radiation passes through or bounces back from the object; pulsed and continuous wave depending on the type of signals, whether discrete or continuous, used

doi:10.1088/978-0-7503-4886-7ch7

in acquiring images; and closed range and standoff categories depending on the distance of the object from the camera. These classifications are explained in tables 7.1–7.3.

Table 7.4 presents a comparison of terahertz imaging with its lower frequency neighbor, millimeter wave imaging.

Table 7.1. Transmission and reflective imaging geometries.

Sl. No.	Transmission imaging	Reflective imaging
1.	Terahertz waves fall on the object under investigation. They penetrate and pass through the object. Emerging out of the object on the opposite side to the terahertz source, they are received and recorded by a detector	Terahertz waves are reflected back from the object under study and received by the detector placed on the same side as the terahertz source
2.	Can be used for thin samples but opacity of thicker specimens may render it unfeasible	Applicable to both thin and thick samples
3.	Small amounts of explosives and illicit materials can be detected because of their transparency to terahertz waves	Larger quantities of explosives and illicit materials are detectable because of absence of the opacity problem

Table 7.2. Pulsed and continuous wave imaging.

Sl. No.	Pulsed imaging	Continuous wave imaging
1.	Uses pulsed laser sources and detectors, e.g., a photoconductive antenna or an optically rectifying non-linear crystal	Uses difference frequency generators and detectors. An example of continuous wave source is the photomixing of radiations from two infrared sources
2.	Suitable for close vicinity range because depth information can be obtained from the difference in propagation time of pulses.	Advantage of higher terahertz power at a specified terahertz frequency
3.	A single pulse of picosecond duration provides wide spectral information (0.1–3 THz) in THz-TDS system	Narrow spectral features are discernible due to spectrally slender continuous wave terahertz beam and the non-requirement of long scanning delay line in THz-CW spectroscopy that will be necessary in a THz-TDS system
4.	Frequency content is spatially non-homogeneous	Frequency content is always spatially uniform
5.	Requires more computational processing. N^2 Fourier transforms are computed when an $N{\times}N$ pixel camera is used.	Uses less intensive computational processing. Only N^2 local maxima or minima need to determined.

Table 7.3. Close-range and stand-off imaging.

Sl. No.	Close-range imaging	Stand-off imaging
1.	Distance less than 3 m	Distance greater than 3 m
2.	Suitable for screening parcels, postal mail, packages and luggage baggages at post offices, courier services, airports, etc, to check for any forbidden content.	Appropriate for long-range screening of people approaching a checkpoint or portal such as airport terminals, railway stations, bus terminuses, shopping malls, etc, to check whether they are carrying any perilous goods.

Table 7.4. THz and millimeter wave imaging technologies.

Sl. No.	THz (0.1–10 THz) imaging	Millimeter wave (0.03–0.3 THz) imaging
1.	Emerging technology	Well-developed technology
2.	10 times superior spatial resolution because the wavelength is 10 times shorter than millimeter waves	10 times lower spatial resolution than terahertz waves
3.	Spectral finger prints of many explosives, chemical and biological agents, and illicit drugs fall in this frequency range	Spectral fingerprints not available in this frequency range

Table 7.5. THz and x-ray imaging technologies.

Sl. No.	THz (0.1–10 THz) imaging	X-rays (10^{16}–10^{19} Hz) imaging
1.	Uses low energy photons	Uses high energy photons
2.	Non ionizing radiation	Ionizing radiation
3.	Biologically benign and non-invasive	Biologically unsafe
4.	Being harmless, can be routinely used for medical imaging and security screening of people	Used sparingly and only when unavoidable

The upper frequency neighbor of terahertz imaging, viz., infrared imaging (0.3–430 THz), known as infrared thermography, provides a representation of heat signature and thermal mapping of an object. It offers a remote temperature sensing technique. Warm-blooded animals such as human beings can be easily seen against the background environment in thermal images, hence very useful for military night vision devices for tracking movements of enemy soldiers. We have already compared terahertz spectroscopy with infrared spectroscopy in section 6.4.

Terahertz imaging is compared with x-ray imaging in table 7.5.

7.2 Close-range imaging techniques

7.2.1 Transmission and reflective pulse imaging by mechanical raster scanning

The object is mounted on a motorized stage which can provide motion in the x- and y-directions (Hu and Nuss 1995). The terahertz radiation is focused to a diffraction-limited spot on the object and the transmitted wave is analyzed. By moving the object, every point of the object is scanned and the resulting transmitted wave is obtained for every point. These points are the pixels of the image.

The experimental assembly contains a femtosecond optical source in the form of a Ti-sapphire laser which supplies 1.0×10^{-13} s duration pulses to excite a photoconductive THz emitter (figure 7.1). The terahertz radiation from the optically gated emitter falls on a pair of paraboloid reflectors, which collimates and then focuses it on the object. A second pair of paraboloids collects and focuses the emerging terahertz radiation from the object on the optically gated terahertz detector, which is a photoconducting dipole antenna. To efficiently couple the terahertz radiation to air, hemispherical silicon lenses are fixed to the backsides of the substrates of the terahertz emitter and detector.

The frequencies of the terahertz waveforms received at the detectors are lowered to kilohertz range using a scanning delay line. The frequency down-converted signal for each pixel is amplified in a current pre-amplifier, digitized by an analog-to-digital converter and fed to a digital signal processor. The spectral information is extracted for each pixel and the image is presented at 12 pixels/second by the display unit. The system is applied to inspection of packages and mapping of chemical composition of biological objects (Hu and Nuss 1995).

Terahertz imaging can be done in the reflective mode by using the setup shown in figure 7.2. The beam of terahertz radiation strikes the object at nearly normal angle of incidence and is focused on the same (Mittleman *et al*1996). The method bears analogy to optical coherence tomography in which light from a source of short coherence length is decomposed into two beams. One of these beams is reflected from the object under test. The second beam is reflected from a reference. Correlation of the two beams provides assessment of the reflectivity of the object.

7.2.2 Reflective pulse imaging by inverse synthetic aperture method

The approach may be understood by referring to radar principles: synthetic-aperture radar is a type of radar using motion of the radar antenna to generate the 2-D image of a target. Inverse synthetic-aperture radar is a method of producing a 2-D image of a target by moving the target.

Here imaging of rotating reflective, opaque target is performed by measurements of electric fields of THz pulses scattered by the target (figure 7.3). Femtosecond duration (8.0×10^{-14} s) pulses from a laser illuminate an optoelectronic THz emitter (McKlatchy *et al* 2001). The THz pulses from the emitter are collected by a silicon lens and collimated by an off-axis paraboloidal mirror. These pulses fall on the target, and undergo scattering. The scattered pulses are received

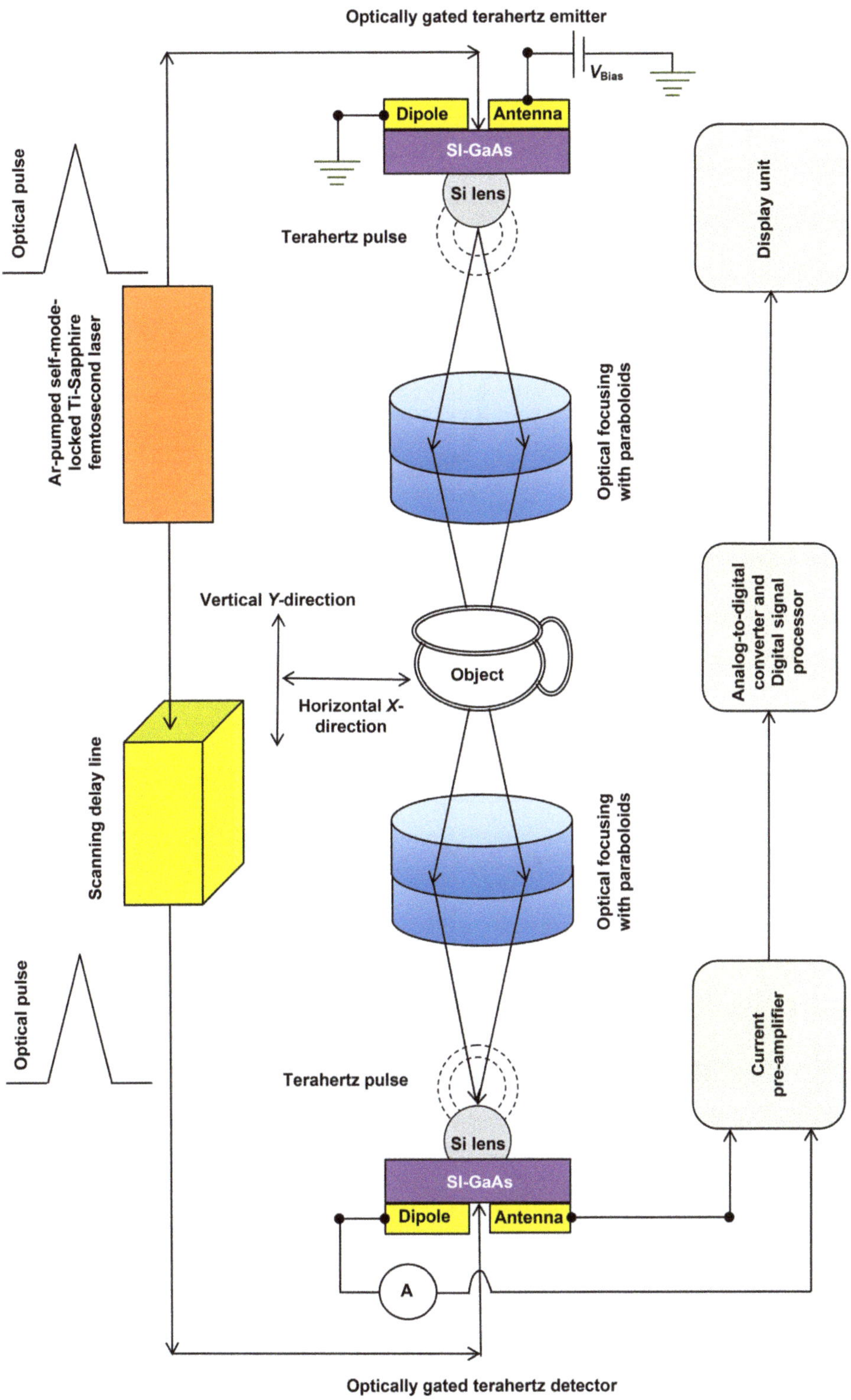

Figure 7.1. Transmission pulse imaging by mechanical raster scanning.

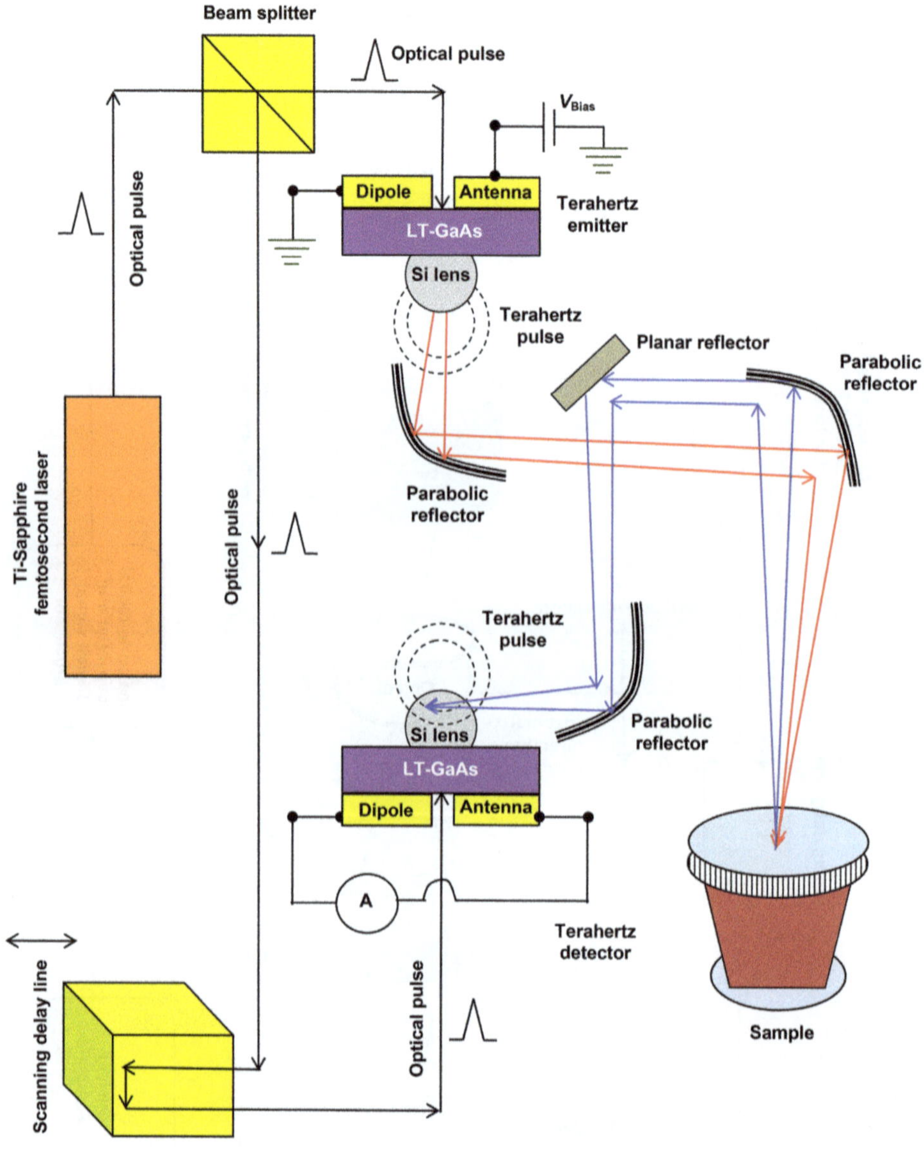

Figure 7.2. Reflective mode terahertz imaging.

by the detector which is coupled to the optical gating pulse through a single-mode fiber. The average current collected by the detector is measured by a current amplifier along with a lock-in amplifier. This current is proportional to the amplitude of the electric field.

The scattered field is measured at different angles θ from 0° to 345° by rotating the target yielding a set of scanning data as a function of angle θ. A 2-D data array is compiled from electric field values as a function of time and angle. The time resolved data are Fourier transformed into the frequency domain. For a particular x position x_0,

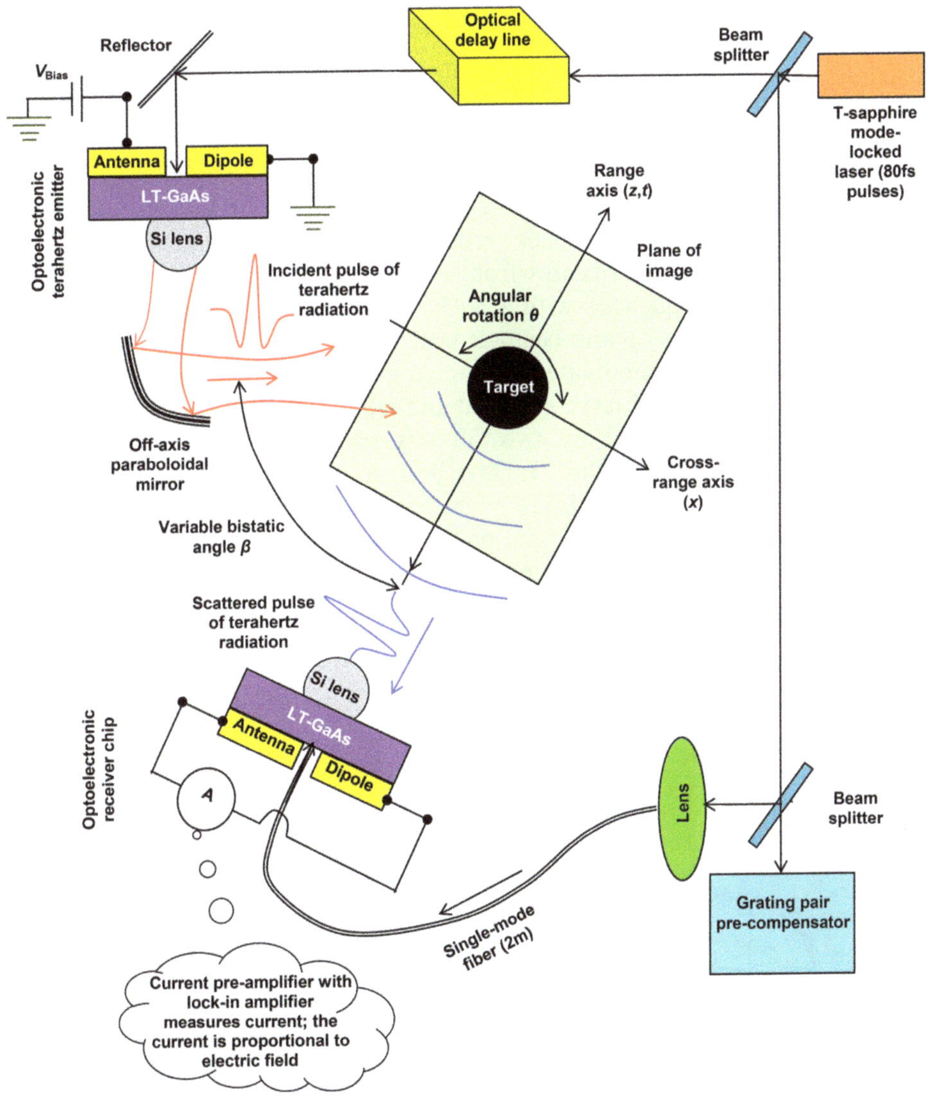

Figure 7.3. Setup for terahertz impulse ranging.

the complex amplitude at every value of angle of rotation θ is multiplied by the phase shift. When the scatterer is present, all the angle-dependent scans are in phase and sum constructively while in the absence of scatterer, they sum destructively. At each value of x_0, all the scans are linearly superimposed. Then by inverse transforming to the time domain, the scattering amplitude is plotted as a function of x_0 and time t.

The image is constructed by treating the target as a superimposition of point scatterers. Using this method, remote imaging of simple and complex opaque objects was demonstrated (McKlatchy *et al* 2001).

7.2.3 2-D imaging of continuous wave terahertz radiation by electro-optic detection

The method entails raising the frequency of terahertz radiation to optical frequency, and this coherent frequency upconversion is followed by optical homodyne (single frequency) detection using an electro-optic crystal (Nahata *et al* 2002).

Continuous-wave terahertz radiation is generated by photomixing in a photoconductive emitter fabricated on low-temperature grown GaAs and coated with antireflection film to prevent back reflection. Difference-frequency mixing is performed by combining the beams from two single-frequency laser diodes tuned to slightly different frequencies with a 50/50 beamsplitter, and driving the photoconductive emitter by this pump beam at beat frequency (figure 7.4).

The terahertz beam is modulated by applying a square wave voltage (amplitude −30 V to +30 V) to the emitter, collimated and refocused on an anti-reflection film-coated <110> ZnTe crystal by two off-axis paraboloidal mirrors. A pellicle beamsplitter is interposed in the terahertz beam path to simultaneously direct an optical probe beam on the crystal. The probe beam is optically biased close to its null point with a compensator in a crossed-polarizer arrangement.

Electro-optic sampling is done using the same two lasers for detection of terahertz radiation.

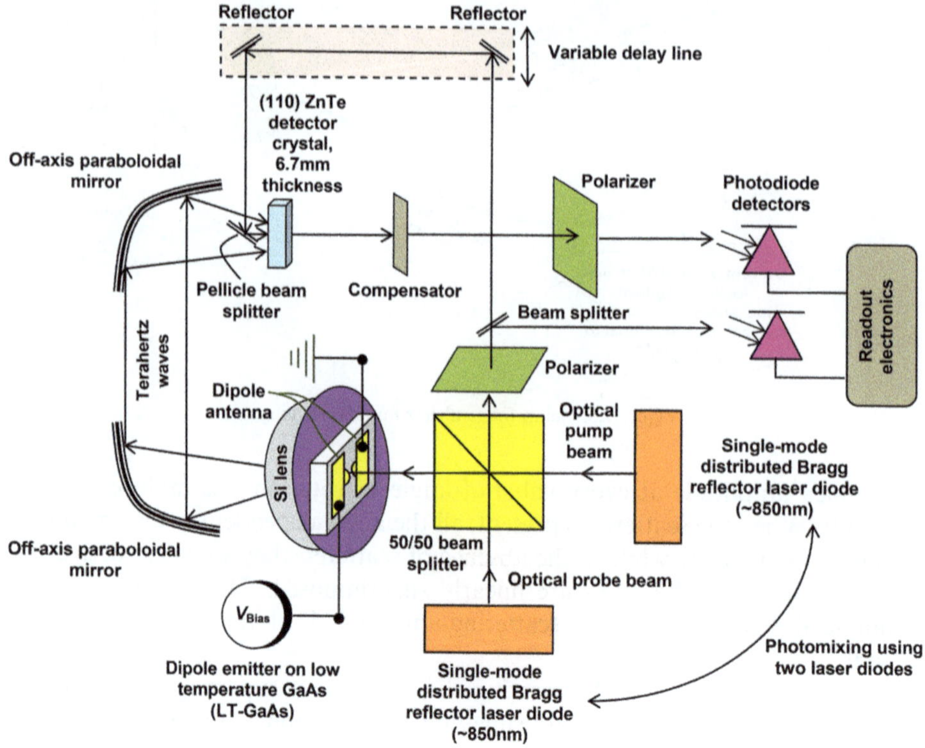

Figure 7.4. Production and detection of continuous wave terahertz radiation.

The intensity of the optical beam $I_{\mathrm{Opt}}\,(t)$ formed by combining the beams from the two lasers is

$$I_{\mathrm{Opt}} = I^0_{\mathrm{Opt}}\{A + \cos(\Omega t)\} \tag{7.1}$$

where I^0_{Opt} is the amplitude of the optical beam, A is a constant and Ω is the beat frequency. The continuous-wave terahertz field $E_{\mathrm{THz}}(t)$ is given by

$$E_{\mathrm{THz}}(t) = E_{\mathrm{THz0}}\cos\left(\Omega t - \phi_1\right) \tag{7.2}$$

where E_{THz0} is the peak value of terahertz field, and ϕ_1 is a phase term linked to the optical path length of the pump arm (d_1) as

$$\phi_1 = \frac{\Omega d_1}{c} \tag{7.3}$$

and c is the velocity of light.

The probe beam is

$$I_{\mathrm{Probe}} = I^0_{\mathrm{Opt}}\left\{A + \cos\left(\Omega t - \phi_2\right)\right\} \tag{7.4}$$

where ϕ_2 is a phase term linked to the optical path length of the probe arm (d_2) as

$$\phi_2 = \frac{\Omega d_2}{c} \tag{7.5}$$

The effective time-varying electro-optic signal $\Delta I_{\mathrm{EO}}(t)$ is proportional to the product of intensity of optical probe beam $I_{\mathrm{Probe}}\,(t)$ and the electric field $E_{\mathrm{THz}}(t)$ of terahertz radiation:

$$\Delta I_{\mathrm{EO}}(t) \propto I_{\mathrm{Probe}}(t)E_{\mathrm{THz}}(t) \tag{7.6}$$

Combining equations (7.2), (7.4) and (7.6) (Nahata *et al*1999)

$$\begin{aligned}
\Delta I_{\mathrm{EO}}(t) &\propto I^0_{\mathrm{Opt}}E_{\mathrm{THz0}}\{A + \cos(\Omega t - \phi_2)\}\cos(\Omega t - \phi_1)\\
&= I^0_{\mathrm{Opt}}E_{\mathrm{THz0}}\{A\cos(\Omega t - \phi_1) + \cos(\Omega t - \phi_2)\cos(\Omega t - \phi_1)\}\\
&= I^0_{\mathrm{Opt}}E_{\mathrm{THz0}}\left\{A\cos(\Omega t - \phi_1) + \frac{\cos(\Omega t - \phi_2 - \Omega t + \phi_1) + \cos(\Omega t - \phi_2 + \Omega t - \phi_1)}{2}\right\}\\
&= AI^0_{\mathrm{Opt}}E_{\mathrm{THz0}}\cos(\Omega t - \phi_1) + I^0_{\mathrm{Opt}}E_{\mathrm{THz0}}\frac{\cos(-\phi_2 + \phi_1)}{2} + I^0_{\mathrm{Opt}}E_{\mathrm{THz0}}\frac{\cos(2\Omega t - \phi_2 - \phi_1)}{2}
\end{aligned} \tag{7.7}$$

since

$$\cos\alpha\cos\beta = \frac{\cos(\alpha - \beta) + \cos(\alpha + \beta)}{2} \tag{7.8}$$

Measurement of the DC component of the time-averaged signal $\langle\Delta I_{\mathrm{EO}}(t)\rangle$ is performed. In equation (7.7), the cosine terms containing the time variable average out to zero. This happens due to the reason that the average value of cosine of an angle over a complete cycle is zero. The average value becomes zero because the positive and negative parts of the cosine wave cancel each other out. So only one term is left in the averaged signal $<\Delta I_{\mathrm{EO}}(t)>$ as:

$$\begin{aligned} < \Delta I_{\mathrm{EO}}(t) > &\propto I^0_{\mathrm{Opt}} E_{\mathrm{THz0}} \frac{\cos(-\phi_2 + \phi_1)}{2} \propto I^0_{\mathrm{Opt}} E_{\mathrm{THz0}} \cos(\phi_2 - \phi_1) \\ &= I^0_{\mathrm{Opt}} E_{\mathrm{THz0}} \cos\left\{\frac{\Omega}{c}(d_2 - d_1)\right\} = I^0_{\mathrm{Opt}} E_{\mathrm{THz0}} \cos\left(\frac{\Omega \Delta d}{c}\right) \end{aligned} \tag{7.9}$$

where the formula

$$\cos(-\theta) = \cos\theta \tag{7.10}$$

is used. Also, equations (7.3) and (7.5) are applied.

The measurement yields both the amplitude and phase of the terahertz field. The amplitude-modulated optical signal is imaged with a CCD camera. The method uses standard optical components.

7.3 Stand-off imaging techniques

7.3.1 Focal plane array (FPA) imaging

Focal plane of a lens/mirror is defined as a plane perpendicular to the axis of the lens/mirror and passing through its focal point. An FPA consists of an array of terahertz detectors. Antenna-coupled microbolometers arranged on the focal plane constitute such an array. In this array, each detector represents a pixel of the image. The object is illuminated with terahertz radiation. The emanating radiation is collected by an objective lens and focused by a hemispherical lens. A transmission-mode terahertz image is formed on the FPA. A complementary metal-oxide-semiconductor (CMOS) readout circuit digitizes the signal from each pixel. A digital image is thus constructed.

A 160 × 120 element microbolometer FPA was demonstrated using quantum cascade lasers (QCLs) (~3 THz) or 10 mW, 2.52 THz far infrared gas laser for illumination at video rate 60 frames/s (Lee and Hu 2005). The bolometer consists of a VO_x thin film on a silicon nitride air bridge (figure 7.5). The FPA is placed behind a germanium camera lens. The noise equivalent power (NEP) is rated at 0.9 pW $Hz^{-1/2}$, and the peak signal-to-noise ratio is 13 dB for a 16 ms acquisition time.

A broadband real-time FPA sensor of 6.4 mm × 8 mm area and 80 × 64 pixel was designed for operation in 0.6–1.2 THz band with 100 frames per second acquisition speed (Trichopoulos *et al* 2013). Broadband planar slot antennas were integrated with heterostructure backward diodes. The FPA was mounted at the center of a silicon hemispherical lens. The antennas captured and directed the terahertz waves on the diodes for detection. When tested with BWO source, the array was found to operate well over 0.6–0.9 THz. The optical sensitivity was 600 V W^{-1} at 0.7 THz and optical NEP was 850 pW $Hz^{-1/2}$.

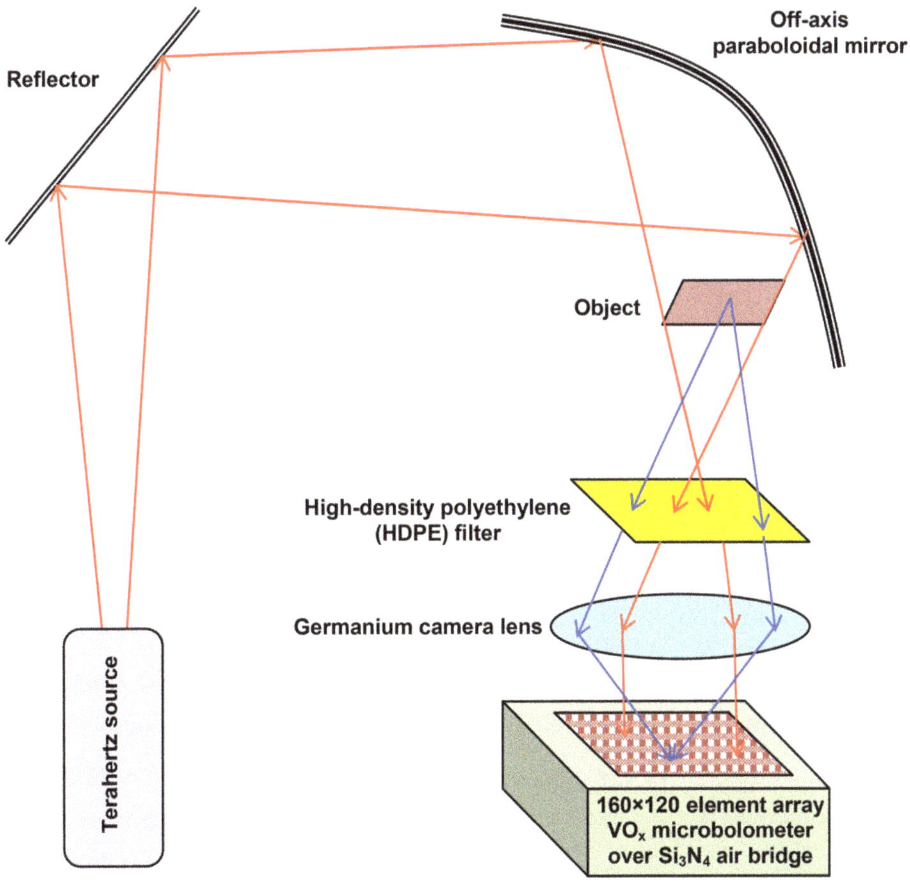

Figure 7.5. Microbolometer focal plane array for continuous wave terahertz imaging.

7.3.2 Interferometric imaging

Interferometry is an investigative technique in which two or more electromagnetic waves are superimposed to create an intensity pattern which is analyzed to extract information about the original state of the waves. In astronomical interferometry, signals from two or more telescopes are combined to produce high-resolution images of celestial objects.

The imaging interferometer works like an astronomical interferometer. It consists of an array of separate pairs of detectors (Federici *et al* 2003). Each detector of the array measures the amplitude and phase of the terahertz radiation received by it. Using the array, the electric fields are determined at several locations of the object. Then the correlated amplitude and phase values of the electric field from various pairs of detectors are applied to produce the sine and cosine components of the brightness distribution. Data from a large number of points are synthesized through Fourier inversion to build the image of the object (figure 7.6).

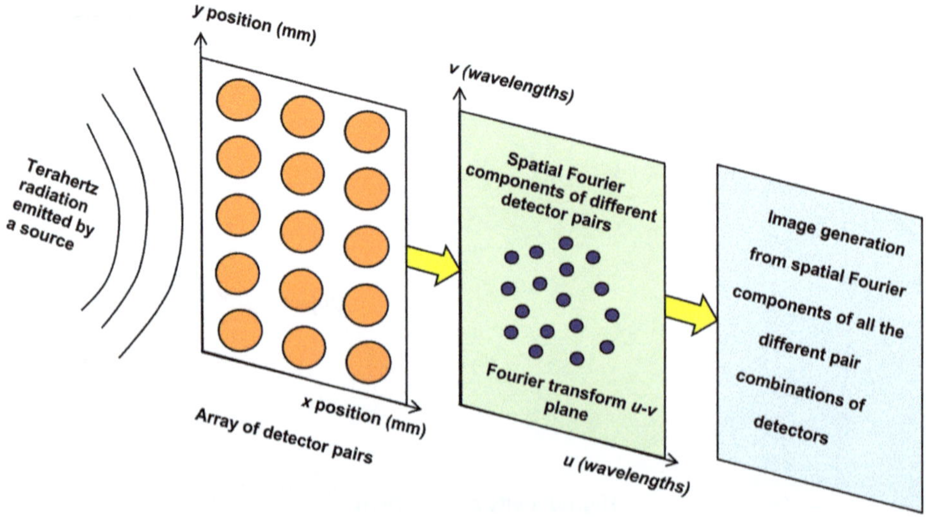

Figure 7.6. Diagrammatic representation of steps in terahertz imaging interferometry.

When the wavefront of the incident terahertz radiation from the object strikes a pair of detectors, one spatial Fourier component of the radiation is measured (Bandyopadhyay *et al* 2006). This spatial Fourier component corresponds to the particular value of distance between the detector pair. The physical separation between detectors in the array is referred to as the baseline. In this way, spatial Fourier components are obtained for various pairs of detectors. These spatial Fourier components are represented as different points in the Fourier transform plane called the *u*-*v* plane. Measurements are performed for $n(n-1)/2$ baseline combinations of *n* detectors. Then the spatial Fourier components of these combinations are used for image generation.

What determines the quality of the image? The *u*-*v* plane is filled with points pertaining to the spatial Fourier components and the methodology adopted in filling decides the image quality. Efficient filling of the *u*-*v* plane depends on the arrangement of detectors during organization of the array. Therefore, the detectors must be spread uniformly over the *u*-*v* plane in accordance with the desired angular resolution. When the number of detectors is small, the *u*-*v* plane can be effectively covered by rotating the array about a fixed axis. If the measurements are done *m* times by rotating an *n*-detector array, the number of base lines is multiplied by *m* so that number of base lines increases to $mn(n-1)/2$, thereby increasing the image quality *m*'fold of that with a small number *n* of detecting elements. However, physical rotation of the array may not be always easy and convenient.

The angular resolution θ of image obtained with an interferometric imaging array in which the separation between the detectors is *d* is expressed as

$$\theta = \frac{\lambda}{d}\text{radians} \tag{7.11}$$

where λ is the wavelength of the terahertz radiation.

The significant advantage gained by using interferometric imaging in place of the conventional methods needs to be elaborated by reference to the analogy with optical or radio telescopes. Angular resolution of a telescope is the parameter whose value tells us the extent to which small details of the viewed object are distinguishable. Taking into account diffraction through a circular aperture, it is given by

$$\theta_{\text{Telescope}} = 1.21966989\frac{\lambda}{D} \approx \frac{\lambda}{D} \tag{7.12}$$

where the symbol D stands for the diameter of the aperture of the lens. How does the factor 1.22 arise? Inclusion of this factor arises from the position of the first dark ring surrounding the Airy disk of the diffraction pattern. Due to this reason, the aperture diameter must be increased to enable discrimination of smaller distances. The practical difficulties faced in building telescopes with very large apertures are easily visualized. Interferometry provides a way out of this difficulty. Consider two telescopes 1 and 2 of apertures D_1 and D_2 located at a distance d apart (figure 7.7). Here the distance d is the baseline of the system. From equation (7.12), the resolutions of the individual telescopes are

$$\theta_{\text{Telescope1}} = \frac{\lambda}{D_1} \tag{7.13}$$

and

$$\theta_{\text{Telescope2}} = \frac{\lambda}{D_2} \tag{7.14}$$

But when the two telescopes are operated in the interferometric configuration, they work as a single giant telescope providing the resolution

$$\theta_{\text{Telescopes1, 2}} = \frac{\lambda}{d} \tag{7.15}$$

By stationing the two participating telescopes at a large distance of separation d, we can lower the $\theta_{\text{Telescopes1,2}}$ by a very large amount without building telescopes with huge apertures. So, this is the way in which interferometry helps us. High angular resolution telescopes can be erected by arranging several telescopes in a two-dimensional array; here, the baseline is the maximum distance between the telescopes.

Example 7.1

(a) Determine the angular resolution in arc seconds achieved with a terahertz interferometric imaging array made with maximum baseline distance between detectors = 2 m. The frequency of the radiation is 3 THz. (b) What is the spatial resolution when the distance of the object from the array is 100 m?

Solution: Applying equation (7.15), the angular resolution is

$$\theta = \frac{\lambda}{d} = \frac{\frac{c}{\nu}}{d} = \frac{c}{\nu} \times \frac{1}{d} = \frac{c}{\nu d} \tag{7.16}$$

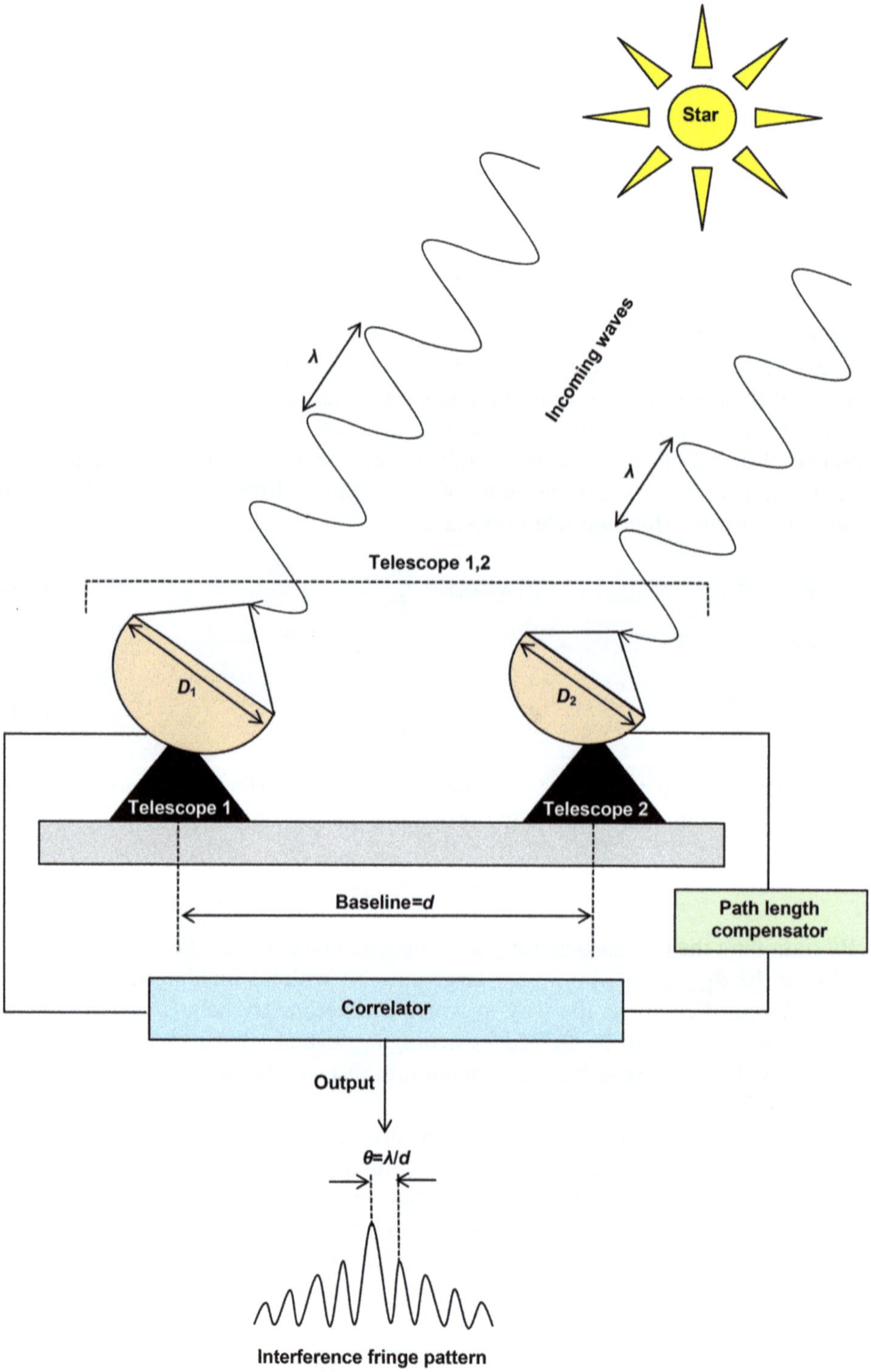

Figure 7.7. Building high angular resolution astronomical telescopes by interferometric imaging principle.

where c is the velocity of waves and ν their frequency. Putting the values in equation (7.16), we get

$$\theta = \frac{c}{\nu d} = \frac{2.9979 \times 10^8}{3 \times 10^{12} \times 2} = 0.49965 \times 10^{8-12} \\ = 0.49965 \times 10^{-4} \text{ radians} \tag{7.17}$$

Since

$$1 \text{ radian} = 57.2958° \tag{7.18}$$

$$\therefore 0.49965 \times 10^{-4} \text{ radians} = 0.49965 \times 10^{-4} \times 57.2958° = 0.00286278° \tag{7.19}$$

Because

$$1° = 3600 \text{ arc second} \tag{7.20}$$

$$\therefore 0.00286278° = 0.00286278° \times 3600 \text{ arc second} = 10.306 \text{ arc second} \tag{7.21}$$

(b)

$$\text{Angle in radians} = \frac{\text{Arc length}}{\text{Radius}} \tag{7.22}$$

or

$$\text{Arc Length} = \text{Angle in radians} \times \text{Radius} \tag{7.23}$$

So, angular resolution is transformed into spatial resolution by multiplying with the distance of the object from the array. Hence, from equations (7.17) and (7.23)

$$\text{Spatial resolution} = \text{Angle in radians} \times \text{Radius} = 0.49965 \times 10^{-4} \times 100 \text{ m} \\ = 0.0049965 \text{ m} = 0.0049965 \text{ m} \times 100 \text{ cm} \\ = 0.49965 \text{ cm} \approx 0.5 \text{ cm} = 5 \text{ mm} \tag{7.24}$$

Example 7.2

What is the baseline required to build a 3.5 THz interferometric imaging setup providing an angular resolution of 4 arc second?

Solution: We know that

$$1 \text{ arc second} = 1/3600° \tag{7.25}$$

Also,

$$180° = \pi \text{ radians} \tag{7.26}$$

or

$$1° = \frac{\pi}{180} \text{ radians} \tag{7.27}$$

From equations (7.25) and (7.27)

$$1 \text{ arc second} = \frac{1}{3600} \times \frac{\pi}{180} \text{ radians} = \frac{\pi}{3600 \times 180} \text{ radians} \tag{7.28}$$

Here,

$$\theta = 4 \text{ arc second} = 4 \times \frac{\pi}{3600 \times 180} \text{radians} = 4 \times \frac{3.14159}{3600 \times 180} \text{ radians} \\ = 4 \times 4.8481 \times 10^{-6} \text{ radians} = 19.39 \times 10^{-6} \text{ radians} \tag{7.29}$$

where equation (7.28) is applied.

From equations (7.16) and (7.29)

$$\theta = \frac{c}{\nu d} \tag{7.30}$$

or

$$d = \frac{c}{\nu\theta} = \frac{2.9979 \times 10^{8}}{3.5 \times 10^{12} \times 19.39 \times 10^{-6}} \\ = \frac{2.9979 \times 10^{8+6-12}}{3.5 \times 19.39} = 0.0442 \times 10^{2} = 4.42 \text{ m} \tag{7.31}$$

Therefore the detectors must be arranged in a 4.42 m × 4.42 m array.

7.4 Security surveillance

Imaging and spectroscopy are intimately associated. They work together hand-in-hand to provide an effective security guard.

7.4.1 Explosives

Vapor pressures of many explosives are extremely low making their detection very difficult by vapor phase sampling, e.g., vapor pressure of RDX (Royal demolition explosive: cyclotrimethylenetrinitramine) at 25 °C is 2.31×10^{-6} Pa, HMX (High melting explosive: cyclotetramethylenetetranitramine) is 8.14×10^{-12} Pa, and PETN (pentaerythritol tetranitrate) is 6.28×10^{-8} Pa (Brady *et al* 2012). Vapor pressure of TNT (trinitrotoluene) at 25 °C is 4.32×10^{-4} Pa (Oxley *et al* 2005, Lee and Weeks 2019) and 1.23×10^{-3} Pa (Lenchitz and Velicky 1970).

THz spectroscopy is a viable tool for detecting low-volatility explosives. Sharp absorption peaks are displayed by many explosives in the terahertz frequency region enabling recognition of their presence from spectral library data. A few examples are: RDX (0.72, 1.26, 1.73 THz); TNT (1.44, 1.91 THz); HMX (1.58, 1.91, 2.21, 2.57 THz); and PETN (1.73, 2.51 THz) (Tribe *et al* 2004). In each case the frequencies of absorption peaks are given in parenthesis. Owing to the smooth, featureless absorption of terahertz radiation by several plastics, clothing, and other materials used for packaging, and moderate weakening of radiation by these packaging materials, these explosives can be easily identified when covered under them.

Example 7.3

The Clausius–Clapeyron equation allows estimation of vapor pressure of a material at a given temperature. The dependence of vapor pressure P of RDX and TNT nanofilms on temperature is expressed as (Hikal *et al* 2011)

$$\ln P(\mathrm{Pa})|_{\mathrm{RDX}} = 39.6 - \frac{15459}{T(K)} \tag{7.32}$$

$$\ln P(\mathrm{Pa})|_{\mathrm{TNT}} = 34.9 - \frac{12058}{T(K)} \tag{7.33}$$

Find the vapor pressures of RDX and TNT at room temperature (25 °C).
Solution: For RDX film, from equation (7.32)

$$\ln P(\mathrm{Pa})|_{\mathrm{RDX}} = 39.6 - \frac{15459}{T(K)} \tag{7.34}$$

or

$$2.303\,\log_{10} P(\mathrm{Pa})|_{\mathrm{RDX}} = 39.6 - \frac{15459}{T(K)} \tag{7.35}$$

or

$$\begin{aligned}\log_{10} P(\mathrm{Pa})|_{\mathrm{RDX}} &= \frac{39.6}{2.303} - \frac{15459}{2.303T} = \frac{39.6}{2.303} - \frac{15459}{2.303 \times 298.15}\\ &= 17.195 - 22.514 = -5.319\end{aligned} \tag{7.36}$$

or

$$P(\mathrm{Pa})|_{\mathrm{RDX}} = 10^{-5.319} = 4.797 \times 10^{-6} \tag{7.37}$$

For TNT, from equation (7.33)

$$\ln P(\mathrm{Pa})|_{\mathrm{TNT}} = 34.9 - \frac{12058}{T(K)} \tag{7.38}$$

or

$$2.303\,\log_{10} P(\mathrm{Pa})|_{\mathrm{TNT}} = 34.9 - \frac{12058}{T(K)} \tag{7.39}$$

or

$$\begin{aligned}\log_{10} P(\mathrm{Pa})|_{\mathrm{TNT}} &= \frac{34.9}{2.303} - \frac{12058}{2.303T} = \frac{34.9}{2.303} - \frac{12058}{2.303 \times 298.15}\\ &= 15.154 - 17.561 = -2.407\end{aligned} \tag{7.40}$$

or

$$P(\mathrm{Pa})|_{\mathrm{TNT}} = 10^{-2.407} = 3.917 \times 10^{-3} \tag{7.41}$$

Example 7.4

Apply the formula for vapor pressure of 2,4,6-trinitrotoluene (TNT) given by Pella (1977) for the temperature range 287.15 to 329.65 K

$$\log_{10}P(\text{torr})|_{\text{TNT}} = 12.31 \pm 0.34 - \frac{5175 \pm 105}{T} \tag{7.42}$$

in the form

$$\log_{10}P(\text{torr})|_{\text{TNT}} = 12.31 - \frac{5175}{T} \tag{7.43}$$

to calculate the vapor pressure in torr and Pascal at (a) 15 °C, (b) 20 °C and (c) 25 °C.

Solution: (a) At 15 °C = 15 + 273.15 K = 288.15 K. From equation (7.43)

$$\begin{aligned}\log_{10}P(\text{torr})|_{\text{TNT}} &= 12.31 - \frac{5175}{T} = 12.31 - \frac{5175}{288.15} \\ &= 12.31 - 17.959 = -5.649\end{aligned} \tag{7.44}$$

$$\therefore P(\text{torr})|_{\text{TNT}} = 10^{-5.649} = 2.24 \times 10^{-6} \tag{7.45}$$

and

$$\therefore P(\text{Pa})|_{\text{TNT}} = 2.24 \times 10^{-6} \times 133.322 = 2.986 \times 10^{-4} \tag{7.46}$$

(b) At 20 °C = 20 + 273.15 K = 293.15 K. From equation (7.43)

$$\begin{aligned}\log_{10}P(\text{torr})|_{\text{TNT}} &= 12.31 - \frac{5175}{T} = 12.31 - \frac{5175}{293.15} \\ &= 12.31 - 17.653 = -5.343\end{aligned} \tag{7.47}$$

$$\therefore P(\text{torr})|_{\text{TNT}} = 10^{-5.343} = 4.539 \times 10^{-6} \tag{7.48}$$

and

$$\therefore P(\text{Pa})|_{\text{TNT}} = 4.539 \times 10^{-6} \times 133.322 = 6.05 \times 10^{-4} \tag{7.49}$$

(c) At 25 °C = 25 + 273.15 K = 298.15 K. From equation (7.43)

$$\begin{aligned}\log_{10}P(\text{torr})|_{\text{TNT}} &= 12.31 - \frac{5175}{T} = 12.31 - \frac{5175}{298.15} \\ &= 12.31 - 17.357 = -5.047\end{aligned} \tag{7.50}$$

$$\therefore P(\text{torr})|_{\text{TNT}} = 10^{-5.047} = 8.974 \times 10^{-6} \tag{7.51}$$

and

$$\therefore P(\text{Pa})|_{\text{TNT}} = 8.974 \times 10^{-6} \times 133.322 = 1.196 \times 10^{-3} \tag{7.52}$$

7.4.2 Illicit drugs

Drugs such as MDMA (3,4-methylenedioxymethamphetamine), methamphetamine and aspirin can be detected at a concentration ~20 mg cm^{-2} in polyethylene bags in an airmail envelope. Absorption spectra of MDMA shows peaks at 1.4, 1.8 THz. For methamphetamine, the peaks are at 1.2, 1.7–1.8 THz. For aspirin, the peak position is at 1.4 THz (Kawase *et al* 2003).

Absorbance spectrum of naproxen sodium shows peaks at 5.2 THz and 6.5 THz. The peak of terfenadine occurs at 3.2 THz and that of acetaminophen is located at 6.5 THz. So these drugs can be identified from THz spectral database (Campbell and Heilweil 2003).

7.5 Pharmaceutical products characterization

7.5.1 Polymorphism in crystalline drug materials

Terahertz time-domain spectroscopy (THz-TDS) is a valuable tool for characterizing polymorphic systems (Markl *et al* 2016). Polymorphic forms of a drug formulation differ in their thermodynamic properties affecting vital parameters such as solubility, hygroscopic nature, melting point, etc. A common failure arises from interconversion between polymorphic forms so that the intended formula is not delivered to the customer. As a result, the efficacy of the drug received by the customer is lower than the prescribed standard. Study of polymorphism in drug content greatly helps in prediction, quantification and confirmation of such transitions and ensuring that product quality is maintained, i.e., the correct polymorphic form of drug is supplied.

7.5.2 Modification of crystalline drug materials by hydration or salt formation

Materials may be modified by hydration. Formation of salt species is another route to material modification. As a consequence of these alterations, differences are observed in the THz-TDS spectra of these materials. Incorporation of new material in the crystalline lattice is responsible for these differences. Detection of occurrence of such phenomena by THz-TDS offers valuable service in quality assurance.

7.5.3 Spontaneous conversion of amorphous drug materials to more stable crystalline form

Materials in the disordered state tend to be less thermodynamically stable leading to higher solubility and therefore ready bioavailability. A drug in amorphous form may undergo transformation to the more stable crystalline state whereby its effectiveness is decreased. The time scales of stability of amorphous drugs must therefore be properly characterized and THz-TDS measurements serve this purpose.

7.5.4 Coating films of tablets

The coating film acts a diffusion barrier in tablets used for modified release. Strict uniformity of film thickness is essential in some applications. Delay time between terahertz pulses reflected from the surface of the tablet and the tablet core is used to determine thickness of the coating. Therefore terahertz imaging is useful for precisely monitoring variability of the thickness of coating films of tablets during manufacturing.

Terahertz imaging is also used for gaining insight into the disintegration and dissolution mechanisms of the tablets because back reflections occur due to creation of subtle differences of the refractive index at interfaces between layers of different materials, e.g., the refractive index is altered when a liquid ingresses into the tablet. So penetration of the liquid into the tablet can be easily tracked from such analysis. On these lines, investigations of the interaction between the tablet and the dissolution medium are applied for designing drug formulations (Markl *et al* 2016).

7.6 Clinical diagnosis

7.6.1 Dentistry

7.6.1.1 Enamel erosion

The tooth has an outer hard, calcified covering layer called the enamel composed of minerals that protect the inner fragile parts against physical and chemical damage. The enamel is reinforced inside by the hard bony tissue known as dentine and then comes the soft pulp. The enamel has an exceptional property of being the hardest tissue in the body. Its hardness is comparable to that of window glass. Due to refractive index differences amongst the tissues constituting the tooth, terahertz radiation is reflected from the dielectric layers in the tooth rendering 3-D imaging of tooth structures possible, thereby exposing subsurface features and mineral changes within the tooth. Terahertz pulse imaging is used for determination of thickness of tooth enamel, and showing its wear and tear (Crawley *et al* 2003, Supriya and Samuel 2020).

7.6.1.2 Dental caries and tooth dimineralization

Caries is detected with great specificity because caries-affected tissue absorbs terahertz radiation to a much greater extent than normal healthy tissue. When terahertz waves strike dental tissue, the intensity of absorption lines of water is higher if the tissue is affected by caries than the corresponding intensity for healthy tissue. Further, the intensity of terahertz radiation reflected by caries-affected enamel is 2.2 times lower while the intensity of reflected radiation from caries-damaged dentine is 4.5 times smaller than the radiation intensity reflected from healthy tissue (Smirnov *et al*2014).

Dental decay causes changes in THz characteristic properties of enamel and dentine (Zinovev *et al*2002). Dimineralization of tooth is indicated by an increased transmission of terahertz signal through tooth tissue. The progress of demineralization is mapped out.

7.6.2 Cancer

7.6.2.1 Basal cell carcinoma (BCC)

Pulsed terahertz images show good contrast between normal and tumor-affected regions of the skin (Wallace *et al*2006). Spectral responses of skin affected by BCC and normal excised human skin to terahertz radiation differ in two primary properties, viz., the absorption coefficient and refractive index. The BCC skin shows a higher absorption coefficient as well as a higher refractive index for terahertz waves than normal skin. Statistically, wide difference is observed between the two skins in the frequency range 0.2–2 THz for absorption coefficient. The difference is maximum at 0.5 THz. In the case of refractive index, the frequency range over which difference is noticed is 0.25–0.9 THz.

In the frequency domain, the implication of higher absorption is that higher frequencies are weakened, and in the time domain, it indicates that the reflected pulses are broadened. The meaning of higher refractive index is that the reflected signal is stronger from BCC skin than normal skin.

Example 7.5

(a) The linear attenuation coefficient of terahertz waves in water is 160 cm^{-1} at 1 THz and 240 cm^{-1} at 2 THz (Zhou *et al* 2019). Find the distances in water at which the intensity of terahertz radiation decreases to one quarter times its initial value. (b) Given that the density of water = 1 g mL^{-1}, what are the mass attenuation coefficients of water at the two frequencies?

Solution: (a) Linear attenuation constant μ, also called linear absorption coefficient, is a constant that describes the fraction of terahertz radiation which is absorbed per unit thickness of the material. Let I_0 denote the intensity of the incident terahertz radiation and I represent the intensity of terahertz radiation after traversing a distance x through the material. Then we can write

$$I = I_0 \exp(-\mu x) \tag{7.53}$$

When natural logarithms of both sides of equation (7.53) are taken, we get

$$\ln I = \ln I_0 - \mu x \tag{7.54}$$

or

$$\ln I_0 - \ln I = \mu x \tag{7.55}$$

or

$$\ln\left(\frac{I_0}{I}\right) = \mu x \tag{7.56}$$

or

$$x = \frac{\ln\left(\frac{I_0}{I}\right)}{\mu} = \frac{1}{\mu}\ln\left(\frac{I_0}{I}\right) \tag{7.57}$$

Here

$$\frac{I_0}{I} = \frac{I_0}{\frac{1}{4}I_0} = 4 \tag{7.58}$$

Putting the value of I_0/I from equation (7.58) in equation (7.57),

$$x = \frac{1}{\mu}\ln\left(\frac{I_0}{I}\right) = \frac{1}{\mu}\ln(4) = \frac{1.3863}{\mu} \tag{7.59}$$

At 1 THz, $\mu = 160\ \text{cm}^{-1}$; hence from equation (7.59)

$$\begin{aligned} x &= \frac{1.3863}{\mu} = \frac{1.3863}{160\ \text{cm}^{-1}} \\ &= 0.00866\text{cm} = 0.00866 \times 10^4\ \mu\text{m} = 86.6\ \mu\text{m} \end{aligned} \tag{7.60}$$

At 2 THz, $\mu = 240\ \text{cm}^{-1}$; therefore from equation (7.59)

$$\begin{aligned} x &= \frac{1.3863}{\mu} = \frac{1.3863}{240\ \text{cm}^{-1}} \\ &= 0.005776\text{cm} = 0.005776 \times 10^4\ \mu\text{m} = 57.76\ \mu\text{m} \end{aligned} \tag{7.61}$$

(b) Mass attenuation coefficient of a material, also called mass absorption coefficient η, describes the fraction of terahertz waves absorbed per unit mass of the material. It is simply the ratio of linear attenuation coefficient μ and the mass density ρ of the material, written as

$$\eta = \frac{\mu}{\rho} \tag{7.62}$$

At 1 THz, $\mu = 160\ \text{cm}^{-1}$; hence from equation (7.62)

$$\eta|_{1\ \text{THz}} = \frac{\mu}{\rho} = \frac{160\ \text{cm}^{-1}}{1\text{g .mL}^{-1}} = 160\ \text{cm}^{-1} \times \frac{1}{\text{g. cm}^{-3}} = 160\ \text{cm}^2\ \text{g}^{-1} \tag{7.63}$$

At 2 THz, $\mu = 240\ \text{cm}^{-1}$; therefore from equation (7.62)

$$\eta|_{2\ \text{THz}} = \frac{\mu}{\rho} = \frac{240\ \text{cm}^{-1}}{1\ \text{g .mL}^{-1}} = 240\ \text{cm}^2\text{g}^{-1} \tag{7.64}$$

7.6.2.2 Oral cancer

Absorption coefficient of ice by terahertz waves is one order of magnitude lower than that of water. So terahertz waves can penetrate deeper into frozen tissues against those in unfrozen state. Lowering of hydration of fresh tissue provides a way for imaging tissues to greater depths. Freeze drying called lyophilization or cryo-dessication is a method of water removal involving three phases: (i) freezing phase carried out in a freezer; (ii) primary drying (sublimation) implemented by lowering the pressure to a few millibars and supplying heat for sublimation of water; and (iii) secondary drying (absorption) phase carried out by raising the temperature to break the bonds between the material and water molecules.

Tissues resected from oral cancer patients are mapped for tumor detection by terahertz waves (0.2–1.2 THz) under two conditions: frozen state at −20 °C and room temperature at 20 °C. In the terahertz images at −20 °C, malignant regions are clearly discriminated from normal mucosa. Similar distinction is not seen in room-temperature images at 20 °C. A reflected pulse is received from a region at a depth of 1.3 mm inside the tumor in the case of the frozen sample at −20 °C. This depth is not reachable in imaging fresh tissue at 20 °C. The study shows the greater penetration of frozen terahertz imaging (Sim *et al* 2013).

As the freezing process sometimes causes tissue necrosis (death of cells), an alternative way to increase the terahertz penetration depth under the skin is to apply a penetration enhancing agent such as glycerol whose absorption coefficient of terahertz radiation is much smaller than water (Oh *et al* 2013).

7.6.2.3 Breast cancer

Cancer is detectable in excised breast tumors with the help of pulsed terahertz imaging and spectroscopy. No contrast agent is required. As we know, the depth of penetration of terahertz waves decreases as we go to higher frequencies. At 0.5 THz, the penetration depth in cancerous tissue is 276 μm (El-Shenawee *et al* 2019). Further, fresh biological tissue containing blood and other body fluids is a highly lossy medium for propagation of terahertz waves. The decreased permeability of terahertz waves restricts imaging to surface tissue. As a result, reflection imaging is more suitable for biological tissues. Nonetheless, transmission imaging is also useful, albeit to a limited extent. The imaging is done in both reflection and transmission modes, and the measured signals are normalized with respect to a chosen standard.

Clear discernibility is found between cancerous and non-cancerous (fibroglandular and fatty) breast tissues. Appreciable contrast is seen between cancerous and healthy fibroglandular tissue. Also, it concurs with pathology (El-Shenawee *et al* 2019).

THz reflection imaging was found to successfully distinguish cancerous tissue from non-cancerous tissue when the tissue was dehydrated, heterogeneous and even for samples of 10 μm thickness (Bowman *et al* 2015).

7.7 Food and agricultural products inspection

7.7.1 Foreign body detection

Foreign objects in food in the form of human hair, egg shells, glass splinters, metal screws, plastic pieces, stones, nails, insects (crickets, maggots, etc), absorb and scatter terahertz waves. The absorption and scattering losses decrease the strength of the signal. Deviations in spectral characteristics (absorption coefficient and refractive index) from normal state indicate the presence of such objects (Jördens *et al* 2006, Hiromoto *et al* 2013).

7.7.2 Toxic substance detection

Aflatoxins (AFB1) have been quantified in peanut oil, soybean oil, corn samples, and maize (Zhao *et al* 2015, Liu *et al* 2019). Presence of melamine, a synthetic nitrogen compound added to food products to raise the protein content, has been found in milk and chocolate powders, and flour. Its peaks have been observed when tested with or without packaging. Revelations have been made of preservatives such as O-phenylphenol and thiabendazole applied to lemons, bananas and grapes; auramine O (AO), a yellow diphenylmethane carcinogenic dye in medical herbs; and talc powder in flour.

7.7.3 Antibiotic detection

Animal-origin foods are vulnerable to contamination by antibiotic residues in feed. Tetracyclines hydrochloride (TCsH) has been found in infant milk powder and food powders; and TCsH and chlortetracycline hydrochloride in rice and chicken (Redo-Sanchez *et al* 2011, Qin *et al* 2015).

7.7.4 Microbial contamination detection

Bacillus subtilis spores are detectable from their absorption band at 1.54 THz (Yu *et al* 2005). Gram-positive and gram-negative bacteria as well as living and dead bacterial cells can be differentiated. Yeast cells, molds and bacteria have been visualized using terahertz slot antennas (Park *et al* 2014).

7.7.5 Moisture content quantification

The absorption coefficient of terahertz radiation by lipid is 1/20 of that of water while that for starch is 1/50 fraction of water absorption coefficient and protein 1/100 part of water absorption coefficient (Qin *et al* 2013). Water strongly absorbs terahertz waves.

7.7.6 Quality assurance

Flaws in food packing such as air-filled and water-filled defects in plastic packages are detected (Morita *et al* 2005). Defects in chocolate bars, crickets in noodle flour, insects embedded in sugar and milk powder, and unwanted objects in food are found (Shin *et al* 2018). Deterioration of meat is assessed and insect infestation of peccans is exposed (Huang *et al* 2015).

7.7.7 Uncovering adulteration

Melamine in milk, glucose or fat powder in dairy products (Baek *et al* 2014), high-fructose corn syrup in honey (Liu *et al* 2016), and yellow chalk powder in turmeric pose serious safety concerns (Nallappan *et al* 2013), which can be resolved from THz spectra of samples.

7.8 Non-destructive testing (NDT)

This is a group of analysis techniques used to evaluate the properties of a material, component, piece of equipment or system without damaging it or impairing its functioning in any way. Common NDT methodologies such as computerized tomography, ultrasonic scanning and infrared thermography suffer from shortcomings, e.g., x-rays used in computed tomography scans have harmful effects, ultrasonic signals are severely attenuated in porous materials making imaging difficult, and thermography becomes ineffective for poor thermal conductivity materials. THz NDT is an upcoming technology which offers several advantages over the prevailing methods. Besides its innocuous and environment-friendly nature, it offers deep penetration, chemical fingerprinting, and low power consumption.

7.8.1 Pulse THz-TDS technique

A pulsed THz-TDS system based on multiple echoes and feature fusion is applied for detection of debonding defects in a five layer ceramic matrix composite (CMC) structure. Good thickness recognition capability is found for 350 μm and 500 μm thick debonding defects (Zhang *et al* 2020).

7.8.2 Reflection 3D THz imaging system

A reflection 3D THz imaging system is used as a frequency modulated continuous wave (FMCW) radar for finding defects in fiber-reinforced CMC (Becker *et al* 2012). Presence of a cavity and/or delamination is confirmed when a region with increased reflection signal intensity is noticed. Exact location of the defect is determined by finding its x, y and z positions.

7.8.3 The FMCW technique

The FMCW technique provides depth-resolved inspection of volume (Ellrich *et al* 2020). The object under examination is showered with linear frequency-modulated terahertz waves from a source (figure 7.8). For a terahertz opaque object, the waves are reflected from the surface of the object. In the case of a terahertz-transparent object, the waves enter the object and undergo reflections at all interfacial regions within the object. The received reflected signals are shifted on the time scale with respect to the sent terahertz

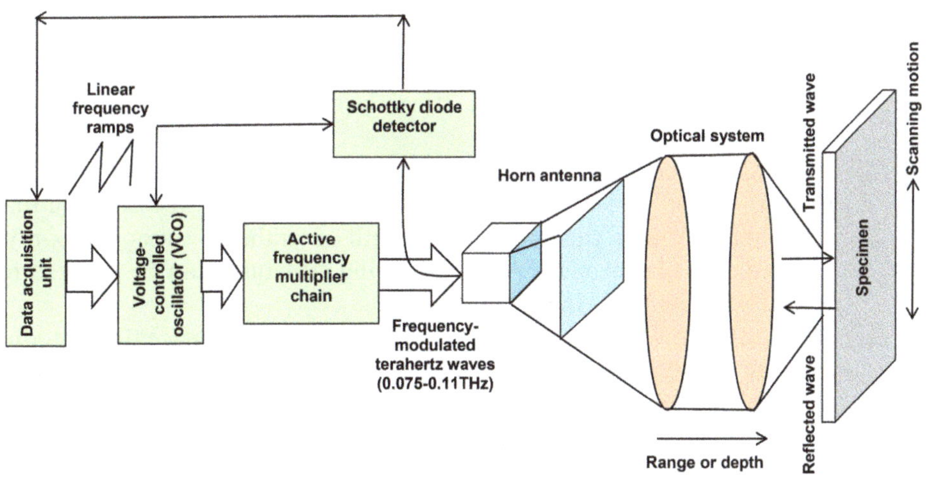

Figure 7.8. Layout of the frequency modulated continuous wave technique.

waves. The terahertz waves sent from the source serve as references from a local oscillator. The received signals are combined with reference signals by a signal processing technique known as heterodyning leading to the creation of a beat frequency signal. A data acquisition unit is used to sample the beat frequency signal. The unit performs Fourier transformation of the signal. It relates the beat frequency to a time, which is interrelated to the distance of the reflecting surface from the source of terahertz waves.

Airbags are vital safety features in modern cars. During an accident, the sudden change in speed is detected by an accelerometer chip, which triggers the airbag circuit. The airbags are hidden either behind the steering wheel of the car or the panel facing the passengers. The airbags are camouflaged by making small perforations on the backsides of the panels. The perforations not only provide an appealing artistic appearance to the car interior but also constitute an essential requirement for airbag inflation. These perforations act as entry points allowing the air to pass through them and fill the airbags. In the unfortunate event of a car mishap, the air must enter hurriedly through them to inflate the airbags fast. Therefore the size of the perforations must be carefully optimized. It cannot be too large otherwise aesthetics of the car interior will be disturbed. It cannot be too small otherwise the air inflow in the airbag will be insufficient. Extremely fast inflation of the airbag during collision is mandatory for safety of the driver/passengers.

The panel material of the car is known as slush skin. Terahertz imaging allows careful inspection of slush skin. Both location and depth of the perforation are measured. The location of the perforation is obtained from the plot of variation of intensity of terahertz waves over the full area of the skin. The depth of the perforation is found from the time-domain data (Ellrich *et al* 2020).

7.9 Discussion and conclusions

(i) THz waves are safe for medical imaging. Being non-ionizing radiation unlike x-rays, THz waves are biologically non-injurious.

(ii) THz radiation can look inside dielectric materials such as paper, cardboard, plastics, wood, masonry, ceramics, pharmaceuticals and clothing to allow nondestructive inspection for revealing cleverly hidden contraband goods, narcotics and explosives. It can replace x-ray scanning of passengers and baggage at airports.

(iii) It can also be considered as an alternative option to x-rays for imaging the interior of solid objects although the images are lower in resolution than x-ray images due to the longer wavelength of terahertz waves than x-rays, and require enhancement. Aeroplane wing structures can be checked for internal defects by THz scanning.

(iv) By comparing with a reference spectrum, a given pharmaceutical product such as a tablet can be analyzed without harming it, enabling close monitoring and quality assurance of manufacturing lines. A comprehensive spectral database of materials in the terahertz frequency range can be referred to for proper chemical identification. Pattern recognition and classification algorithms can play useful supportive roles.

(v) Many biomolecules like proteins have distinct absorption lines in the 0.1 to 5 THz frequency range serving as spectral fingerprints for their identification and quantification. Malignant growths can be detected early using THz waves.
(vi) The high attenuation of THz waves by absorption in water can hinder the widespread use of THz imaging over long distances in atmospheric humidity conditions. Necessary systems operating in THz communication windows at desired power levels are needed for such applications, taking into account the signal loss incurred.
(vii) The high absorption sensitivity of terahertz waves is also a boon in some applications. It has been applied for investigation of water diffusion in wood. Water transport in wood weakens the structures and also affects glued joints. Industrialists involved in fabrication of wooden furniture and those in house construction want non-intrusive visualization of water diffusion in wood. THz radiation can show the water transport in wood during drying and re-humidification cycles (Bensalem *et al* 2017).

Every story has an end. But the end marks a new beginning in the same way as a full stop indicates the end of a sentence, but we can start a new sentence immediately after it. So it is the beginning of a new sentence. Many new and interesting developments are awaited over the horizons in the realm of terahertz electronics in the coming years. The future is bright!

References

Baek S H, Lim H B and Chun H S 2014 Detection of melamine in foods using terahertz time-domain spectroscopy *J. Agric. Food Chem.* **62** 5403–7

Bandyopadhyay A, Stepanov A, Schulkin B, Federici M D, Sengupta A, Gary D and Federici J F 2006 Terahertz interferometric and synthetic aperture imaging *J. Opt. Soc. Am.* **23** 1168–78

Becker S, Ullmann T and Busse G 2012 3D terahertz imaging of hidden defects in oxide fiber reinforced ceramic composites *4th Int. Symp. on NDT in Aerospace 2012 – Th.4.A.2, AeroNDT2012 (Augsburg, Germany, November 13–14)* pp 1–6

Bensalem M, Mindeguia J C, Sommier A, Batsale J C and Pradere C 2017 Measurement of water content in a wood sample by terahertz imaging *Poromechanics VI: Proc. of the Sixth Biot Conf. on Poromechanics (Paris, July 9–13, 2017)* M Vandamme, P Dangla, J-M Pereira and S Ghabezloo (eds.) (Reston, VA: American Society of Civil Engineers (ASCE)) pp 847–54

Brady J E, Smith J L, Hart C E and Oxley J 2012 Estimating ambient vapor pressures of low volatility explosives by rising-temperature thermogravimetry *Propellants Explos. Pyrotech.* **37** 215–22

Bowman T C, El-Shenawee M and Campbell L K 2015 Terahertz imaging of excised breast tumor tissue on paraffin sections *IEEE Trans. Antennas Propag.* **63** 2088–97

Campbell M B and Heilweil E J 2003 Non-invasive detection of weapons of mass destruction using THz radiation *Proc. SPIE* **5070** 38–43

Crawley D, Longbottom C, Wallace V P, Arnone B C D and Pepper M 2003 Three-dimensional terahertz pulse imaging of dental tissue *J. Biomed. Opt.* **8** 303–7

Ellrich F, Bauer M, Schreiner N, Keil A, Pfeiffer T, Klier J, Weber S, Jonuscheit J, Friederich F and Molter D 2020 Terahertz quality inspection for automotive and aviation industries *J. Infrared Millim. Terahertz Waves* **41** 470–89

El-Shenawee M, Vohra N, Bowman T and Bailey K 2019 Cancer detection in excised breast tumors using terahertz imaging and spectroscopy *Biomed. Spectrosc. Imaging* **8** 1–9

Federici J F, Gary D, Schulkin B, Huang F, Altan H, Barat R and Zimdars D 2003 Terahertz imaging using an interferometric array *Appl. Phys. Lett.* **83** 2477–9

Hikal W M, Paden J T and Weeks B L 2011 Thermo-optical determination of vapor pressures of TNT and RDX nanofilms *Talanta* **87** 290–4

Hiromoto N, Shiba N and Yamamoto K 2013 Detection of a human hair with polarization-dependent THz-time domain spectroscopy 2013 *38th Int. Conf. on Infrared, Millimeter, and Terahertz Waves (IRMMW-THz) (Mainz, Germany, 1–6 September 2013)* (New York: IEEE) pp 1–2

Huang X B, Hou D B, Huang P J, Ma Y H, Li X and Zhang G X 2015 The meat freshness detection based on terahertz wave *Proc. SPIE* **9795** 97953D

Hu B B and Nuss M C 1995 Imaging with terahertz waves *Opt. Lett.* **20** 1716

Liu W, Zhang Y and Han D 2016 Feasibility study of determination of high-fructose syrup content of Acacia honey by terahertz technique *Proc. SPIE* **10030** 100300J1–6

Jördens C, Rutz F and Koch M 2006 Quality assurance of chocolate products with terahertz imaging (Poster 67, NDT.net Issue 2006) *9th European Conf. on NDT, September 2006 (ECNDT 2006) (Berlin)* pp 1–8

Kawase K, Ogawa Y and Watanabe Y 2003 Non-destructive terahertz imaging of illicit drugs using spectral fingerprints *Opt. Express* **11** 2549–54

Lee A W M and Hu Q 2005 Real-time, continuous-wave terahertz imaging by use of a microbolometer focal-plane array *Opt. Lett.* **30** 2563–65

Lee Y J and Weeks B L 2019 Investigation of size-dependent sublimation kinetics of 2,4,6-trinitrotoluene (TNT) micro-islands using *in situ* atomic force microscopy *Molecules* **24** 1895

Lenchitz C and Velicky R W 1970 Vapor pressure and heat of sublimation of three nitrotoluenes *J. Chem. Eng. Data* **15** 401–3

Liu W, Zhao P, Wu C, Liu C, Yang J and Zheng L 2019 Rapid determination of aflatoxin B_1 concentration in soybean oil using terahertz spectroscopy with chemometric methods *Food Chem.* **293** 213–9

Markl D, Ruggiero M T and Zeitler J A 2016 Pharmaceutical applications of terahertz spectroscopy and imaging *Eur. Pharm. Rev.* **21** 45–50

McKlatchy K, Reiten M T and Cheville R A 2001 Time resolved synthetic aperture terahertz impulse imaging *Appl. Phys. Lett.* **79** 4485–7

Mittleman D M, Jacobsen R H and Nuss M C 1996 T-ray imaging *IEEE J. Sel. Top. Quantum Electron.* **2** 679–92

Morita Y, Dobroiu A, Otani C and Kawase K 2005 A real-time inspection system using a terahertz technique to detect microleak defects in the seal of flexible plastic packages *J. Food Prot.* **68** 833–7

Nahata A, Yardley J T and Heinz T F 1999 Free-space electro-optic detection of continuous-wave terahertz radiation *Appl. Phys. Lett.* **75** 2524–26

Nahata A, Yardley J T and Heinz T F 2002 Two-dimensional imaging of continuous-wave terahertz radiation using electro-optic detection *Appl. Phys. Lett.* **81** 963–65

Nallappan K, Dash J, Ray S and Pesala B 2013 Identification of adulterants in turmeric powder using terahertz spectroscopy 2013 *38th Int. Conf. on Infrared, Millimeter, and Terahertz Waves (IRMMW-THz) (Mainz, Germany, 1–6 September 2013)* (New York: IEEE) pp 1–2

Oh S J, Kim S-H, Jeong K, Park Y, Huh Y-M, Son J-H and Suh J-S 2013 Measurement depth enhancement in terahertz imaging of biological tissues *Opt. Express* **21** 21299–305

Oxley J C, Smith J L, Shinde K and Moran J 2005 Determination of the vapor density of triacetone triperoxide (TATP) using a gas chromatography headspace technique *Propellants Explos. Pyrotech.* **30** 127–30

Park S J, Son B H, Choi S J, Kim H S and Ahn Y H 2014 Sensitive detection of yeast using terahertz slot antennas *Opt. Express* **22** 30467–72

Pella P A 1977 Measurement of the vapor pressures of TNT, 2,4-DNT, 2,6-DNT, and EGDN *J. Chem. Thermodyn.* **9** 301–5

Qin J, Ying Y and Xie L 2013 The detection of agricultural products and food using terahertz spectroscopy: a review *Appl. Spectrosc. Rev.* **48** 439–57

Qin J, Xie L and Ying Y 2015 Determination of tetracycline hydrochloride by terahertz spectroscopy with PLSR model *Food Chem.* **170** 415–22

Redo-Sanchez A, Salvatella G, Galceran R, Roldós E, García-Reguero J A, Castellari M and Tejada J 2011 Assessment of terahertz spectroscopy to detect antibiotic residues in food and feed matrices *Analyst* **136** 1733–8

Shin H J, Choi S-W and Ok G 2018 Qualitative identification of food materials by complex refractive index mapping in the terahertz range *Food Chem.* **245** 282–88

Sim Y C, Park J Y, Ahn K-M, Park C and Son J-H 2013 Terahertz imaging of excised oral cancer at frozen temperature *Biomed. Opt. Express* **4** 1413–21

Smirnov S V, Grachev Y V, Tsypkin A N and Bespalov V G 2014 Experimental studies of the possibilities of diagnosing caries in the solid tissues of a tooth by means of terahertz radiation *J. Opt. Technol.* **81** 464–67

Supriya M and Samuel A V 2020 Terahertz technology: a review on dental perspective *J. Pharm. Sci. Res.* **12** 1320–5

Tribe W R, Newnham D A, Taday P F and Kemp M C 2004 Hidden object detection: security applications of terahertz technology *Proc. SPIE* **5354** 168–76

Trichopoulos G C, Mosbacker H L, Burdette D and Sertel K 2013 A broadband focal plane array camera for real-time THz imaging applications *IEEE Trans. Antennas Propag.* **61** 1733–40

Wallace V P, Fitzgerald A J, Pickwell E, Pye R J, Taday P F, Flanagan N and Ha T 2006 Terahertz pulsed spectroscopy of human basal cell carcinoma *Appl. Spectrosc.* **60** 1127–33

Yu B, Alimova A, Katz A and Alfano R R 2005 THz absorption spectrum of Bacillus subtilis spores *Proc. SPIE* **5727** 4

Zhang J-Y, Ren J-J, Li L-J, Gu J and Zhang D-D 2020 THz imaging technique for nondestructive analysis of debonding defects in ceramic matrix composites based on multiple echoes and feature fusion *Opt. Express* 1990**28**

Zhao Z, Lian F and Zhang Y 2015 Aflatoxin B1 detected by terahertz time-domain spectroscopy 2015 *8th Int. Congress on Image and Signal Processing (CISP) (Shenyang, China, 14–16 October 2015)* (New York: IEEE) pp 1225–30

Zhou J, Rao X, Liu X, Li T, Zhou L, Zheng Y and Zhu Z 2019 Temperature dependent optical and dielectric properties of liquid water studied by terahertz time-domain spectroscopy *AIP Adv.* **9** 035346-1–-7

Zinovev N N, Fitzgerald A F, Strafford S M, Wood D J, Carmichael F A, Miles R E, Smith M A and Chamberlain J M 2002 Identification of tooth decay using terahertz imaging and spectroscopy *Twenty Seventh Int. Conf. on Infrared and Millimeter Waves (San Diego, CA, 26 September 2002)* (New York: IEEE) pp 13–4

www.ingramcontent.com/pod-product-compliance
Ingram Content Group UK Ltd.
Pitfield, Milton Keynes, MK11 3LW, UK
UKHW051041120926
8556IPUK00030B/673